Sydney's Food Landscapes

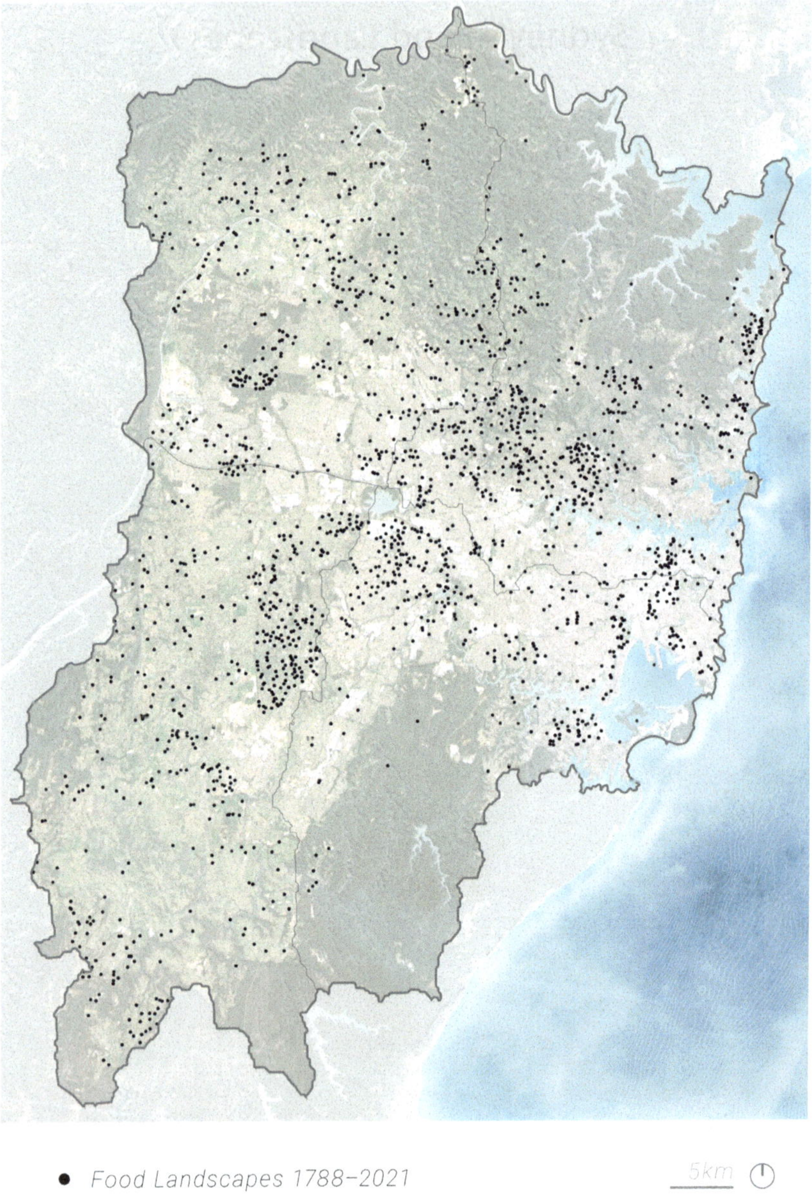

Food Landscapes Map of Sydney, 1788–2021. (Source: Joshua Zeunert and Josh Gowers)

Joshua Zeunert • Alys Daroy

Sydney's Food Landscapes

Agriculture, Planning, Sustainability

palgrave
macmillan

Joshua Zeunert
Arts, Design and Architecture
UNSW Sydney
Sydney, NSW, Australia

Alys Daroy
English and Theatre
Murdoch University
Murdoch, Australia

ISBN 978-981-96-0709-9 ISBN 978-981-96-0710-5 (eBook)
https://doi.org/10.1007/978-981-96-0710-5

Cover illustration: Composed by Joshua Zeunert. Sources: Edward Dayes, Museums of History NSW; Mitchell Library, State Library of New South Wales; Western Sydney University Archives; Nearmap.

This Palgrave Macmillan imprint is published by the registered company Springer Nature Singapore Pte Ltd.
The registered company address is: 152 Beach Road, #21-01/04 Gateway East, Singapore 189721, Singapore

Acknowledgements

While writing this book over multiple years we have worked and lived on the lands of the Bedegal, Dharug, Tharawal, Kaurna and Whadjuk Noongar Nations. We pay particular respect to the custodians of the ancestral lands that constitute the area of this book's focus. We acknowledge that sovereignty was never ceded, and pay our respect to Elders past, present and emerging. We recognise the deep and enduring care for Country over tens of thousands of years.

Despite a keen prior interest in metropolitan agriculture, we had only encountered a few disparate fragments of Sydney's publicised food histories after colonial invasion: failed early colonial undertakings, early grand estate kitchen gardens and orchards, La Perouse and Kyeemagh Chinese market gardens and ethnically diverse food production on the ever-shifting peri-urban fringes. We little suspected just how many lost histories would emerge. Synthesising Sydney's commercial agricultural history proved to be a gargantuan task. "Just an overview", we reassured ourselves time and again as we become lost in the city's complex, incomplete or competing narratives. Summarising many thousands of sources with rigour has taken significant mental perspiration over six years—many longer than originally anticipated. Despite best efforts, gaps and uncertainties inevitably remain given the breadth and scope of the undertaking. We have encountered many wonderful people on this long journey, to whom we are most grateful for their kind assistance and support.

Special thanks go to Palgrave Macmillan, especially Marion Duval for her faith in the proposal and turning the pitch into a reality and for her patience and understanding, and to Springer, especially Lavanya Devgun for all of her help, support and crucial assistance with preparation, peer review and delivery production processes (and upbeat tolerance of the stream of queries!).

This research was catalysed by a University of New South Wales (UNSW) Faculty/School of Built Environment grant. It has been further supported and enabled through UNSW's Scientia Program and an Australian Research Council DECRA Fellowship. These allowed the sporadic employment of research assistants from 2018 to 2024. Without this assistance, this book would not have been possible. We thank our home institutions, including the leadership, research teams and staff in the UNSW Built Environment and Scientia Program and Murdoch University's School of Humanities, Arts and Social Sciences. Special thanks go to Michael Ostwald and to Robert Freestone, specifically for the latter's invaluable input into *Periods Four* and *Five*. Thanks also to Philip Oldfield, Evelyn Mike, Rebekah Yee, Uttra Benton, Maria Aloe and all the staff in the UNSW Landscape Architecture programme, including Sara Padgett Kjaersgaard, for their support, given that this undertaking took far longer than originally anticipated.

The initial faculty grant allowed us to hire historian Sandra Pullman to assist with archival research. Sandi—thank you so much for your dedication to the project and your invaluable assistance, cheery and upbeat disposition, and can-do attitude. Your contribution has been absolutely invaluable. It took longer than hoped, but better late than never!

Huge gratitude is extended to two integral research assistants, Josh Gowers and Stephanie Stankiewicz. SS and JG, thank you for your brilliant work. Together we've developed rough sketches and vague musings into draft drawings, countless iterations (sorry!), and too many drawing inventories seeking to keep track of a sprawling figures assemblage. We've also (hopefully) managed to merge distinct graphic styles into a consistent and legible package on a small budget. Thanks to you both, the end results are brilliantly evocative graphics, beautiful maps and information that well exceeded original aspirational visions. Despite all not making it into the book's figure and size constraints, each and every one is appreciated.

Thanks also to Elisabeth Granville for her research assistance with image auditing and navigating copyright clearance. Cheers, Beth! A thank you also to Matt Stanley for editing several chapters of the draft manuscript. We also thank Stuart Read at NSW Environment and Heritage for his effusive generosity throughout the research process, along with his considerable knowledge that has substantially enriched the book's content. Stuart, it was a pleasure!

We would like to acknowledge and thank *Nearmap*, particularly Alex Aitken and Natalie Fisher for granting us use of their 2009–2021 high-resolution aerial imagery in the book. This was indispensable in illuminating the visual sociology and environmental change in the story of Sydney's lost food landscapes. We greatly appreciate it. Similarly, we gratefully acknowledge Spatial Services NSW and their Creative Commons Historical Imagery as indispensable to generating knowledge and visually evidencing the narrative.

Thanks to all the contributors of primary source material and artwork included in the figures. Many historical works are unattributed, and we have acknowledged original creators wherever possible. This gratitude is extended to all the people and staff who have provided digitised, online and physical material, archives, records, books, publications, imagery and the like. Specific thanks are given to: The State Library of NSW (Isabelle Andersson and all the excellent librarians) and its Mitchell Library and Dixson Galleries; the National Library Australia (Piers Robinson); Museums of History NSW (Bonnie Wildie, Stephanie Bailey, Anne Johnson), State Archives of NSW and Caroline Simpson Library; Western Sydney University archives (Cathie Lester); Hornsby Shire Recollects (Sue Miller); State Library of Victoria; Museum of Applied Arts and Sciences (Kathy Hackett); National Archives of Australia (Georgina Jansen); National Museum of Australia; Taylor and Francis (Karin Beesley); University of Sydney archives (Vipasha Mukherjee); The National Trust (David Burdon, James Bosanquet and Heath Quint); Bayside Council Library (Kirsten Broderick); City of Canada Bay Local Studies Collection (John Johnson); Northern Beaches Council Library (Jason Slattery); Royal Australian Historical Society; Australian War

Memorial; Georges River Libraries Local Studies Collection (Steph Gildaly, David Clarke); Elisabeth Dowle; Judy Horton; The Dural and District Historical Society (Michael Bell) and Knox Geelan; Randwick and District Historical Society; Justin Cahill; Trove and Pittwater Online News. A special thanks to Catherine Warne from Kingsclear Books for her invaluable legacy of local history publishing in Sydney.

A heartfelt thanks to all the people that have generously donated their time, knowledge, contacts, suggestions and photographs. This is not limited to: Jo Russell-Clarke; David S. Jones; David Martin; Benjamin Driver; Philip Thalis; Anna Richards; Sarah James; Laura Wynne; Colleen Morris; Mary Casey; Angie Thomas; Lorraine Stacker; Nicole Rawson; Angela Phippen; Rod Simpson; Christine Edney; Sarah Luke; Ros Berryman; Graham Wilson; Naomi Crago; Daphne Lowe Kelley; Peter Moore; Karla Whitmore; Tamara Prentice; Stan and Rachel Shore; Chris O'Brien; Jason Slattery; Aleem Aleemullah and Danièle Hromek. Our sincere apologies if we have missed anyone in this list over the multiple years of research.

Warm thanks to our three peer reviewers of the book proposal and the peer reviewer of the manuscript. This is a huge undertaking and we very much appreciate the effort and helpful and constructive feedback provided. You enriched and enabled the book to be realised.

Last but certainly not least, we extend ongoing thanks and love to both our families for their support and generosity.

August 2024

Dr Joshua Zeunert
Dr Alys Daroy

The Prickly Pear (*Opuntia* spp.) embodies complex intersections of colonisation, sustenance and survival shaping Sydney's foodscapes. (Source: © Elisabeth Dowle)

A 3D Food Landscapes Map of Sydney, 1788–2021. (Source: Joshua Zeunert and Josh Gowers)

Contents

List of Figures

Frontispiece

Introduction

Site and Topic

Period One: 1788–1850

Period Two: 1851–1900

Period Three: 1901–1945

Period Four: 1946–1999

Period Five: 2000–2020

Then and Now

2021+

List of Tables

List of Boxes

1

Introduction: Marginal or Marginalised? Sydney's Metropolitan Food Landscapes from 1788

From the moment the Union Jack's flagpole drove down into the resistant sandstone lands of the Gadigal Nation, colonial foodways have inscribed themselves upon the Australian continent (Fig. 1.1).[1] Sydney's broader agricultural story, from invasion to global metropolis, is one of marginality and erasure.[2] It is also one of resilience, ingenuity and vibrant metropolitan agriculture now buried beneath the city's built footprint. This is an untold history of Sydney's lost commercial food production since colonisation in 1788. It demonstrates the city's collective agricultural successes and diverse socioeconomic and cultural food production over 233 years. Now one of the world's fastest growing cities, Sydney is also a 'canary in the coal mine' for climate change in the Anthropocene—the geological epoch in which human activities have altered planetary stability.[3] The city is sustained by a complex web of local, state, national and international logistics chains extending well beyond its metropolitan region, placing it in a potentially fraught position for future food supply. This book traces the arc of Sydney's localised agriculture and planning, mapping its past so that we might more sustainably, and equitably, determine its future.

J. Zeunert, A. Daroy, *Sydney's Food Landscapes*,
https://doi.org/10.1007/978-981-96-0710-5_1

Fig. 1.1 Sydney's cultivated agricultural lands, 1798/1817. Source: State Library of NSW

In January 1788, Governor Arthur Phillip selected Warrane at the mouth of the waters flowing through Wogganmugule, Pannerong and Booragy towards Burramatta for Sydney's establishment. The wider region was home to First Nations people from around 29 different Dharug, Kuringgai and Dharawal language groups, with deep connections to land and complex Indigenous knowledge systems governing food. The area was renamed Port Jackson—now also known as Sydney Harbour—and it became the first site of the British penal colony of New South Wales. Britain was eager for the colony to establish self-sufficiency, a task set in motion through replicating British and Irish diets and Anglo-European foodways. The founding colony acted as a testbed for an array of farming practices and industries applied more widely across a country now spatially dominated by agriculture.[4] Sydney's colonial food production thus became Australia's dominant agricultural blueprint and the city's agricultural story parallels that of many other colonised countries worldwide.

Prior to colonisation, First Nations Australians managed the landscape for millennia. An onslaught of land clearance, 'slash-and-burn' agriculture, the mass deployment of exotic species and the subsequent carving up of land parcels for development occurred *after* the uninvited arrival of the British. As such, this book picks up at the point of colonisation in 1788 to assess Sydney's commercial food trajectory towards a fragile food future. Colonisation had a colossal and continuing impact upon First Nations Aboriginal communities and food systems. Colonial/settler and Indigenous value systems and food practices were at times irreconcilable but at others, more nuanced or even enmeshed. Stories of resistance, resilience and influence exist alongside conflicts.[5] Investigating Sydney's metropolitan agriculture inevitably invokes these complexities and exchanges.

We recommend the following important publications that explore Indigenous food and land management practices as companion works: Rupert Gerritsen's *Australia and the Origins of Agriculture* (2008); Bill Gammage's *The Biggest Estate on Earth* (2011); Bruce Pascoe's *Dark Emu: Aboriginal Australia and the Birth of Agriculture* (2014)[6] and Sutton and Walshe's *Farmers or Hunter-Gatherers? The Dark Emu Debate* (2021); Zane Ma Rhea's *Frontiers of Taste: Food Sovereignty, Sustainability and*

Indigenous–Settler Relations in Australia (2016); Billy Griffiths' *Deep Time Dreaming: Uncovering Ancient Australia* (2018); Tyson Yunkaporta's *Sand Talk* (2019) and *Right Story, Wrong Story* (2023); and David S. Jones and Darryl Low Choy's *Indigenous Knowledge Systems and Yurlendj-nganjin* (2021).[7]

Further understanding food landscapes prior to invasion requires Indigenous-led discussion and knowledge transmission. Australia's 121 Local Aboriginal Land Councils (LALCs) maintain Indigenous knowledge systems and landscape management practices across Country and are crucial to further discussion and future planning. Within the New South Wales Aboriginal Land Council (NSWALC), the Dharug Custodian Aboriginal Corporation[8] is critical to providing Indigenous perspectives in relation to this book's spatial boundary. We note the role of Indigenous-owned and Indigenous-managed organisations and educational and cultural hubs for expanding awareness and improving future planning.[9] We further echo Ma Rhea in identifying the need for further research to,

> [E]xamine the interface of Indigenous Australian and settler foodways up to the present as a way of thinking through how the production of endogenous foods known to Indigenous Australian people might be re-examined as the basis for a national Australian cuisine that could be scaled up but that also addressed Indigenous Australian food sovereignty and enduring food security needs, be ecologically sustainable ('clean and green'), drawing on the best of Indigenous Australian and newcomer knowledge.[10]

The Sydney region originally contained at least 200 indigenous plants yielding edible seeds, fruits, tubers/roots/rhizomes, leaves, flowers and nectar, although none have been at production and consumption scales examined by this book (see *Site and Topic*).[11] The only two commercially produced native (as opposed to indigenous)[12] species within Sydney's boundary are the Sydney Rock Oyster (*Saccostrea glomerata*),[13] devastated by a virus in 1994,[14] and the Macadamia nut (*Macadamia integrifolia*), not local to Sydney nor widely planted in the region.[15] Any idealism pertaining to Sydney diets' integration of bush foods requires calibrations based on population numbers supported by spatial-energy yields and interrogation of the reality of consumer demand and value systems. The fact is Sydney's 5 million+ residents solely, or at least chiefly, rely on exotic plant and animal species subject to extensive breeding over millennia by cultures from across the globe.

Western productivist agriculture is akin to 'mining' the soil to extract nutrients. Trends are geared to intensification rather than planned degrowth or intergenerationally ecologically sustainable land management. High-yield crops are prioritised, which mirror consumers' environmentally intensive dietary choices.[16] This work therefore focuses on uncovering Sydney's lost food landscapes since colonisation in light of future food security. A key part of this mission revolves around one central, dominant and persistent narrative—that the region's land is inherently 'marginal'. 'Marginal' land refers to agricultural land of insufficient inherent productive capacity to warrant protection or ongoing use for food production. Whether due to colonial ignorance of foreign conditions in New South Wales' early days or convenient economic expediency in the face of skyrocketing property values in Sydney's latter period, the city's metropolitan agriculture has been characterised as a tale of struggle. Writing in 1829, Edward Wakefield, for instance, termed New South Wales' soil "not particularly fertile", adding, "hundreds of years must elapse before land, here, will be cultivated as well as in the most barbarous countries of Europe".[17] This book challenges the veracity of this claim, finding that the great majority of the city's agricultural pursuits were surprisingly productive. We argue instead that Sydney's colonial story has been dominated by perennial cycles of economic speculation, greed and short-term planning influencing the way in which the city's productive land has been viewed and treated. As is evident from the stories that follow, top-down cultural factors have shaped and continue to influence Sydney's agriculture as much, if not more, than the geographic.

In the colony's early days, emancipated convict farmer James Ruse and others extracted yields despite very limited infrastructure, saving the colony from starvation.[18] The fertile Green Hills (now Windsor) and nearby Pitt Town Bottoms spurred the fledgling colony's growth and support the longest (through barely) continuing farms in Australia.[19] Sydney catalysed Australia's pastoralism, becoming the foundation of the national economy until surpassed by twentieth-century manufacturing.[20] Australia's first agricultural irrigation district commenced in Liverpool in 1856 and its archetype continues to undergird horticultural industries (irrigated cropping, grains, fruits and, nuts) across regional locations. Sydney was also integral to Australian viticulture. Despite devastation wrought by the *Phylloxera* mite, the city continued to produce a notable proportion of New South Wales' grapes into the mid-twentieth century, with Queen Elizabeth II served a Sydney-produced sparkling wine in 1954. Geometric patterns of thriving stone fruit and citrus orchards once draped across undulating topography, particularly in greater Parramatta, the north-west and the Hills District.

Fruit production persisted commercially into the mid-twentieth century, with a few remnants remaining at the time of writing.[21] Dairying was particularly active, with over 500 dairies across all Sydney's regions in 1901[22] of which a few still operate, including one of Australia's largest.[23] Local abattoirs once dumped their extensive waste streams into the Harbour, attracting sharks,[24] while livestock production saw animals regularly roam across commons and streets until displaced by privatised motor vehicles' critical mass. Despite varying motives and both positive and problematic outcomes, Sydney's agricultural productivity was quite extraordinary. This is not to overstate the area's fertility, which varies and encompasses challenging terrain including its aforementioned sandstone, as detailed in *Site and Topic*. Rather, it is to bring to light the kaleidoscopic variety of agricultural pursuits sustained by Sydney's lands since lost under the stomp of its built footprint.

Agriculture was also once Sydney's most spatially dominant metropolitan land use (Figs. 1.2 and 2.10 bottom). Most food landscape successes have vanished or fragmented across disparate sources. The city's past cumulative agricultural scale and production diversity may come as a surprise, as it did to us during the research process. Operations spanned macro-to-micro enterprises, mirroring polar ends of the socioeconomic spectrum. 'Bunyip aristocracy' pastoralists, often the beneficiaries of land grants,[25] contrasted culturally marginalised market garden labourers frequently living below the poverty line on comparatively postage-stamp-sized lessee plots.[26] The agricultural middle ground is more challenging to grasp. Australian history tends to exclude small-scale colonial farmers

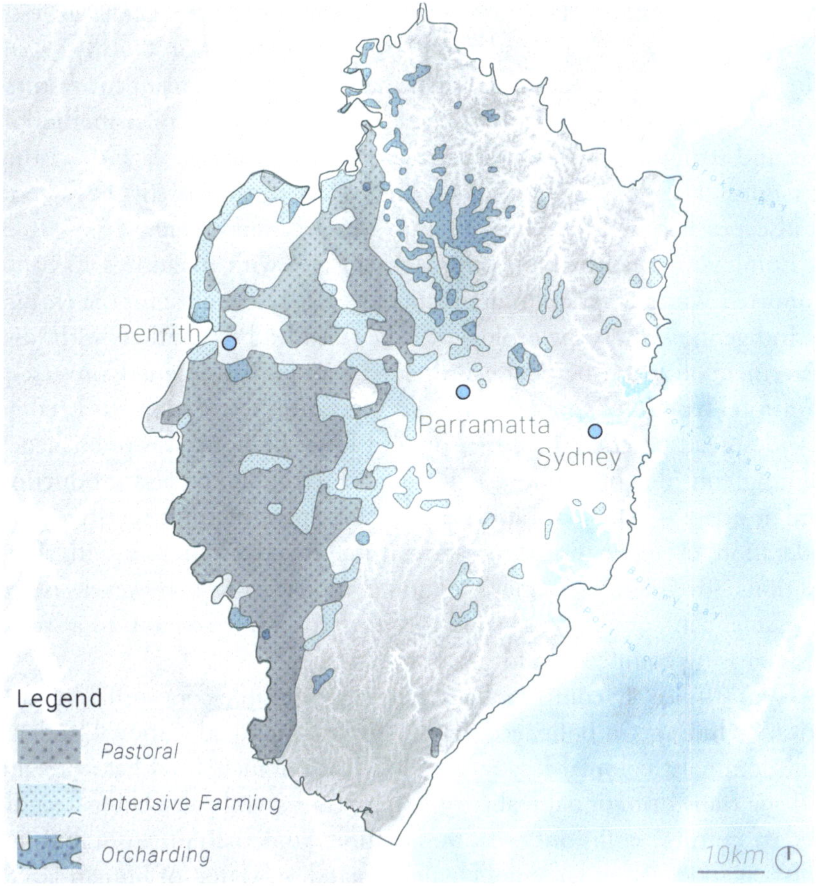

Fig. 1.2 In the mid-twentieth century, agriculture was still Sydney's most spatially dominant land use. Source: Stephanie Stankiewicz and Joshua Zeunert (Map adapted from Winston, 1957: 10)

from grand notions of nation-building, giving preference to the stories of pastoralists, swagmen, Bushmen and miners.[27] These stories are collected and synthesised across social strata and agricultural typologies to reveal a bigger and more nuanced picture.

This book was inspired by Doug Benson and Jocelyn Howell's *Taken for Granted: The Bushland of Sydney and Its Suburbs* (1990). Benson and Howell fastidiously investigate and compile historic information about

Sydney's indigenous plants and plant communities to generate an estimated overview prior to, and appraisal after, colonisation (1788–1990). Their work provided a foundational interpretative blueprint for restoration of indigenous plant assemblages.[28] We employ a similar methodology and structure, in this case canvassing historic materials and existing agricultural remnants to generate an overall picture of Sydney's food landscapes after 1788, seeking to improve agri-food literacy.

Employing a comparable approach engages with paradoxes. Because imported Western agricultural practices were a key catalyst for the erasure of Indigenous landscape ecologies,[29] they also lead to conflicts with, displacement of, and atrocities towards, First Nations custodians. Conversely, non-food based indigenous plant and vegetation restoration has become a widespread practice since Benson and Howell's book was published,[30] gaining considerably more traction than metropolitan food production and at often at odds with it. The contested space requires sensitive consideration of restoration and agricultural practices in line with First Nations sovereignty. Sydney's ongoing food security, however, needs assessment in context of its trajectory towards a megaregion, in a world of increasing competition for natural resources.

Food studies are complex due to crossing multiple conceptual boundaries, including symbolic, ecological, structural and educational under a still dominant colonial legacy.[31] As Ma Rhea argues, food has "literally fed the transformational reality of colonization, while also being a signifier of cultural collisions, accommodations and incommensurabilities, and a marker of fundamental change of land, of water, of human labor, in fact of entire ecosystems".[32] Food landscapes therefore reveal the values underpinning agricultural decisions and practices, providing insights into the ethics, or their lack thereof, influencing urban development and land use. A lack of value for metropolitan agriculture accords with a history that, since its founding on unceded lands, has been largely shaped by self-interest and short-term land-based speculations.[33] This book is founded on respect for Indigenous knowledge systems and the need to think and act long term, with care, and to recognise multi-species kinship. To think from an Earth-first perspective is to acknowledge the ecological limits of a finite planet. Increasing metropolitan agriculture can link consumption and production, support local producers and economies and reduce the heavy environmental toll of long supply chains and industrialised agribusiness.

A number of voices have expressed concern at the loss of Sydney's agriculture.[34] How does this work differ? Prior publications tend to focus on truncated time periods, specific sites, districts, cultural groups or practices, such as Chinese market gardeners, early western Sydney pastoralism, or Hills District orcharding. None have attempted to document, interrogate and synthesise Sydney's overarching agricultural narrative. Although some generalisations are inevitable, a 'big-picture' view reveals over-arching patterns and trends. Previous accounts lack comprehensive visual accompaniments, such as mappings, illustrations and photographs. These are crucial for illuminating landscape perception, preferences, the spatial makeup of activities, experiences and changes over time. We therefore include dedicated mappings of Sydney's food landscapes within 110 figures containing 270 individual images. It is our hope that this symbiosis of narrative and image will illuminate agricultural land as fundamentally important for sustainable city regions.

Research was propelled in part by Yi-Fu Tuan's notion of "topophilia", the love of place and landscape,[35] for Sydney's captivating context, dramatic coastline, vibrant and diverse city-region, unique biodiversity and productive plains. Moreover, this book is sustained by a hope to contribute to an ongoing, albeit declining, campaign for bolstering local food systems. This involves meaningful engagement with Indigenous land custodianship, preserving agricultural land, reclaiming the commons, supporting viable local agri-economics and increasing living-wage local food systems. Our central concern is ultimately the long-term integrity and viability of local food production at a time of increasing environmental precarity. Metropolitan Sydney is here the focus but we invite you to question how the city's lost metropolitan agriculture might serve as a case study for wider application.

Although a geographic and spatial landscape history of Sydney, this book therefore explores the potential for sustainable futures more broadly. It calls into question "subjunctive historiographies", that though seemingly engaged with the past, provide models of agency for the present.[36] Sydney's lost food landscapes offer models of food production and highlight the impressive range of edible species that can be grown in a region too often dismissed as unreceptive, unsuited or challenging to justify commitment to metropolitan agriculture.[37] While the majority of the book's content is directed at unearthing past food landscapes and practices, it ponders future possibilities. What role should food-producing landscapes have in Sydney's and Australia's metropolitan areas? How might varying agricultures be planned and executed with respect to species, techniques and landscapes? Bolstering metropolitan agriculture can spur new practices in contextually appropriate landscapes and cultural contexts.

This book hence focuses on Sydney's commercial-scale agriculture for its landscape impact, in line with Wendell Berry's notion that "eating is an agricultural act".[38] The invisible processes of industrial, global food chains invoke Val Plumwood's concept of "shadow places", namely the "multiple disregarded places of economic and ecological support, a split between our idealised homeplace and the places delineated by our ecological footprint".[39] Local and tangible food systems, on the other hand, are relational, place-based and more accountable to communities and bioregions. It is easy to feel helpless to create change when a single individual among 5 million urban inhabitants, within a nation of around 27 million, in a world of over 8 billion. For those with the privilege of choice, what we eat and where we source it from, offers an opportunity to shape social and ecological futures, not just for Sydney but for the planet.

Notes

1. It is doubtful if the large flagpole as depicted in Algernon Talmage's recreation could actually have been inserted into the sandstone land, or if it did marginally penetrate the surface, whether the shallow or non-existent 'topsoil' would actually support the large pole and its Union Jack. See: https://viewer.slv.vic.gov.au/?entity=IE547849&mode=browse.

   For the challenge of determining varying First Nations names including 'Eora', see: https://darugcorporation.com.au/the-eora-dilemma/ and https://www.sl.nsw.gov.au/stories/eora.

2. The following chapter, *Site and Topic*, defines the book's boundary area for 'Sydney'.
3. Crutzen and Stoermer, 2000.
4. ABARES, 2022; Zeunert, 2023, 2024a, b.
5. Ma Rhea, 2017.
6. See Pascoe (2014) for a contested counter history about First Nations agriculture. Pascoe drew extensively from Gerritsen (2008). Pascoe's work is heavily critiqued by Sutton and Walshe (2021).
7. Though specific to a non-Sydney area, we also note reports such as *Aboriginal People, Bush Foods Knowledge and Products from Central Australia* by Merne Altyerre-ipenhe Reference Group, Josie Douglas and Fiona Walsh (2011). See also Jones et al., 2018; Jones and Clarke, 2018.
8. Also spelled 'Darug'.
9. See, for example, Deadly Story's "Food and Agriculture" resource portal (a partnership between VACCA, SNAICC, The Koorie Heritage Trust, the Federation of Victorian Owners Corporation, Brightlabs and the Department of Families, Fairness and Housing) (https://deadlystory.com/page/culture/Life_Lore/Food), or in the Hunter Valley, the Wonnarua Nation Aboriginal Corporation's Language and Culture Archive (https://wonnarua.org.au/), amidst numerous other resources.
10. Ma Rhea, 2017: 8.
11. Hunwick, 2018: 10. See Jones & Clarke, 2018, who provide a list of species in commercial production (i.e. the macadamia) and further reading lists.
12. *Native* plants usually denote species occurring in a country before the arrival of and/or significant disturbance by humans, e.g. within Australia. *Indigenous* plant species usually denotes plants found in an area of greater focus, such as the Sydney region, or a bioregion (Fig. 2.9). *Endemic* denotes species occurring only within a specific area and not found elsewhere.
13. Which is also native to parts of New Zealand.
14. See '*AQ*' in *Period Four*. The subsequent recovery was hard-fought and ongoing. Also see: https://www.youtube.com/watch?time_continue=353&v=jCQlxgG903Y&feature=emb_logo.
15. South Africa produced more macadamia nuts than Australia (in 2018).
16. Zeunert, 2024a. Also see Pimentel & Pimentel, 2008; Oppenlander, 2013; Poore & Nemecek, 2018; Springmann et al., 2018; Zeunert, 2018; Scarborough et al., 2023.

17. Wakefield, 1829.
18. Agricultural success saved Sydney as the indefinite deployment of food supply ships was not an economic or logistical liability British rulers would continually incur.
19. Albeit, are being severely eroded by urban encroachment, see Fig. 8.13.
20. Hoorn, 2007.
21. See for example McClymont, 2013.
22. Lewis, 1980.
23. Leppington Pastoral Company, see Fig. 7.2. They sold part of their land to encompass part of the (under construction) Western Sydney Airport amidst controversy from the price paid by the government, see *Period Five.*
24. See Homebush State Abattoir Complex in *'IL'* in *Period Three* and *Four.*
25. A term coined by Deniehy in 1853, see *Period 2: 1851–1900.*
26. Christie, 1988; James, 2016a, 2016b.
27. McIntyre, 2009.
28. See e.g. SER, 2004; Zeunert, 2013.
29. See Ma Rhea, 2017: 133–134.
30. While bush regeneration's roots extend earlier than 1990, Benson and Howell's text helped to increase Sydney's and Australia's ecological literacy, particularly in its methodology for reading and interpreting place-specific past assemblages of vegetation. Moreover, it provided guidance about likely locations, species and vegetation community assemblages across the overall metropolitan context and its landscape patchwork.
31. Ma Rhea, 2017: 1–2.
32. Ma Rhea, 2017: 1–2.
33. Lines, 1991.
34. Such as Parker & Jarecki, 2003; Sinclair et al., 2004; Mason & Knowd, 2010; Merson et al., 2010; James et al., 2010; James, 2014, 2016a, 2016b; Sinclair, 2015a, 2015b; James & O'Neill, 2016; Carey & James, 2018; Lehmann, 2020; Wynne et al., 2020.
35. Tuan, 1974.
36. Quayson, 2000: 65, cited in Ma Rhea, 2017: 8–9.
37. Quayson, 2000: 65, cited in Ma Rhea, 2017: 8–9.
38. Berry, 2010: 146.
39. Plumwood, 2008, np.

References

ABARES. (2022). *Snapshot of Australian Agriculture 2022.* Australian Bureau of Agricultural and Resource Economics and Sciences: Australian Government Department of Agriculture, Fisheries and Forestry. Accessed May 1, 2022. https://www.awe.gov.au/abares/products/insights/snapshot-of-australian-agriculture-2022

Berry, W. (2010). *What are people for?* Counterpoint Press.

Carey, R., & James, S. (2018). Peri-urban agriculture in Australia: Pressure on the urban fringe. In J. Zeunert & T. Waterman (Eds.), *Routledge handbook of landscape and food* (pp. 213–227). Routledge.

Christie, M. (1988). *The Sydney markets 1788–1988.* Sydney Market Authority.

Crutzen, P. J., & Stoermer, E. F. (2000). The "Anthropocene". *Global Change Newsletter, 41*, 17.

Hoorn, J. (2007). *Australian pastoral: The making of a white landscape.* Fremantle Press.

Hunwick, H. (2018). *The food and drink of Sydney: A history.* Rowman & Littlefield.

James, S., Dimeski, B., Crabtree, L., & O'Neill, P. (2010). *Sydney's agricultural lands: An analysis.* Parramatta, N.S.W.: University of Western Sydney. https://researchdirect.westernsydney.edu.au/islandora/object/uws:11823/datastream/PDF/view

James, S. (2014). Protecting Sydney's Peri-Urban agriculture: Moving beyond a housing/farming dichotomy. *Geographical Research, 52*, 377–386.

James, S. (2016a). *Farming on the Fringe: Peri-urban agriculture, cultural diversity and sustainability in Sydney.* Springer.

James, S. (2016b). Beyond 'Local' food: How supermarkets and consumer choice affect the economic viability of small-scale family farms in Sydney, Australia. *Area, 48*, 103–110.

James, S., & O'Neill, P. (2016). Planning for Peri-urban Agriculture—a geographically-specific, evidence based approach from Sydney. *Australian Geographer, 47*, 173–194.

Jones, D., & Clarke, P. (2018). Aboriginal culture and food-landscape relationships in Australia: Indigenous knowledge for country and landscape. In J. Zeunert & T. Waterman (Eds.), *Routledge handbook of landscape and food* (pp. 41–60). Routledge.

Jones, D. S., Choy, D. L., Tucker, R., Heyes, S., Revell, G., & Bird, S. (2018). *Indigenous knowledge in the built environment: A guide for tertiary educators.* Sydney: Dept of Education and Training.

Lehmann, M. (2020). Farming on Sydney's Fringe; can it survive urban sprawl? *The Weekend Australian Magazinee*. https://www.theaustralian.com.au/weekend-australian-magazine/farming-on-sydneys-fringe-can-it-survive-urban-sprawl/news-story/048b857628c86161f482319238dc9bad
Lewis, M. (1980). Milk, mothers, and infant welfare. In J. Roe (Ed.), *Twentieth century Sydney: Studies in urban & social history* (pp. 193–207). Southwood Press.
Lines, W. (1991). *Taming the great south land: A history of the conquest of nature in Australia*. University of Georgia Press.
Ma Rhea, Z. (2017). *Frontiers of taste: Food sovereignty, sustainability and Indigenous-settler relations in Australia*. Springer.
Mason, D., & Knowd, I. (2010). The emergence of urban agriculture: Sydney, Australia. *International Journal of Agricultural Sustainability, 8*(1–2), 62–71.
McClymont, J. (2013). *Pictorial history: The Hills District*. NSW: Kingsclear Books.
McIntyre, J. (2009). Not rich and not British: Philip Schaeffer, 'failed' colonial farmer [online]. *Journal of Australian Colonial History, 11*, 1–20. https://search.informit.com.au/documentSummary;dn=997044012583652;res=IELHSS
Merson, J., Attwater, R., Ampt, P., Wildman, H., & Chapple, R. (2010). The challenges to urban agriculture in the Sydney Basin and lower Blue Mountains region of Australia. *International Journal of Agricultural Sustainability, 8*, 72–85.
Oppenlander, R. A. (2013). *Food choice and sustainability: Why buying local eating less meat and taking baby steps won't work*. Langdon Street Press.
Parker, F., & Jarecki, S. (2003). *Peri-urban agriculture and public policy: Sydney's forgotten farmers, paper presented at social inequality and the knowledge society conference, 12 November 2003*. Macquarie University.
Pascoe, B. (2014). *Dark Emu*. Magabala Books.
Pimentel, D., & Pimentel, M. (2008). *Food, energy, and society*. CRC Press.
Plumwood, V. (2008). Shadow places and the politics of dwelling. *Australian Humanities Review, 44*. https://australianhumanitiesreview.org/2008/03/01/shadow-places-and-the-politics-of-dwelling/
Poore, J., & Nemecek, T. (2018). Reducing Food's environmental impacts through producers and consumers. *Science, 360*, 987–992.
Scarborough, P., Clark, M., Cobiac, L., et al. (2023). Vegans, vegetarians, fish-eaters and meat-eaters in the UK show discrepant environmental impacts. *Nature Food*. https://doi.org/10.1038/s43016-023-00795-w
SER. (2004). *The SER international primer on ecological restoration, version 2*. Society for Ecological Restoration.

Sinclair, I., Docking, A., Jarecki, S., Parker, F., & Saville, L. (2004). From the outside looking. In *The future of Sydney's rural land: Background issues and workshop outcomes*. University of Western Sydney.

Sinclair, I. (2015a). *Growing food in a residential landscape*. Paper presented at the State of Australian Cities Conference 2015, 9–11 December 2015, The Gold Coast, Queensland.

Sinclair, I. (2015b). *Sydney Peri-urban network issues paper*. Edge Land Planning.

Springmann, M., Clark, M., Mason-D'Croz, D., et al. (2018). Options for keeping the food system within environmental limits. *Nature, 562*, 519–525. https://doi.org/10.1038/s41586-018-0594-0

Sutton, P., & Walshe, K. (2021). *Farmers or hunter-gatherers?: The Dark Emu debate*. Melbourne University Press.

Tuan, Y.-F. (1974). *Topophilia: A study on environmental perception, attitudes and values*. Prentice-Hall.

Wakefield, E. (1829) *A letter from Sydney: The principal town of Australasia*. Accessed February 11, 2020. http://gutenberg.net.au/ebooks13/1306631h.html

Winston, D. (1957). *Sydney's great experiment: The progress of the Cumberland County plan*. Halstead Press.

Wynne, L., Ruoso, L.-E., Cordell, D., & Jacobs, B. (2020). 'Locationally disadvantaged': Planning governmentalities and peri-urban agricultural futures. *Australian Geographer, 51*, 377–397.

Zeunert, J. (2013). Challenging assumptions in urban restoration ecology. *Landscape Journal, 32*(2), 231–242.

Zeunert, J. (2018). Challenges in agricultural sustainability and resilience: Towards regenerative practice. In J. Zeunert & T. Waterman (Eds.), *Routledge handbook of landscape and food* (pp. 231–252). Routledge.

Zeunert, J. (2023). *Food | landscapes | Australia*, https://www.foodlandscapes.com.au/. Doi: https://doi.org/10.26190/s0kk-8x57

Zeunert, J. (2024a). Bridging rural and urban disconnections: Spatial graphic explorations of Australia's livestock landscapes. In R. Pernice & B. Chen (Eds.), *Urban regeneration and rural revitalization in Australia and China* (pp. 110–143). Routledge.

Zeunert, J. (2024b). States of Australia's agri-environment: Visual extractions from degraded landscapes. *Visual Studies, 39*(4), 609–621. https://doi.org/10.1080/1472586X.2023.2294004

2

Site and Topic: How to Read This Book

Beneath Sydney's glittering towers and hardscapes lies land that once supported a complex commercial food history. Sydney's lost agricultural landscapes were shaped over two centuries of metropolitan food production and planning since colonial invasion in 1788. This chapter orients the book's historical sweep, diverse food landscape typologies, maps and spatial boundary (Fig. 2.1). Cartographic representations encompass the broader geographic, morphological and sociocultural milieu. Figure 2.2 overviews each chapter's topic, timeframe, formats and '11 typologies', and Figure 2.2 spatially contextualises changing understandings of colony, state and country.

J. Zeunert, A. Daroy, *Sydney's Food Landscapes*,
https://doi.org/10.1007/978-981-96-0710-5_2

Fig. 2.1 Sydney's encircling topographic basin and waterways greatly limit the spatial area for metropolitan agriculture. This coincides with this book's focus and boundary area (grey outline). Source: Joshua Zeunert and Josh Gowers

Fig. 2.2 This diagram conveys the book's topic, timeframe and time periods, presentation formats and each chapter's 11 typologies. Source: Joshua Zeunert and Stephanie Stankiewicz

Fig. 2.3 New South Wales' shifting boundaries and areas by year (Note this is graphically indicative only. First Nations sovereignty has never been ceded. Tasmania was invaded in 1803 and obtained self-governance in 1856). Source: Stephanie Stankiewicz, Josh Gowers and Joshua Zeunert

Timeframe: 1788–Present

The harbour city's commercial food production is divided into five periods. *Period One: 1788–1850* (P1) introduces us to the uninvited colonists and their struggles to establish a self-sustaining colony based on British foodways. *Period Two: 1851–1900* (P2) traverses the gold rush, pastoral boom and the need to nourish a burgeoning population. *Period Three: 1901–1945* (P3) includes Australia's Federation, World War I, the bubonic plague outbreak and the transformational shift from agrarian to mechanical and chemical industrialised food systems. *Period Four: 1946–1999* (P4) discusses the impact of post-World War II immigration, the advent of formalised planning and the immense loss of local food production due to the growth of Sydney's population and urban form, creating a tipping point for local food security. *Period Five: 2000–2020* (P5) catapults us into the new millennium and into untrammelled metropolitan expansion through urban development. As *Period Five* magnifies the two most recent decades, it is intentionally shorter. It is followed by a brief visual essay, *Then and Now*, juxtaposing key historic agricultural districts with their metamorphoses into urban landscapes. Finally, *2021+* peers across the horizon to contemplate Sydney's resilience amidst climate precarity and uncertain food futures.

Classification of Productive Landscapes: '11 Typologies'

Distinct 'typologies', or food archetypes, emerged from our initial analysis of Sydney's diverse agricultural practices. Like the book's timeframes, classifications overlap and their broad structure accounts for temporal shifts. They serve to help navigate the city's evolving metropolitan food landscapes, facilitate narrative consistency across periods, correlate the Map Database and categorise individual types of commercial food production (Box 2.1).

Box 2.1 11 Food Landscape Typologies

- ***Agricultural Facilities and Education*** **(AF)**: Schools, training facilities, universities, asylums, prisons or hospitals with food production. Also encompassing field or applied research facilities and other government agricultural facilities and initiatives.
- ***Mixed Farming*** **(MF)**: Mixed estate farms, agricultural operations, adjacent land parcels, and/or localised districts with varying food production.
- ***Market Gardening*** **(MG)**: Commercial market gardening, intensive vegetable production, greenhouses, large estates' kitchen gardens and mushroom production.[1]
- ***Fruit*** **(FR)**: Orchards, groves and agglomerations of fruit trees and table fruits.
- ***Cropping*** **(CR)**: Field crops and irrigated cropping, frequently for animal fodder.[2]
- ***Intensive Livestock*** **(IL)**: Dairies, piggeries, poultry pens and sheds, feedlots, irrigated pastures with (for example) adjacent dairy herds.[3]
- ***Grazing*** **(GZ)**: Pastoralism and grazing livestock (usually cattle or sheep).
- ***Vineyard Wineries*** **(VW)**: Wineries producing their own grapes for fortified and table wines.
- ***Agricultural Infrastructure*** **(IN)**: Processing facilities, transportation hubs, storage infrastructure such as grain mills and silos, abattoirs, markets and wharfs, dairy depots, fertiliser plants, agricultural pesticide factories, salt pans and factories processing local agricultural outputs.
- ***Aquaculture*** **(AQ)**: Seafood production.
- ***Miscellaneous*** **(MC)**: Nurseries (productive not ornamental), agricultural society exhibitions and shows, city farms, community gardens and public urban agriculture.

Mapping

Food Landscape Maps

The heart of the book's research is a suite of dedicated food landscape summary maps chronicling Sydney's food production from 1788 to 2021. The overarching master map spans the five historical periods (the frontispiece and Figs. 9.4, 9.5, and 9.6). Figure 2.4 presents a visual overview of the map assemblage. Each chapter, and the book itself, conclude with a 3D exploded-axonometric map and plan revealing commercial food landscapes' spatial distribution and typological composition.[4]

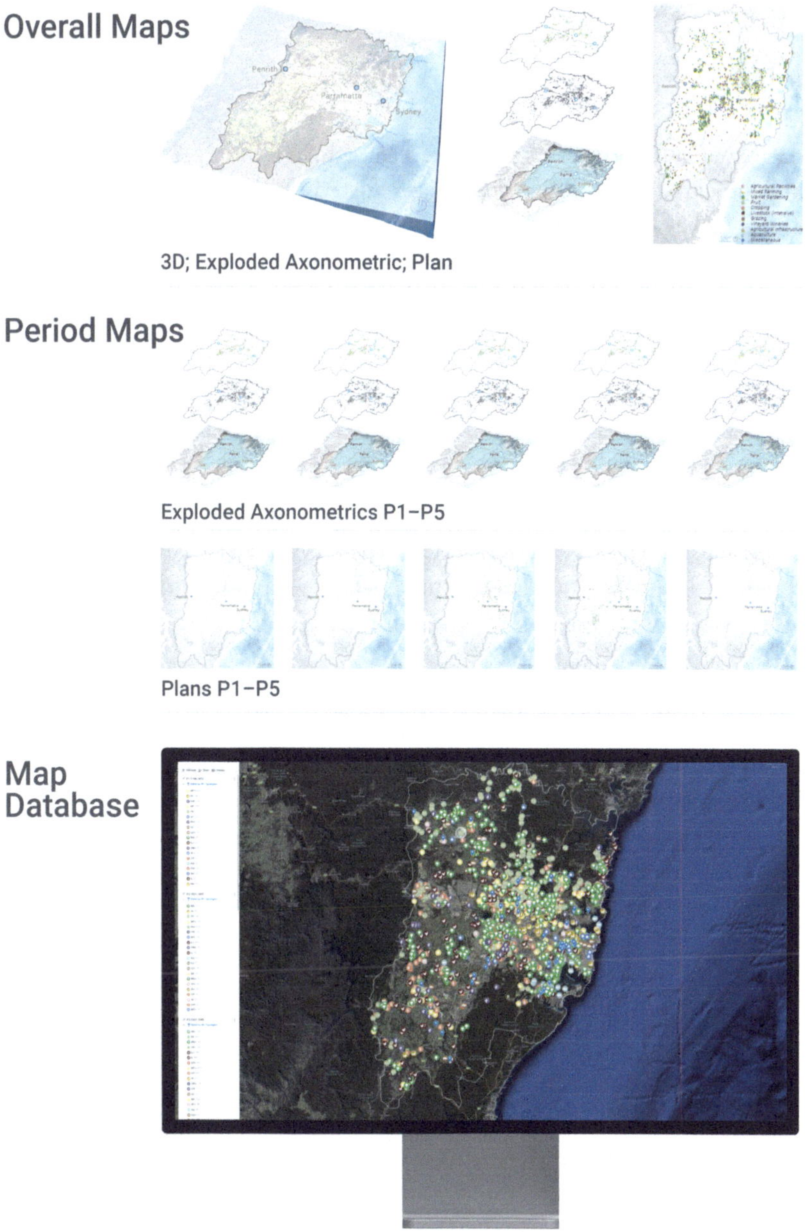

Fig. 2.4 This diagram summarises the book's maps overall, using 3D, exploded axonometric and plan drawings across the five time periods. The maps were compiled from the online Map Database (Box 2.2). Source: Joshua Zeunert, Josh Gowers and Stephanie Stankiewicz

Online Database

The maps included in this book are a selection of drawings from an original geolocated Master Map Database hosted online in Google My Maps (Box 2.2, Fig. 2.4). This database offers an information-rich interactive format enabling the user to pan out and zoom in to pinpoint precise food landscape locations, as well as provides summary information and links.[5]

Box 2.2 Online Google Master Map Database

https://www.google.com/maps/d/u/0/edit?mid=1Bkr5I0gMxNtY8WASlev-ArSqOik-XtDI&ll=-33.658444299163435%2C150.66546108070236&z=17

The Master Map Database contains over 1400 entries. It was compiled 2018–2021, and acknowledges that Sydney's ongoing urbanisation creates constant change since it was undertaken. A few notes to consider when exploring the database:

- When multiple food landscapes appear in proximity they are grouped under single entries. In other words, the 1400 entries represent several thousand individual food landscapes.
- Like the book's maps and focus chapters, database entries are organised into the five time periods frames as layers, which can be turned on or off to reveal changing patterns over time.
- Individual entries are classified according to the 11 typologies. If more than one typology was present in a location, additional classifications were added into the heading, listed in (approximate) hierarchical order from most to least prevalent (multiple entries are nested under the most prevalent typology only).[6]
- Variations to the maps' digital versus static display required reflexivity. Many entries cross different time periods; for example, a dairy (intensive livestock: *IL*) spanning *Periods Two* to *Four* is listed only in *Period Two* in the online database. The online database hence only contains entries based on their first time period.[7] The maps showcased within the book, however, reveal every period in which a particular food landscape was found, organised according to primary typology. For

example, if an entry was 'P2–P4: IL, MG', it is included only in *Period Two* online, while in the book, *IL* is mapped (maroon-brown) in *Periods Two*, *Three* and *Four* and denoted as an 'entry containing multiple typologies' (see map legends). It does not appear in *Market Gardening* in either case.
- Both the online database and maps in the book use a variant in *Period Five* denoted as 'current'. This means that at the time an entry was recorded it appeared operational.[8]
- Classifications are the best indicative estimates possible within permitted access to sites and aerial overviews.[9]

Both the book's static maps and the online database have benefits and limitations—their strength therefore lies in their combined use.

Mapping Methodology

Despite dedicating several years to data collection, the task was immense. In 1921, for example, Sydney had 427 registered metropolitan dairymen,[10] and in 2009, 1052 verified vegetable growers.[11] To manage the abundant individual operations over a long timeframe, we have grouped and hierarchically emphasised data to present metropolitan temporal and spatial patterns. Notably, we prioritised food production activities lost to urban expansion over active operations at the time of data capture. Many thousands of sources were used and the research process cross-referenced multiple resources for verification. These varied sources, ranging from written histories to maps, illustrations and photographic evidence,[12] each contributed to our understanding of Sydney's agricultural patterns. High-resolution aerial photography played a pivotal role in both the database creation and spatial analysis, helping to create a fuller understanding of Sydney's metropolitan agricultural spatial extent and patterning.

Figure 2.5 offers examples of each typology visible in aerial photographs to provide a sense of the method undertaken and its challenges. Aerial imagery, and typologies changing over time, are presented from left to right. Changes in practices are evident in, for instance, commercial poultry-keeping, which largely shifted from outdoor pens into sheds from the mid-twentieth century.

Fig. 2.5 Aerial imagery showing examples of each typology, moving from left (oldest) to right (newest). Source: Joshua Zeunert

The earliest reliable aerial imagery available during the compilation dates to 1943. Consequently, there are a greater number of Map Database entries from the end of *Period Three*. Initial access to aerial images was limited, centring on 1943 and 1951/53 snapshots, Google Maps' recent captures and Nearmap imagery (2009–current). In 2020, a critical resource emerged: the NSW Historical Imagery Viewer portal.[13] However, its dynamic state during the latter research phase (2020–2021) led to fluctuating and inconsistent imagery. As a result, some irregularities are unavoidable.

Figure 2.6 illustrates the years of aerial photography consulted, emphasising spatial coverage within the book's boundary area. Notably, certain regions, particularly outer agricultural districts, lack historical aerial photographic documentation. Undoubtedly, countless additional food landscape sites existed, regrettably eluding capture in our database and mapping.

Limitations and Caveats

There are seven key limitations stemming from the mapping methodology: Availability, singularity, capture, identification, classification, specificity and sources. These are briefly summarised below. Firstly, as mentioned above, workable-resolution aerial imagery became available from 1943 (late *Period Three*) onward. Coverage during the 1960s to 2000s was less prevalent, smaller and lower in resolution compared to imagery from 1943 and 1951/53 (Box 2.3). While Nearmap's high-resolution maps (2009–2020) enhanced classification accuracy, their focus on urban areas resulted in outer metropolitan agricultural regions having limited detailed resolution and infrequent coverage. Consequently, identifying entries beyond these coverage extents and years posed challenges, resulting in inevitable omissions.

Secondly, aerial imagery provides a snapshot frozen in (day)time. In 1943, for example, despite the high demand for fresh food production, the total farmed area was affected by labour deployment to war efforts. Additionally, fields might have been fallow, seasonal or devoid of livestock at the point of image capture. Dairy herds could be concealed in

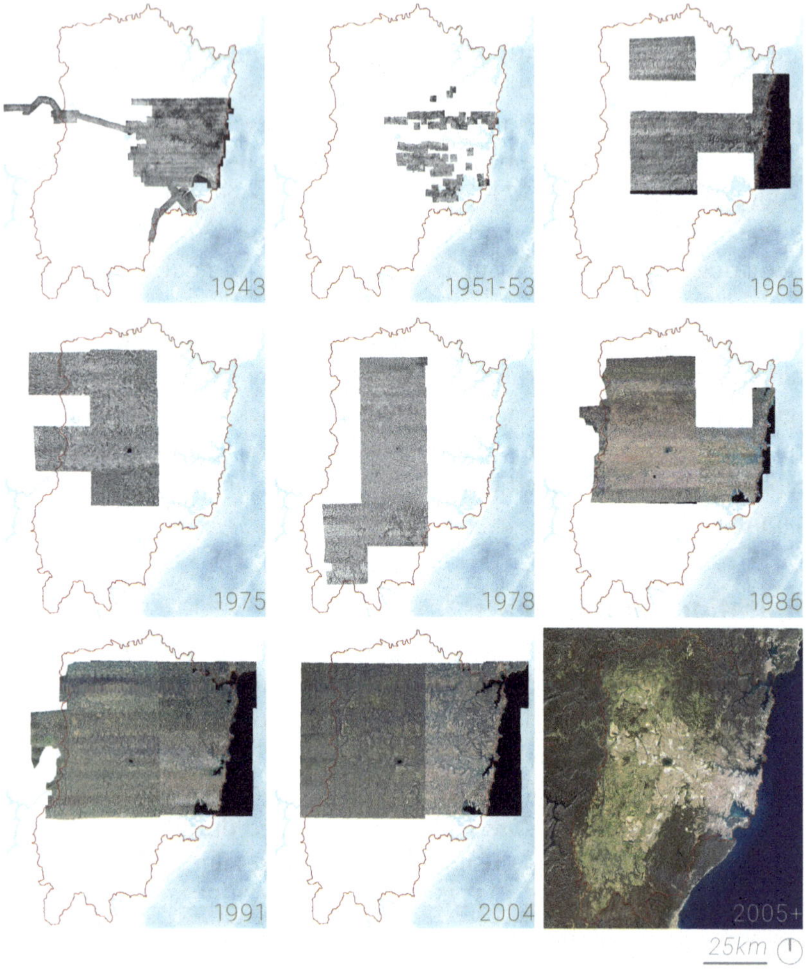

Fig. 2.6 Aerial image spatial coverage by year of source (Extents shown are approximate and show coverage at the end of the research period (2021). Actual coverage available at the time of survey (late 2018–2021) was smaller for some sources. Coverage extents for some years were in flux over the research period as new information was being uploaded into the Historical Imagery Viewer.). Source: Josh Gowers and Joshua Zeunert

milking sheds or crops difficult to discern within grainy aerial imagery. Classification therefore leaned towards conservatism rather than assumption.

Thirdly, despite considerable efforts to minimise errors, the manual survey process is inherently prone to human oversight (refer to 'Errata' in this chapter). Surveying up to 5325-square kilometres across all available years of aerial photography coverage involved meticulously combing historic imagery. An automated computer-based approach was ruled out due to complexities, cost constraints and the need for manual validation, given varying resolution, image quality and evolving agricultural typologies over time.

Fourthly, distinguishing between various typologies can be challenging. Differentiating smaller irrigated crop fields from market gardens, or irrigated pastures from cropped areas, can pose difficulties. Similarly, discerning dairy fields from those used for beef cattle requires careful observation. Considering additional landscape features helped to increase accuracy: Market gardens typically exhibit orderly rows between plants, cropping areas tend to have larger geometries (Fig. 2.5), dairies are identifiable by milking sheds and worn tracks to and from these, while cattle or sheep grazing often involves larger field sizes and farm dams. Image clarity and lack of colour continued to present hurdles, as well as activities such as poultry-keeping becoming obscured within sheds.

Fifthly, land use classifications do not necessarily directly correlate with food. Vineyards and fruit trees may produce table fruits or wine and cider (though statistics confirm that in the case of the former, wine was dominant). Grazing fields might be for recreational equine[14] activities and not for cattle or sheep.[15] The complexity of land use extends beyond straightforward food production, and these nuances made distinct identification challenging at times.

Sixthly, it difficult to ascertain an end date for many of the database entries, unless definitive information was available, or clear incremental imagery showed its obliteration through urban development and/or land-use change. Research tackled these hurdles through synthesising a range of sources as evidence, offering nuanced descriptions of land use and sociocultural complexities and openly acknowledging limitations. The book therefore aims for accuracy while recognising the complexities inherent in historical survey and land use classification.

Finally, the scope of this research is necessarily broad in order to consider multidimensional nested elements rather than focusing on a narrow topic area and timeframe, which presents numerous challenges. The work inevitably draws upon a wide array of secondary sources. Sources have been cross-checked wherever possible and any seemingly questionable sources have been rigorously investigated and discarded where appropriate. While every care has been taken during the research process, it is likely that there are unintentional errors, for which we apologise. Please see this work as starting point for further research into the more specific topic areas. This book's intended strength lies in its overview and 'big picture' provided from synthesising many its thousands of sources.

Box 2.3 Aerial Imagery Sources

Five key sources of aerial imagery were used to populate the Map Database:

- Sixmaps 1943 aerial imagery[16];
- High-resolution 1951/53 government utilities and military imagery[17];
- NSW Imagery's 'Historical Imagery Viewer' portal (1955, 1965, 1966, 1975, 1978, 1986, 1991, 2004 and 2005 (other years and wider coverage subsequently became available after the research process)[18];
- Google Maps satellite imagery[19];
- Nearmap's 'MapBrowser' imagery 2009–2021.[20]

Focus

This book focuses on the dynamic nexus between the urban environment and metropolitan agricultural landscapes. To emphasise productive agricultural landscapes lost under Sydney's expanding footprint since 1788, focus was firstly prioritised on built up regions, secondarily on peri-urban, with tertiary focus on metropolitan rural regions[21] (noting these spatially shifted over time). Figure 2.7 ('Built Form') depicts Sydney's built footprint at the end of each of the five time periods correlating with the city's population as graphed (top).

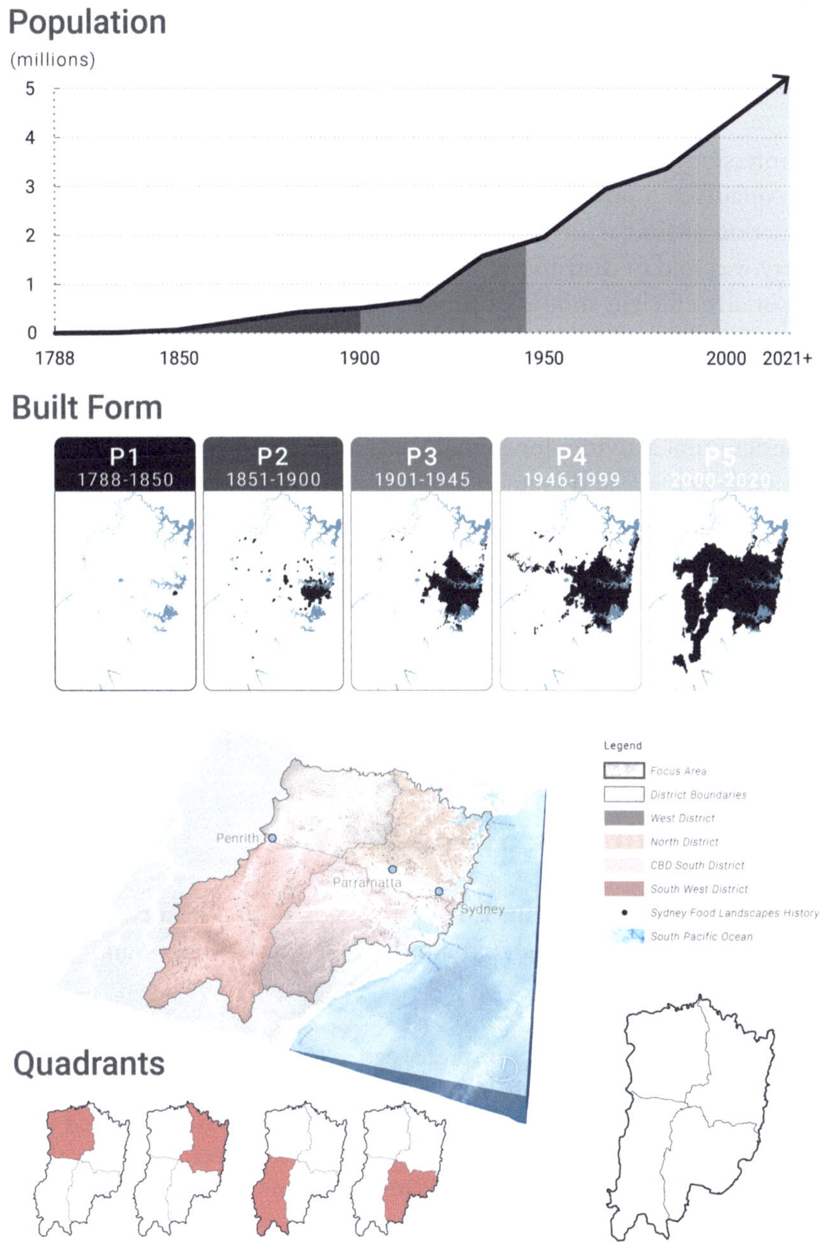

Fig. 2.7 Population and urban footprint over time during *Periods One* to *Five* (top, middle). Bottom: Sydney boundary divided into four quadrants, in 3D and key plans (showing focus quadrants in red). Source: Stephanie Stankiewicz, Josh Gowers and Joshua Zeunert

The research focus was further influenced by prioritising:

- Professional activities, namely, where commercial thresholds were met to qualify as a primary producer, activities in areas zoned as rural production or similar, or where evidence indicated food grown on a property was sold or distributed[22];
- Spatially efficient modes of production, for example, market gardening rather than grazing on pastures[23];
- Production of fresh and perishable foods for Sydney-based consumption[24];
- Agricultural activities for human consumption wherever possible such as table grapes[25] and grain.[26]

Part of the book's rationale is to emphasise relevant future practices for bolstering Sydney's future food security as a case study of similarly industrialised urban regions. Discussion asks how agriculture can coexist in metropolitan areas given the need to periodically relocate, establish quickly and deliver fast economic returns and ongoing viability, while navigating urban encroachment and increasing land prices.

What Constitutes 'Sydney'?

Sydney's topographical basin is key to its agricultural history. The city's geography and topography is bound by sea, river valleys and ranges (Figs. 2.1 and 2.10).

> The Sydney metropolitan area spreads across a natural, oval-shaped basin roughly four thousand square kilometers in area, some sixty-five kilometers from east to west, and eighty-five kilometers north to south, bounded by the Pacific Ocean in the east, the Hawkesbury and Nepean Rivers in the north and west, and the Woronora Plateau in the south.[27]

Unlike cities situated on plains or deltas, Sydney is constrained within a finite terrestrial basin. Given these natural borders, we might assume that there is relative agreement about what constitutes the city's limits. Instead, numerical estimates vary wildly across policies and publications.

Figure 2.8 overlays six existing boundaries and terms commonly used to demarcate Sydney's overall area, as elucidated in Box 2.4.[28]

Box 2.4 Existing Boundaries of Note

- **Sydney Basin Bioregion:** This was also eliminated due to its focus on larger ecological assemblages which do not correlate with this book's investigation of metropolitan agriculture nor the social primacy of food production.
- **Sydney Geological Basin:** The expansive 'Sydney basin' relates to Sydney's *geological* region,[29] but is also contradictorily used as a loose term to refer to varying and inconsistent conceptions of the city's area.[30] The geological boundary was ruled out due to extending well beyond immediate Sydney[31] and its topographic basin.
- **Sydney Food Futures:** This boundary relates to a 2016 study examined in *Period Five* and relating to the 'SPUN' boundary area (following).
- **Sydney Peri-Urban Network:** The 'Sydney Peri-Urban Network' ('SPUN')[32] stretches approximately 300 kilometres north to south to include 12 regional Councils beyond metropolitan Sydney. The SPUN indicates pressures exerted on Sydney's peri-urban hinterlands and also suggests the city's infrastructural supply tentacles' stretch further beyond city limits for water, energy and other supply lines. The SPUN boundary exceeds metropolitan Sydney's immediate agricultural landscapes[33] and exaggerates the spatial conception of 'peri-urban', problematically suggesting that this extensive area is land waiting for development.
- **Greater Sydney Region:** In terms of human-defined jurisdictions and boundaries, the Greater Sydney Commission's (GSC) (2015–2022) 'Greater Sydney Region'[34] is spatially dominated by the expansive Local Government Areas (LGAs) of Hawkesbury, Blue Mountains and Wollondilly.
- **County of Cumberland**: Of all the eight boundaries shown in Fig. 2.8, this is the closest to this book's boundary. The County's historic metropolitan area provides a neat and clear boundary. It excludes key parts of Sydney and agricultural lands to the west of the Hawkesbury Nepean to the slopes of the Blue Mountains, such as Camden, Penrith and Richmond.[35] It can also be confused with the County Cumberland Council,[36] adding to a legacy of confusing historic sub-nomenclature of hundreds, districts and parishes, which do not correlate with current Local Government Area (LGA) boundaries.
- **Metropolitan Rural Area:** The GSC's 'Metropolitan Rural Area' (MRA)[37] is useful for an agricultural focus and thus shown in Fig. 2.8. It is not, however, strictly a boundary but rather a patchwork of areas within the GSC's regional boundary. Much of the MRA also extends too far west into the Blue Mountains to be considered part of immediate Sydney.
- **Urban Area:** The GSC's urban area boundary relates to the existing built footprint prior to the organisation publishing its 2018 metropolitan strategy for Sydney.

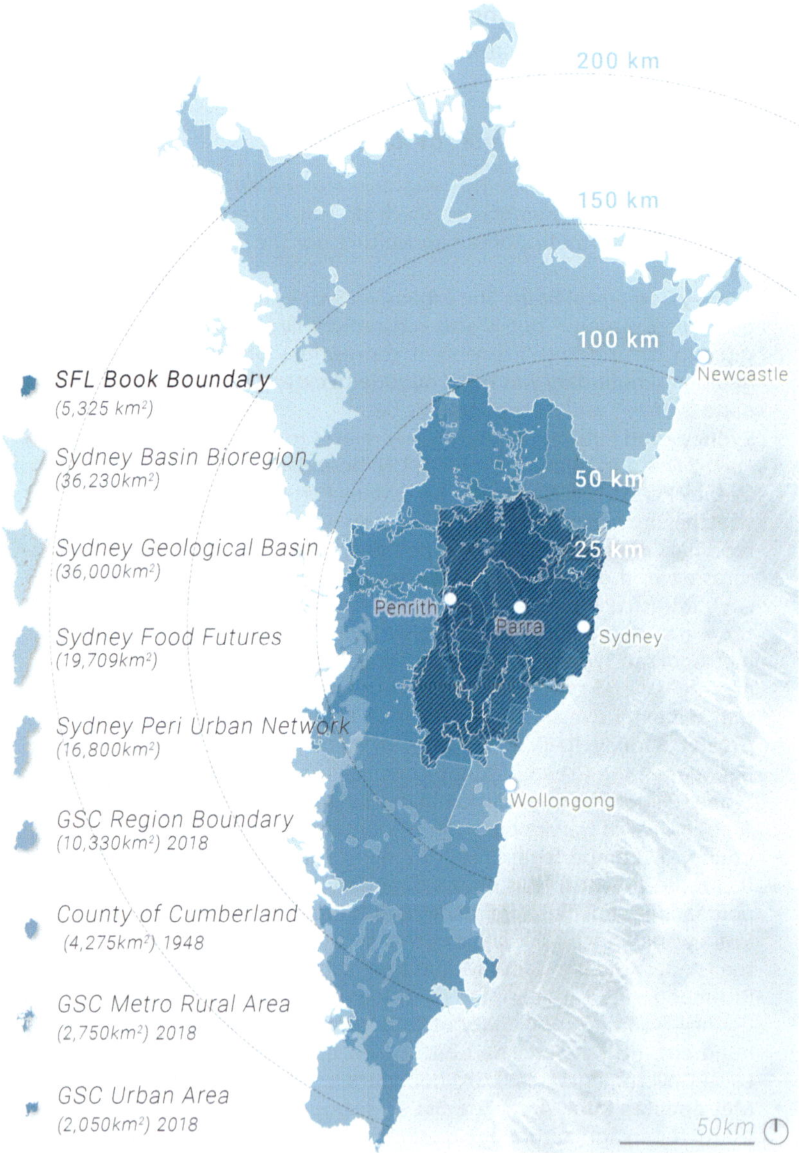

Fig. 2.8 The book's boundary as compared with eight others encountered in the research and literature. Vastly different spatial areas are commonly termed 'Sydney' without explicit reference. Source: Joshua Zeunert, Stephanie Stankiewicz and Josh Gowers

Determining a boundary for Sydney's metropolitan agriculture given the diverse spatial variations (Box 2.4 and Fig. 2.8) is compounded by the 233-year time frame in which commercial metropolitan agriculture has taken place. Hunwick's aforementioned quote offers valuable clues about the area's immovable geological and geographical edges. To achieve greater spatial precision, a meticulous process of sieve mapping and thematic comparison analysed the broader contexts and boundary areas.[38] This considered a wide array of factors, including First Nations' Countries, topography, geology, ecological bioregions, vegetation, river catchments, rainfall patterns, land and soil capabilities, land use, population density, population fluctuations, temporal footprint, spatial annual mean temperature and spatial annual mean rainfall (Fig. 2.9).

After a protracted process of sieve mapping, comparison and analysis,[39] we settled on Sydney's *topographic basin*, namely, the "broad lowland [basin] east of the Blue Mountains".[40] Water catchment boundaries, rivers and the Pacific Ocean were key to determining the city's edges. Figures 2.1, 2.7, 2.8, 2.9, and 2.10 show the boundary and area included in the book. Its total area is 5325 km^2. We further subdivided the boundary area into four quadrants to help identify individual food landscapes through a key map system with each highlighted in red, as shown in Fig. 2.7 (bottom).[41] These largely correlate with Sydney's four major river systems: north-east (Parramatta), north-west (Hawkesbury), south-east (Georges) and south-west (Nepean) (Fig. 2.11, top).

Context Analysis

Topography

Sydney's encircling geographic rim and large population results in its "foodbowl"[42] extending further afield. This starts at (just under) 100 km away by road with the modestly sized Central Coast Plateau and further distance again to the west, south or north. Importing food from outside this agricultural basin involves overcoming significant distances across challenging topography and geology (Figs. 2.1 and 2.10). Whether by road, rail, ship, or plane (or yet to be proven alternatives),[43] these

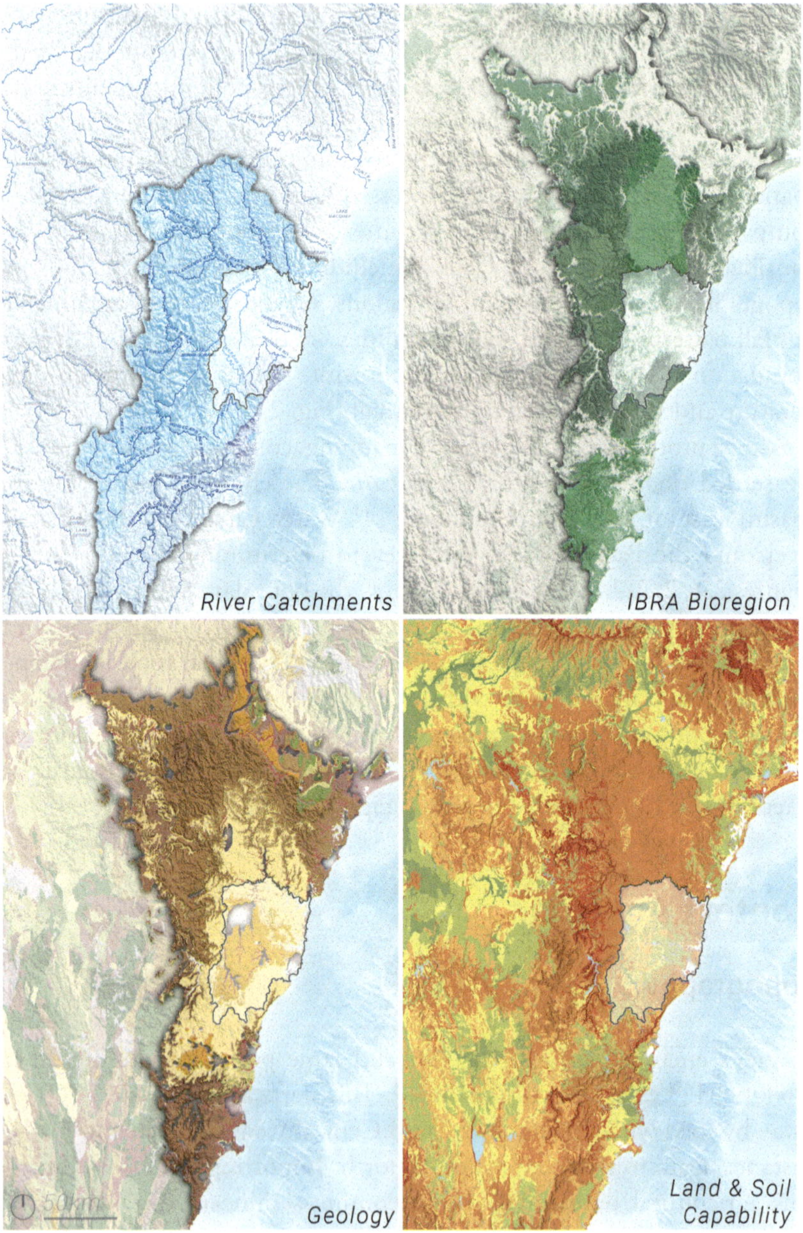

Fig. 2.9 Analysis maps of Sydney's wider area: river catchments, bioregion, geology, land and soil capability relating to the book's boundary area. Source: Josh Gowers, Joshua Zeunert and Stephanie Stankiewicz

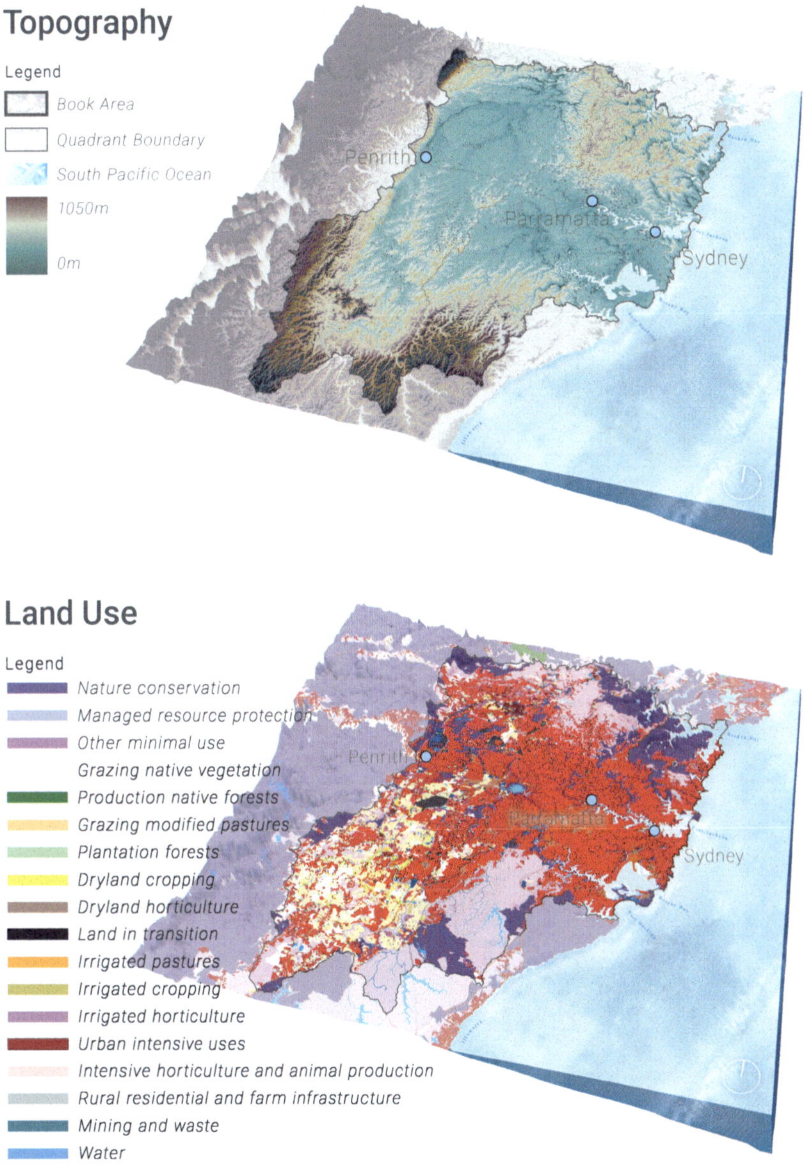

Fig. 2.10 3D analysis maps of boundary area: topography (top), and Sydney's land use in 2020 (bottom) (CLUM, 2020). Source: Josh Gowers, Joshua Zeunert and Stephanie Stankiewicz

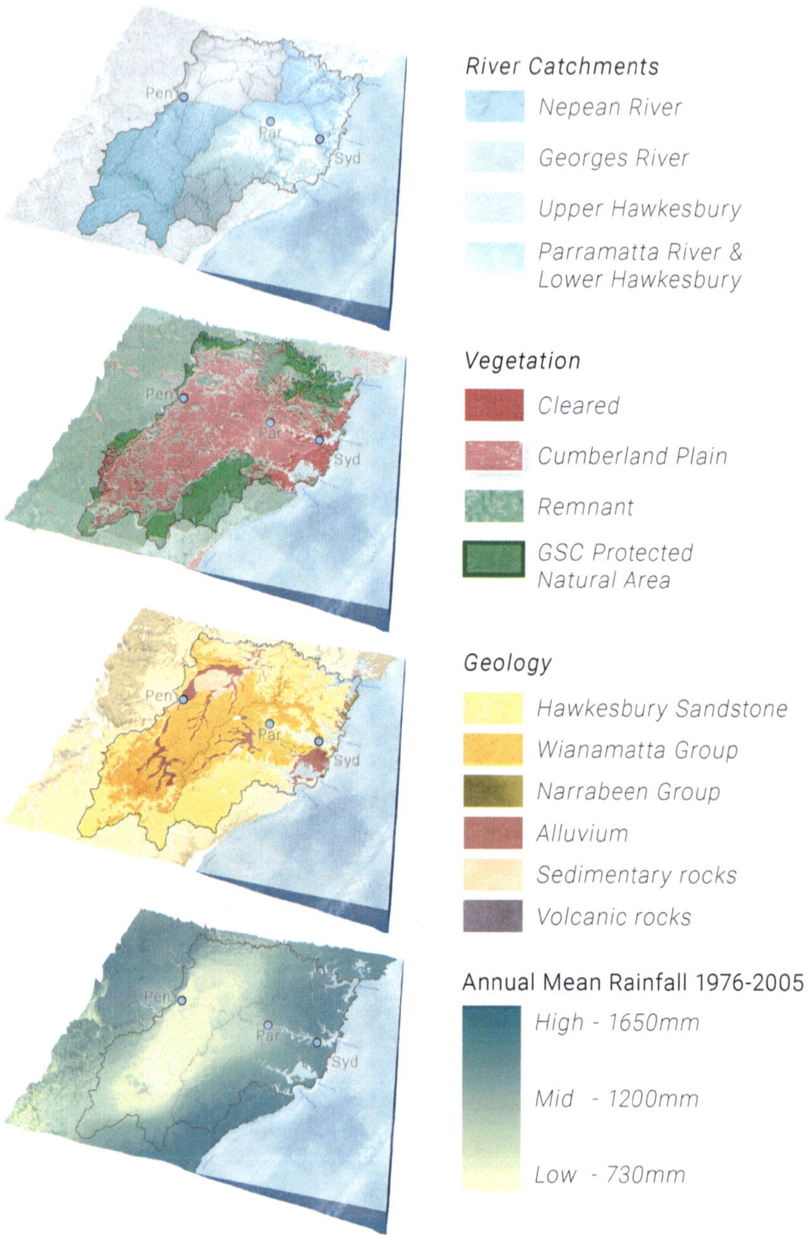

Fig. 2.11 3D analysis maps of boundary area: river catchments; vegetation; geology and rainfall (The Bureau of Meteorology (BOM) do not provide metropolitan maps or spatial mapping datasets showing nuanced rainfall data across the city, and thus 1976–2005 rainfall database was used. Source: adapted from State Government of NSW and NSW Department of Climate Change, Energy, the Environment and Water 2020). Source: Josh Gowers, Joshua Zeunert and Stephanie Stankiewicz

transportation systems necessarily defy the basin's terrestrial and topographic limitations,[44] which takes and embodies considerable fossil energy.[45]

Climate and Microclimate

The New South Wales government classifies Sydney's climate as warm temperate, sub-humid without a dry season.[46] It is crucial to note that metropolitan Sydney includes a variety of microclimates that range notably in temperature and rainfall (see 'rainfall' below and Fig. 2.12). It is challenging to gain a clear spatial understanding of this metropolitan variation. The Bureau of Meteorology, while an invaluable source for current and historic data, no longer provides annual rainfall or temperature maps to assist with determining climatic nuances across the Sydney region.[47]

Climate and microclimate are major determinants of edible species' success: The influential Diggers Club, for instance, states that "ninety percent of gardening problems" are caused by "not understanding the climate needs".[48] The club provides three climatic and growing maps, revealing that Sydney is in a 'warm'/'warm coastal'/'subtropical' zone experiencing 30 to 60 days above 30 degrees and 240 'growing days' of over 15 degrees annually.[49] Though useful in situating Sydney nationally, the maps are not detailed enough to canvass Sydney's microclimatic nuances that vary considerably across the city's sprawling footprint. Temperatures along the city's harbour and coastal fringes differ from those inland, which typically experience greater variations between minimum and maximum temperatures.

Figure 2.12 shows how average temperatures spatially vary across the metropolitan area.[50] Despite lacking data after 2005, it reveals that higher temperatures correlate with the city's urban footprint (Fig. 2.10, bottom red area). This phenomenon is known as the 'urban heat island effect'[51] and largely results from hard materials (roads, buildings) absorbing and reflecting heat, where previously soft landscapes supported cooler overall temperatures. The urban heat island effect means that wider climatic systems are temporally affected by urban microclimates and particularly the presence and agglomeration of built form and hard surfaces.[52]

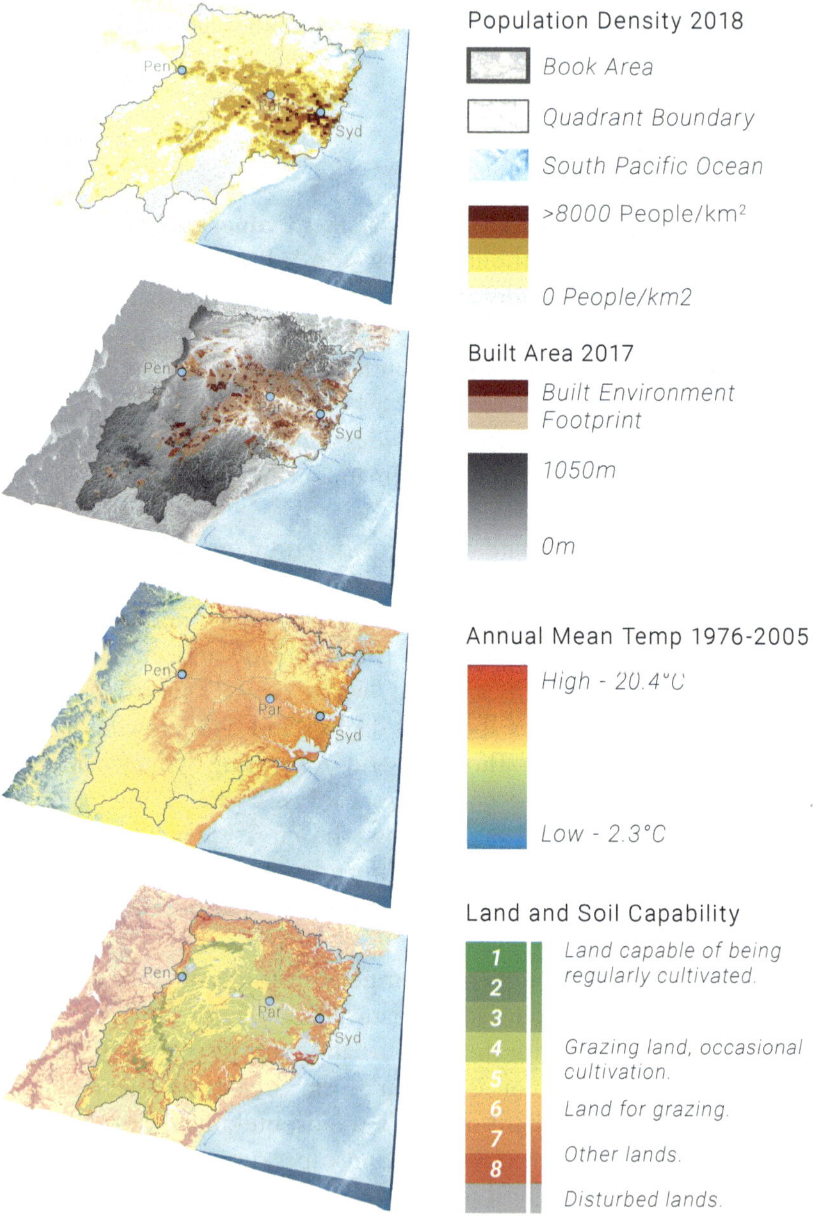

Fig. 2.12 3D analysis maps of boundary area: population density (2018); built area 2017; average temperature; and land and soil capability. Source: Josh Gowers, Joshua Zeunert and Stephanie Stankiewicz

Sydney's microclimates are therefore constantly changing, particularly in developing areas. Historic data offers a guide rather than an accurate indication of the future. Higher built densities increase the urban heat island effect and thus change the microclimates and conditions suited to particular edible species. The extent to which the urban heat island effect has altered Sydney's climate and microclimates over time is evident, for example, in a quote from an 1883 article regarding Gelding's 'Victoria' Nursery at (now) Dulwich Hill (Fig. 4.1),

> The nursery is well exposed on all sides, and the winter frosts are somewhat severe. These conditions would seem to secure perfect hardiness in the stock raised, for if plants are "coddled" by close protection in the nursery they speedily succumb to even ordinary conditions when sent out to customers.[53]

Dulwich Hill is now considered a relatively 'inner' suburb and its climate is certainly much milder today than in 1883, largely due to the urban heat island effect. More widely across urban Sydney, tall individual buildings create high wind velocities that can be destructive to sometimes fragile edible species, while overshadowing hinders growth of edible plants, particularly those developing tubers and fruits. Orientation and solar access are important for edible species, where the standard rule of thumb is 6 hours of direct sunlight per day. Global climate change creates localised impacts that make it challenging to precisely forecast future conditions. The consensus is that existing and ongoing greenhouse emissions will create increasing climatic variability and volatility, along with increased temperatures and reduced overall chill factor.[54] Thus, Sydney's microclimates shape which edible plants and varieties can be successfully grown across the city. Climatic averages and generic species growing guides for Sydney can be misleading or too generic to meaningfully inform optimum specific species selection for a given location in the metropolitan area. Anecdotal sources, such as blogs can be misleading unless location and microclimatic particulars are provided.

Chill Factor

Some fruit and nut species require cooler temperatures to set fruit. There are different measures to quantify chill including chill portions, chill units and most commonly, chilling hours. Chill hours are the number of annual hours needed by a plant under seven degrees Celsius. They relate to plant dormancy and flowering requirements for some species (such as some nuts, berries, pome and stone fruits—citrus do not require chill). Plant breeding has adapted species for varying climates and 'low chill' conditions; apples, for example, can vary between 300 and 1800 chill hours required depending on specific varieties.[55] Outside of professional growers and especially in metropolitan areas, chill hours are not well understood,

> [K]nowing how much winter chill different crops and cultivars need for healthy flowering and strong yields is essential for growers when deciding what to plant in their region. Failure to suitably match the chilling requirement of a crop or cultivar with the local winter chill will result in reduced productivity.[56]

No nuanced spatial data or mapping were identified during the research process to show the considerable variation in chill hours between Sydney's cooler western inland regions—where frost can be experienced—versus the milder overnight temperatures near the coast. Accordingly, PlantNet's classification of Sydney as 'medium' for chill hours at 620 is a rough indication but would generally be lower along the coast and in more built-up areas experiencing the urban heat island effect.[57] See here[58] for additional information and here[59] for a calculator tool.

Given the ongoing advance of the built environment and hard landscapes in Sydney, combined with warming global temperatures, as a general rule, chill hours are reducing across Sydney. Over time this continues to lend growing conditions as better suiting species with low or no chill needs.

Rainfall

Like its microclimates, Sydney's rainfall significantly varies from the flatter regions in the dryer west, such as Camden at 782 mm annually, to the wetter coastal regions. Sydney's CBD averages 1210 mm. More rainfall is received in the north and south in elevated areas such as Royal National Park. This has been mapped in Fig. 2.11. This figure includes rainfall contours adapted from Benson and Howell's 1990 book which show that annual rainfall varied by as much as 900 mm across metropolitan Sydney in the space of around 30 km from Camden at c. 700 mm to c. 1600 mm 30 km east to the coastline.[60] This amount of rainfall variation is more than the total rainfall of each Australian capital city except Darwin and Brisbane (the latter is just over 1000 mm). Even Camden's annual rainfall, although appearing low for Sydney, is higher than all of southern Australia's capitals: Perth, 730 mm; Melbourne, 663 mm; Canberra, 620 mm; Hobart, 614 mm and Adelaide, 530 mm.[61] The annual rainfall distribution is also relevant to successfully growing edible species—Sydney is blessed with a relatively consistent spread of rainfall across the year having no dry season[62] in contrast to most other Australian capital cities that typically suffer extended dry periods and can require more irrigation.

Box 2.5 Fruit Fly

Like many agricultural regions in Australia, Sydney suffers from the fruit fly pest, first becoming susceptible three decades after founding (see *Period One: 1788–1850*). Propensity for fruit fly influences species selection. Green Harvest suggest two solutions, one is to grow fruit fly resistant fruits including: avocado (avoiding thin-skinned *Fuerte* and *Rincon* varieties); bananas; citrus (avoid thin-skinned varieties such as *Meyer* lemon); custard apple (winter ripening varieties are best); grumichama; ice cream bean; jaboticaba; kiwifruit; longan; lychee; macadamia; passionfruit; pawpaw; persimmon (avoid early varieties like *Fuyu*); pineapples and pomegranate. 'Sometimes affected' fruit-fly species are noted as including blueberries, grapes, mulberries and nashi pears. Green Harvest further suggest keeping trees compact so that fruit or the entire tree can be covered/netted to eliminate fruit fly and also bird problems.[63]

Land and Soil Capacity for Agriculture

As mentioned in *Introduction*, arguments marginalising agriculture in Sydney are regularly based on the claim that inherent land capacity is poor, and/or that Sydney lacks highly ranked productive land classifications.[64] The Sydney region provides a range of soil types that vary in their suitability for different agricultural activities and approaches. Some of Sydney's lands are effectively unsuited to agriculture, primarily its iconic sandstone regions (see Geology, Figs. 2.9 and 2.11). Figures 2.9 and 2.12 show Sydney's land and soil capability' (LSC).[65] As a general rule, the soils in built-up areas of metropolitan Sydney have been modified, degraded and polluted from urban development.

Box 2.6 Nomenclature

This book employs and combines several terms (such as agriculture, landscape and food) which can be inconsistently defined and/or contradictory in and of themselves, yet alone in unison. Their use and intent are clarified as follows, along with other key terms and concepts:

Agriculture: In *The Oxford Companion to Australian Gardens*, Aitken and Looker define agriculture as the "science and practice of cultivating the soil and rearing animals...primarily based on production and market-driven outcomes".[66] This book applies this definition, focusing on commercial production within Sydney's metropolitan area defined by the book's boundary. Emphasis is on the generation of food surpluses, rather than hobbyist[67] and self-sufficiency scales, as the latter are well trodden by numerous other works. Nonetheless, the "close relationship between agriculture, horticulture, and gardening"[68] is acknowledged. Research excludes non-edible, land-based activities and primary production such as equine and agroforestry.

Landscape: We use 'landscape' to characterise the reciprocal relationships between human cultures and environments shaped by our practices and activities. In other words, landscapes may be considered the combined works of 'nature' and humans over time. Waterman and Zeunert explain that, "landscape describes places as relationships and processes", creating people's identities and guiding their actions and their preferences, "while at the same time are shaped by the actions and forces of people".[69] Accordingly, 'landscape' moves well beyond scenic and decorative notions.

Food Landscape: As suggested above, "food and landscape are mutually constituted",[70] whereby "food, as currency, medium, and sustenance, is a fundamental part of... landscape relationships".[71] The combination of *food* and *landscape* implies the reciprocity between consumption and production—food preferences shape production landscapes and place-based

(*continued*)

Box 2.6 (continued)

production constraints determine what can be consumed freshly and locally. Accordingly, food sourced from outside the defined boundary area of the metropolitan Sydney region is considered imported from outside the local food landscape.

Landscape Context: The book's specific landscape context is shown in Fig. 2.10 (top) and encompasses Sydney's built-up areas (Fig. 2.12) as well as its topographic landscape basin. Focus is placed on terrestrial landscapes rather than aquatic, although some of these are included where relevant, chiefly in inland waters.

British: Refers to the combined geographic British Isles including the Republic of Ireland rather than Great Britain and the United Kingdom, reflecting the early colonists' dominant ethnic and food heritage. This is intentionally differentiated from the commonly deployed, yet overly homogenous, 'European'.

Urban: Refers to Sydney's 'built-up' areas forming its physical city footprint.

Peri-urban: Land on the urban fringe occupying the zone between rural and urban, which often transitions to built-up urban conditions during urban growth.

Metropolitan: Countries, states and municipalities use differing measures. Here, it includes the zones beyond the immediate urban areas, namely peri-urban and metropolitan-fringe regions affected by their proximity to the metropolis.[72] This correlates with the book's boundary area (Figs. 2.1 and 2.10).

Rural: Areas outside of the urban, metropolitan and peri-urban regions, which are largely excluded from this work. Given the 233-year timeframe, it is important to note that much of Sydney's now urban and peri-urban area was once rural but are now captured within the book's metropolitan boundary.

Cumberland Plain: The largest single spatial area of Sydney if using Sydney Basin Bioregion and Subregion classifications (Figs. 2.9 and 2.11) and Sydney's primary agricultural production region under British methods.

Country: 1. The term and concept used by First Nations Aboriginal Australians encompassing their place in the world and the lands, waterways and seas to which they are connected.[73] 2. An English often idealised notion of rural land outside of urbanised areas implying a dualistic notion of 'town' and 'country'. 3. An internationally recognised nation state.

Horticulture: Encompassing intensive plant production of food and non-food-producing plants, such as turf growing, cut flowers and nurseries cultivating non-agricultural/edible species (usually ornamental or native plants).

Scale: The physical dimensions of an agricultural landholding, crop or commodity, individually or collectively.

Foodways: A broad concept encompassing many dimensions of food's production and consumption. Waterman offers an overview not limited to "the sets of collective and individual practices associated with cuisine" and "global and transnational food and landscape practices and cultures" as well as "local, regional, and heritage practices".[74]

Land Use

Figure 2.10 (bottom) shows 2020 land use government classifications.[75] The dominance of the red 'urban' area shows land zoned as urban or residential and gives a sense of the creeping built environment filling the finite topographic basin with hard landscapes and marginalising other land uses.

Five chapters present five different time periods structured as follows:

Key Points

- Relevant wider global and national events;
- Relevant Sydney events;
- 11 typologies (Agricultural Facilities & Education; Mixed Farming; Market Gardening; Fruit; Cropping; Intensive Livestock; Grazing; Vineyard Wineries; Agricultural Infrastructure; Aquaculture; Miscellaneous);
- Time period food landscape maps.

We begin with *Period One* (1788–1850), opening with a painting of a denuded ecology cleared to allow agricultural incision and grazing livestock at one of Sydney's Government Farms (Fig. 3.1).

Notes

1. Mushrooms are usually produced in sheds and thus could be classified into their own category, however, we sought to reduce the overall number of 'typologies'. Although tomatoes, for example, are technically a fruit, they are classified under market gardening (dominantly vegetables).
2. Note that it is at times difficult to discern between field cropping and vegetable production in aerial imagery. Also, irrigated cropping often takes place for other uses such as feeding dairy cattle.
3. This can at times be difficult to differentiate from irrigated field crops.
4. Each time period map was further broken into four quadrant maps (Fig. 2.7) offering more spatial detail: north-east; north-west; south-east; and south-west. Unfortunately due to space restrictions, these maps are not included.

5. Whenever this information could be ascertained—if an exact location is not known this has been noted in an entry's description.
6. Early mixed farms, for example, regularly had a diverse range of typologies present, but as a single property are entered under their (estimated) primary typology.
7. Due to limitations in the Google My Maps software used. This was an unavoidable shortcoming at the time the database was compiled.
8. In aerial photography, Google Street View and or/on its website if available.
9. Google Street View was used if available but this imagery was often several years old at the time.
10. Codrington, cited in Wilkinson, 1999: 3.
11. Malcolm & Fahd, 2009: 3.
12. Photography became available in Sydney from mid-late P2 onwards.
13. https://portal.spatial.nsw.gov.au/portal/apps/webappviewer/index.html?id=f7c215b873864d44bccddda8075238cb.
14. Horse meat (chevaline) is not consumed in Australia (unlike much of Europe) and thus Australia is one of the largest exporters of horse meat. Consequently, horse agistment was excluded where discernible due to likelihood of its recreational use rather than a specific food-production landscape.
15. As discussed, grazing was not a focus for recording in the Map Database given both its vast spatial extents outside or urban areas and its limited scope as an economically viable metropolitan land use looking ahead.
16. https://maps.six.nsw.gov.au. This became available on the NSW Historical Imagery Viewer in 2020.
17. Sourced from prints taken for the military, the utilities and government departments were scanned at high resolution by Andy Brill and made available online at Flickr.com: https://www.flickr.com/photos/angeljim46/albums/72157632418208341/with/10183323943
18. https://portal.spatial.nsw.gov.au/portal/apps/webappviewer/index.html?id=f7c215b873864d44bccddda8075238cb. Please note that partway through the research process, NSW Historical Imagery became available and was used to confirm database entries. Halfway through using this imagery, however, the database changed from listing image layers as decades (1940s/1950s/1960s/ 1970s/1980s/1990s/2000s) to the exact years noted (1955/1965/1975/1978/1986/1991/2004/2005, while also removing the 1940s imagery). The years available changed over the course of verifying entries. Also, the website address link to access NSW

Historical Image Portal changed over the course of this research and may again change from the time of publishing. Focus of aerial coverage is usually on already built-up areas at the time of capture rather than peri-urban regions. Some periods had very limited extents of coverage at the time of research (e.g. 1950s, 1970s, 1980s) (Fig. 2.6). The tool is very useful to observe the transition of specific sites over time when multiple coverage years are available.

19. https://www.google.com/maps. This displays the most recent capture imagery of a given area.
20. https://apps.nearmap.com/maps/.
21. Some of these are well covered in existing works, for example, the Hawkesbury—see Proudfoot, 1987/2017; Boon, 2017.
22. This is due to seeking to feed the maximum proportion of the population as locally as possible. Some hobby farming is inevitably included through aerial photographic surveys. Inclusion of a domestic component was incidental and not prioritised—see Gaynor, 2006 (concentrating on Perth but also accommodating some Sydney activities).
23. Such activities primarily take place well beyond metropolitan regions due to lower land values.
24. One of the very reasons for local food production is to reduce time from harvest to sale to maximise nutritional content and freshness.
25. Noting that the two are very difficult to differentiate in aerial imagery alone. Wine/alcohol does not adequately supplement nutritional needs, despite perpetually heroic attempts by many of Sydney's residents to prove otherwise.
26. This was often difficult to ascertain but was applied as a general principle.
27. Hunwick, 2018: 1.
28. This is suggestive of spatial-geographic illiteracy, as elucidated in *Period Five* and *2021+*.
29. See for example Martyn, 2018: 27; Danis et al., 2011; Geoscience Australia quantify it as 64,000 square kilometres of onshore and offshore land, see: https://www.ga.gov.au/scientific-topics/energy/province-sedimentary-basin-geology/petroleum/offshore-eastern-australia/sydney.
30. An example further discussed in *2021+* (SFF, n.d.), uses the term 'Sydney Basin' to connote the 'Sydney Peri-Urban Network' boundary area as outlined.
31. Including, for example, Newcastle and the Hunter valley. See Danis et al., 2011: 519.

32. See for example, https://www.wollondilly.nsw.gov.au/shire-projects/growth-in-wollondilly/sydney-peri-uban-network-spun/.
33. See *2021+* chapter for a further discussion on this, including a key reference (SFF, n.d.) that uses this SPUN area.
34. GSC, 2018: 25, 157.
35. Winston commented that the areas west of the Hawkesbury-Nepean rivers but within the lowlands should have been included in the remit of the metropolitan town planning body that was the County Cumberland Council operating from 1945 to 1964. See Winston, 1957: 4.
36. See *Period Four.*
37. GSC, 2018: 144, 157.
38. See McHarg, 1969.
39. See McHarg, 1969.
40. See Martyn, 2018: 26, and Nolan & Benson in Karskens, 2010: 20.
41. Quadrants were also intended to provide more zoomed views of the food landscape maps but these were deleted due to figure limitations.
42. See Carey & James, 2018.
43. See Kunstler, 2006.
44. The gentlest topographic route to depart Sydney by road is the frequently congested M31 to the south-west. As explored in *Period Five*, The Greater Sydney Commission (responsible for producing Sydney's 2018 metropolitan planning strategy) explicitly acknowledge this: "The geography of the Sydney Basin including its waterways, ridges and the ring of national parks that frame the Basin, limit the spatial extent of Greater Sydney. The ridges and ravines in particular, impede the development of roads and rail and in some areas significantly constrain accessibility" (GSC, 2018: 8).
45. Zeunert, 2018.
46. Environment NSW, n.d. Also see: https://greenharvest.com.au/SeedOrganic/ClimateZones.html.
47. Contact was made three times with the BOM seeking climatic mapping or summaries for Sydney, who confirmed that these no longer exist.
48. https://www.diggers.com.au/pages/my-garden-path.
49. Along with other areas as diverse as Brisbane to Melbourne, Adelaide, Longreach and much of central Australia also in this classification. See Blazey & Varkulevicius, 2006: 20, and https://www.diggers.com.au/pages/my-garden-path.

50. Albeit this figure does not adequately convey variations between high and low temperatures found inland from the coast.
51. See for example Mohajerani et al., 2017.
52. For Sydney specific material see Santamouris et al., 2017 and Irger, 2014.
53. https://trove.nla.gov.au/newspaper/article/64034792.
54. See for example the 'Details' tab 'How will climate change impact winter chill?': https://hort-science.shinyapps.io/ChillCalculator/.
55. Blazey & Varkulevicius, 2006: 94.
56. https://hort-science.shinyapps.io/ChillCalculator/.
57. https://www.plantnet.com.au/plant-care/chill-hours-guide/.
58. https://greenharvest.com.au/GreenGardenNotes/FruitTreesForSmallGardens.html#ChillFact.
59. http://grf-smartfarm.daf.qld.gov.au:3838/apps/chillcalculator/.
60. Noting that this base data is over 30 years old and which the Bureau of Meteorology advised no recent equivalent was available. See Benson & Howell, 1990: 10.
61. https://www.abc.net.au/news/2018-04-24/perth-rainfall-higher-than-melbourne-hobart-and-london/9688142.
62. Sydney Basin—climate, NSW Office of Environment and Heritage, Accessed 16 May 2019. https://www.environment.nsw.gov.au/bioregions/sydneybasin-climate.htm.
63. https://greenharvest.com.au/PestControlOrganic/Information/FruitFlyControl.html.
64. See for example Clarke, 2017. Pers. Comm. Rod Simpson, 2018. See *2021+* and Zeunert et al., 2025.
65. Available: https://geo.seed.nsw.gov.au/Public_Viewer/index.html?viewer=Public_Viewer&locale=en-AU&runWorkflow=AppendLayerCatalog&CatalogLayer=SEED_Catalog.111 and here: (click 'show on seed map') https://datasets.seed.nsw.gov.au/dataset/land-and-soil-capability-mapping-for-nsw4bc12.
66. Aitken and Looker, 2002: 12–13.
67. Given the expansive timescale of this book and the constantly shifting dynamics of the city and its peri-urban fringes, any inclusion of hobby farming in the Map Database is incidental rather than targeted.
68. Ibid.
69. Waterman & Zeunert, 2018: i.
70. Potteiger, 2013: 261.
71. Waterman & Zeunert, 2018: i.

72. Agricultural activity is significant when conceived and measured within the metropolitan context. In 1989, over a third of the USA's agricultural production by value occurred within Metropolitan Statistical Areas (Gorgolewski et al., 2011: 232, citing Smit et al., 2001, citing Heimlich, 1989). Over two decades later, US metropolitan counties still accounted for 38.7 per cent of total agricultural sales' value (USDA, 2012; T. Parker, USDA, pers. comm. 20 April 2017). This value is primarily generated in peri-urban regions from, for example, irrigated vegetable production, dairying, and pig and poultry raising.
73. See Jones et al., 2018: 14.
74. Waterman, 2018: 520–521.
75. CLUM, 2020.

References

Aitken, R., & Looker, M. (2002). *The Oxford companion to Australian gardens*. Oxford University Press.

Benson, D., & Howell, J. (1990). *Taken for granted: The bushland of Sydney and its suburbs*. Kangaroo Press.

Blazey, C., & Varkulevicius, J. (2006). *The Australian fruit and vegetable garden: Grow the best fruit and vegetables for good health and flavour*. Digger's Club.

Boon, P. (2017). *The Hawkesbury River: A social and natural history*. CSIRO Publishing.

Carey, R., & James, S. (2018). Peri-urban agriculture in Australia: Pressure on the urban fringe. In J. Zeunert & T. Waterman (Eds.), *Routledge handbook of landscape and food* (pp. 213–227). Routledge.

Clarke, M. (2017). *Values of the metropolitan rural area of the greater Sydney region*, AgEconPlus Consulting. Accessed 10 December 2018. https://gsc-public-1.s3.amazonaws.com/s3fs-public/Values_of_the_Metropolitan_Rural_Area_of_the_Greater_Sydney_Region_(Ag_Econ_Plus).pdf

CLUM. (2020). State Government of NSW and Department of Planning and Environment.

Danis, C., O'Neill, C., Lackie, M., Twigg, L., & Danis, A. (2011). Deep 3D structure of the Sydney Basin using gravity modelling. *Australian Journal of Earth Sciences, 58*(5), 517–542. https://doi.org/10.1080/08120099.2011.565802

Environment NSW. (n.d.) *The Sydney Basin Bioregion*. https://www.environment.nsw.gov.au/-/media/OEH/Corporate-Site/Documents/Animals-and-plants/Bioregions/bioregions-sydney-basin.pdf

Gaynor, A. (2006). *Harvest of the suburbs: An environmental history of growing food in Australian cities*. UWA Press.

Gorgolewski, M., Komisar, J., & Nasr, J. (2011). *Carrot city*. Monacelli.

GSC. (2018). Greater Sydney region plan: A metropolis of three cities, Greater Sydney Commission: NSW Government. https://www.greater.sydney/publications

Hunwick, H. (2018). *The food and drink of Sydney: A history*. Rowman & Littlefield.

Irger, M. (2014). The effect of urban form on urban microclimate, PhD thesis, The University of New South Wales.

Jones, D. S., Choy, D. L., Tucker, R., Heyes, S., Revell, G., & Bird, S. (2018). *Indigenous knowledge in the built environment: A guide for tertiary educators*. Department of Education and Training. https://researchsystem.canberra.edu.au/ws/portalfiles/portal/24156189/ID12_2418_Deakin_Jones_2018_Guide.pdf

Karskens, G. (2010). *The Colony: A history of early Sydney*. Allen & Unwin.

Kunstler, J. (2006). *The long emergency: Surviving the converging catastrophes of the 21st century*. Atlantic Books.

Malcolm, P., & Fahd, R. (2009). *Ground truthing of Sydney vegetable industry in 2008*. Horticulture Australia Limited, NSW Department of Primary Industries. https://www.futuredirections.org.au/wp-content/uploads/2015/05/VG07073_Groundtruthingsurveyofvegindustrystats.pdf

Martyn, J. (2018). Rocks and trees: A photographic journey through the rich and varied geology, scenery and flora of the Sydney region, STEP: Turramurra, NSW.

McHarg, I. (1969). *Design with nature*. Natural History Press.

Mohajerani, A., Bakaric, J., & Jeffrey-Bailey, T. (2017). The urban heat Island effect, its causes, and mitigation, with reference to the thermal properties of asphalt concrete. *Journal of Environmental Management, 197*, 522–538.

Pers. Comm's. Rod Simpson, 2018.

Potteiger, M. (2013). Eating places: Food systems, narratives, networks, and spaces. In W. I. Madison (Ed.), *Landscape Journal 32–2* (pp. 261–275). University of Wisconsin.

Proudfoot, H. (1987/2017) *The Hawkesbury: A thematic history, Hawkesbury City Council* (2017 edition), Accessed 21 Aug 2020. https://www.hawkesbury.nsw.gov.au/__data/assets/pdf_file/0004/119911/The-Hawkesbury-A-Thematic-History-2017-February.pdf

Santamouris, M., Haddad, S., Fiorito, F., Osmond, P., Ding, L., Prasad, D., Zhai, X., & Wang, R. (2017). Urban Heat Island and overheating characteristics in Sydney, Australia: An analysis of multiyear measurements. *Sustainability, 9*(5), 712. https://doi.org/10.3390/su9050712

USDA (2012) Census of Agriculture, United States Department of Agriculture. Accessed 15 May 2015. https://www.agcensus.usda.gov/Publications/2012/

Waterman, T. (2018). Taste, foodways, and everyday life. In J. Zeunert & T. Waterman (Eds.), *Routledge handbook of landscape and food* (pp. 517–530). Routledge.

Waterman, T., & Zeunert, J. (2018). Introduction. In J. Zeunert & T. Waterman (Eds.), *Routledge handbook of landscape and food* (p. 5). Routledge.

Wilkinson, J. (1999) *Dairy industry in NSW: Past and present*, Briefing Paper No 23/99, NSW Parliamentary Library Research Service. Accessed 15 July 2020. https://www.parliament.nsw.gov.au/researchpapers/Documents/dairy-industry-in-nsw-past-and-present/23-99.pdf

Winston, D. (1957). *Sydney's great experiment: The progress of the Cumberland County Plan.* Halstead Press.

Zeunert, J. (2018). Challenges in agricultural sustainability and resilience: Towards regenerative practice. In J. Zeunert & T. Waterman (Eds.), *Routledge handbook of landscape and food* (pp. 231–252). Routledge.

Zeunert, J., Hawken, S., & Gowers, J. (2025). Visualising and valuing urban agriculture for land use planning: A critical GIS analysis of Sydney and neighbouring regions. *Land, 14*(4), 854. (1–28). https://doi.org/10.3390/land14040854

3

Period One: 1788–1850

Box 3.1 Population 1788 and 1850

Population 1788: c. 1,420
Population 1850: c. 50,000
= ~35-fold increase

J. Zeunert, A. Daroy, *Sydney's Food Landscapes*,
https://doi.org/10.1007/978-981-96-0710-5_3

Fig. 3.1 Castle Hill Government Farm, 1806. Source: John Lewin, Mitchell Library, State Library of NSW

Box 3.2 Key Events 1788–1850

- British invasion and occupation of First Nations Aboriginal lands and the establishment of Sydney and New South Wales in 1788;
- Conflicts between First Nations people and British occupiers;
- Imposition and implementation of British foodways and agricultural techniques;
- Establishment of multiple colonies across the continent and subsequent changes in boundaries (Fig. 2.3);
- The 1840s economic depression.

Box 3.3 Key Events for Sydney 1788–1850

- Tensions between civic-minded and self-interested settlers;
- Discovery of the Cowpastures and pastoral lands;
- Crossing the Blue Mountains in 1813;
- Importation of ice from 1839.

Box 3.4 Sydney's Metropolitan Agriculture Trends 1788–1850

- Identifying suitable lands for agriculture, initially in Burramatta (Parramatta) then Bulyayorang (at the Hawkesbury River) renamed Green Hills (now Windsor);
- Establishment of Government Farms;
- Land grants and convict labour provided to freeholders and emancipists;
- Agricultural experimentation, failures and successes;
- Reliance on rudimentary farming techniques;
- Lack of agricultural inputs, infrastructure, knowledge and skills;
- Enforced or unwilling agricultural labourers;
- Initial lack of livestock and breeding delays;
- Pastoralism's establishment and expansion beyond Sydney;
- Fledgling agricultural-related industries such as milling;
- Smallholders and crops occupying riverbanks and more fertile pockets of land;
- Grazing on arable lands, in larger landholdings and on uncleared land;
- The founding of five Hawkesbury River farming towns (Wilberforce, Pitt Town, Windsor,[1] Richmond and Castlereagh) in 1810;
- Regular flooding of the Hawkesbury River;
- Cessation of convict labour to New South Wales in 1850.

This chapter contains the names of deceased Aboriginal people.

Prior to colonisation, the Koorie Nations across the broader region managed food systems for millennia, as discussed in the *Introduction*. The cultural and spiritual significance of Country carried across generations through deep attunement with land, seasons and species, and was maintained within complex Indigenous knowledge systems. Under colonisation's scythe, traditional foodways were increasingly cut away as transported food production practices expanded. Vegetation along the Dyarubbin, or Deerubbin (Hawkesbury River), for example, was cleared for imported crops, removing the staple food of darug (yams). When Dhurag and Darkingjun people were further restricted from river access or forced off the land attempting to harvest new crops, British troops were deployed in punitive expeditions.[2] Stories of massacre, conflict, oppression, exploitation, but also survival, resistance and resilience, are therefore entwined within colonial agriculture. Some people, like Bidjigal man Pemulwuy, fought back, "by burning crops and raiding food stores".[3] Others faced malnutrition, starvation, disease or forced dependency on purported 'colonial benevolence'.[4] While the violent dispossession of traditional hunting and harvesting lands destroyed Indigenous food systems and access, colonists' early attempts to transplant exotic species and British farming practices also led to near starvation for convicts, free settlers and officials alike. This chapter traces Australia's current dominant commercial and industrialised agriculture back to its colonial inception between 1788 and 1850. It is thus particularly concerned with the colonists' lack of planning and agricultural place-based literacy. Successful harvests and accompanying economic and population growth incrementally grew as dispossession and degradative land practices concurrently increased across Sydney, and over time, the continent.

Relevant Global and National Events Pre-1788

The Undertaking

Struck by the "breathtaking audacity" of Britain's colonisation of the unfamiliar Australian continent, Tim Flannery terms the venture "every bit as monumental as the mission to land a man on the moon".[5] Eleven sailing ships in the 'First Fleet' transported 775 convicts, 645 free settlers and enough supplies for 2 years across the world on a 36-week journey.[6] The imperative was only partially a response to overbrimming prison hulks on London's River Thames and the newly independent United States of America's refusal to continue to accept British convicts in 1776.[7] The "arks, bearing the spores of an agricultural colony"[8] also sought to utilise English and Irish convict labour to develop a "self-sufficient farming settlement" and yield economic and strategic benefits.[9]

The First Fleet were remarkably underprepared for an agricultural endeavour.[10] As Robert Hughes observes, "never had a colony been founded so far from its parent state, or in such ignorance of the land it occupied".[11] There was a near total lack of reconnaissance undertaken prior to settlement. 'Survey' consisted of just over a week in the Kamay region (renamed 'Botany Bay') during the Royal Society's South Sea expedition led by then-Lieutenant James Cook on the *Endeavour* (1768–1771). The expedition's focus was heavily influenced by the aristocratic 25-year-old "gentleman of large fortune"[12] Joseph Banks and his accompanying coterie, who were focused on botanical survey and specimen collection.[13] Banks' interest reflected British Enlightenment fashions for procuring the novel and 'exotic', not only as aesthetic curiosities but also in the name of the Empire's scientific advancement and possible commercial gains.[14] Banks' passion for natural history spurred an extensive collection of seeds, plants, insects, molluscs, reptiles, fish, birds and marsupials. This diverse array reflected scientific and social interest rather than the practical needs of food production in a faraway colony with unknown conditions.

It is no surprise that imperial arrogance in an age of Western expansionism, slavery and hierarchical notions of race also translated to botany and food-producing plants. Rather than Indigenous names, species were largely given Latin or English common names. Some anthropological interest in Aboriginal names and species belied a greater dismissal in a blank colonial canvass deemed to be *Terra nullius*, or 'land belonging to no one', denying First Nations' sovereignty. As Ma Rhea notes, the "cultivation of the soil for food was not simply an act of securing food needs", rather, British styles of agriculture "from earliest colonial times in Australia proved ownership".[15] The colony's landscape was viewed through a lens of class-based elitism and compared to the wide vistas of English parklands and gentlemen's estates as popularised by landscape architects of the era such as Lancelot 'Capability' Brown.[16] The colonists saw the new lands as ripe for transplanting prestigious pastoralism, namely large-scale cattle and sheep grazing for meat and wool production. They failed to recognise that the supposedly untouched wilderness'[17] parkland character was the result of First Nations' long-term active land management and burning practices—Bill Gammage's notion of the "Biggest Estate on Earth".[18]

In 1779, the then-34-year-old Banks testified to the House of Commons' Bunbury Committee[19] that "a colony of felons could support itself within a year" based on apparent good climate, arable soil, an abundance of fish, pasture, fresh water and wood.[20] Banks' words carried particular weight, given that he had led Australia's only terrestrial reconnaissance.[21] He again testified in 1785 before the Beauchamp Committee,[22] the sceptical but time-pressured Home Secretary Thomas Townshend Lord Viscount Sydney agreed to move the convicts to Botany Bay.[23] His scepticism regarding the site's agricultural potential was suggested by his directive of providing 2 years of provisions in spite of Banks' assertion that 1 year's supplies would be enough to establish self-sufficiency.[24] Despite Sydney's hesitancy, Britain raised the stakes. The Official Charter required the colony to not only reach self-sufficiency but to provide returns on investment without delay.[25]

Relevant Sydney Events: 1788–1850

From the First Fleet's arrival in January 1788, it was immediately clear that that landscape was not as Banks had suggested.[26] His descriptions of floral abundance and species diversity were indeed true but had little correlation with the fertility required for British agriculture best suited to deep and fertile loams. Banks' account was in fact misleading to the extent that Captain Arthur Phillip discarded Botany Bay as the earmarked site after 5 days of reconnaissance.[27] Instead, the lands of the Gadigal, Cammeraygal and Wangal Nations at the harbour's mouth were deemed more favourable. Renamed 'Port Jackson', the site had the benefits of larger trees (initially thought to indicate greater soil fertility), harbour access, a summer-flowing fresh water source (the 'Tank Stream', now tunnelled within Sydney's CBD) and sheltered anchorage.[28] The bay of

Warrang/Warrane was declared Sydney Cove (now Circular Quay) and claimed as British territory in a flag raising ceremony in the name of King George III on 26 January 1788.[29]

While an improvement on Botany Bay, Sydney Cove's challenges soon started to show. The land "looked fertile and lovely, but it proved arid, reluctant, incomprehensible",[30] with no "significant fruits, roots or berries growing amidst the botanical profusion".[31] The "cantankerous and duplicitous"[32] Lieutenant-Governor Robert Ross pronounced that "in the whole world there is not a worse country than what we have yet seen of this",[33]

> All that is contiguous to us is so very barren and forbidding that it may with truth be said here nature is reversed; and if not so she is nearly worn out, for almost all the seeds we have put into the ground has rotted, and I have no doubt but will, like the wood of this vile country when burned or rotten, turn to sand.[34]

Unlike Britain's younger, deeper soils and softwood trees, Sydney's landscape and vegetation appeared aged, gnarled and tough. Soils near Sydney Cove were thin and sandy, punctuated by protruding sandstone outcrops. It was rapidly apparent that the region's significant trees proved not to indicate soil fertility suitable for British species, the vegetation frequently growing out of narrow, rocky cracks in sandstone geology. The colonists began the arduous task of clearing the land through forced convict labour to sow imported crops, graze introduced animals and strip the land of its timber to build shelter.

An Antipodean Colony

The colonists set about inscribing rigid geometric patterns across the land.[35] Roads were carved into the landscape and imposed over existing Dharug and Koorie tracks. Crops were sown in straight rows in ploughed fields (Fig. 3.2). Settler livestock, alien to the Australian subcontinent, were shackled, shepherded, bred and eventually penned in boxed runs.[36] Cartesian geometry demarcated the boundaries of Crown and private lands and emphatic notions of legal ownership. Over time, parcels of forcefully occupied land would be granted to favoured Crown subjects,

Fig. 3.2 Sydney Cove April 1788 (top) and Government Farm at Rose Hill c. 1791 (bottom). Sources: Francis Fowkes, National Library Australia Bib ID: 1084748; Dixson Galleries, State Library of NSW

or more commonly, leased or sold. The perceived wilderness was tamed into orderliness and subjugated.

Colonial Sydney's landscape management philosophy was typical of a Western anthropocentric (human-centred) nature/culture dualism, counter to First Nations care for Country. 'Nature' was viewed as first and foremost a human resource to be controlled by human mastery, as exemplified since classical antiquity through Judeo-Christian dominion over the Earth,[37] Plato and Aristotle's hierarchical Great Chain of Being, or Roman orator Cicero (106–143 BCE),

> We are absolute masters of what the earth produces. We enjoy the mountains and the plains. The rivers are ours. We sow the seeds and plant the trees. We fertilise the earth. We stop, direct, and turn the rivers; in short, by our hands and various operations in this world we endeavour to make it as it were another nature.[38]

If the colonists sought to create 'another nature' in the likeness of a Western ideal, they continued to be confronted by the mirage that the abundant vegetation provided little familiar sustenance, remaining a "wet desert" that left them "starving in a visual garden of Eden".[39] This, inevitably, gave rise to what has become known as the 'hungry years'.

The Hungry Years

Australia was the first continent to be colonised in which "there was no tradition of agriculture and gardening that the colonists recognised, no native gardeners to be employed and no gardens in which to plant the imported plants and seeds".[40] Rather, colonists rejected the unfamiliar endogenous foods, establishing familiar foodways as quickly as possible and "bringing their exogenous foods to new lands and waterways, a pattern that was to recur as the colonial frontier expanded across the landmass of Australia".[41] Predominantly British land management practices often proved ill adapted to Australia's climatic "fickleness" and the "poor quality of Australian soils".[42] Broadly speaking, these soils were subject to millions of years of geological stability and weathering, and many millennia of cultural burning, both considerably diminishing suitability to Western agriculture. Few deliberate attempts were made to devise specific farming approaches for Australian conditions.[43] Moreover, poor agricultural planning and execution deserves greater emphasis in the historical record. While imported practices were imposed, widespread incompetence, desperation and frequent disdain for the 'new' land heightened degradative agricultural and land management outcomes. Unsurprisingly, failure ensued.

The first years of the colony (1788–1792) became known as the 'hungry years', a time of "near starvation and hardship"[44] in which "all lived for five years on the bleak edge of starvation".[45] Despite the First Fleet's rations designed to last 2 years, the colony was dependent on additional imported food from distant corners of the British Empire. In October 1788, a ship was deployed to Cape Town to obtain supplies, mercifully returning with provisions in May 1789 but wrecked in 1790 on its subsequent run to Caton. In 1789, supply ships never arrived and the fledgling colony almost starved, "smallpox ravaged the First Nations population and misery was everywhere".[46] Convicts and soldiers alike faced reduced rations. An anonymous letter, possibly by Marine officer Watkin Tench, dated 14 April 1790 details the "dread of perishing by famine" that "stares us in the face",

> Our present allowance is a short one: two pounds of pork (which was cured four years ago, and shrinks to nothing if boiled), two pounds and a half of flour, a pound of rice and a pint of pea[s] [...] we use every means to get fish [...] Were the ground good, our gardens would be found of infinite use to us in these days of scarcity, but with all our efforts we cannot draw much from them.[47]

The story, however, is nuanced. There are various accounts of colonists eating bush foods from the immediate environment to supplement government rations, including trading with First Nations people for game.[48] Foods included wild plants—generally those that appeared similar to European species—such as purslane, wild sage, celery, spinach (Warrigal greens), samphire and the herb *Smilax glyciphylla* used as tea. Meat included kangaroo, possums, fish, eels, stingrays, turtles, birds (and their eggs) and occasionally reptiles and insects. According to Captain John Hunter, the "pot or spit received everything we could catch or kill, and the common crow was relished here".[49]

Claims that "settlers added native foods to their diet, not simply because of hunger [...] they were valued for their nutritive and medicinal qualities, and for variety and freshness",[50] however, should not be overstated. The "acceptance of native produce was grudging [. . .] and temporary",[51] with the majority of settlers "clinging as closely as they could to the dietary pattern to which they had been accustomed",[52] and remaining "intent to reestablish their familiar foods from Britain and Europe".[53] British cultural influences therefore "utterly dominated from the outset, and in fact did so until well after the Second World War", with colonists relying on food deliveries from 'home' until they could "duplicate their foodways locally".[54] The colonisers' foreign, processed foods similarly held little appeal for local people who were, according to one account, "very little interested in our biscuit and fresh bread", and "took it and mimicked us pretending to eat it, but as soon as one left them, they threw it away".[55] There was therefore very little integration between British and Aboriginal diets, their associated foodways, and therefore, mutual ways of approaching land management.

The colony remained on tight rations at risk of starvation, despite provisions returning with the HMS Sirius in May 1789. By the end of 1789, every colonist, including the Governor, was on two-thirds rations except for those responsible for procuring food, such as gamekeepers and fisherman.[56] The situation worsened with the arrival of the Second Fleet in June 1790. Operated by a private contractor, it became known as the 'Death Fleet' following 37 deaths en route and delivering 127 sick and famished convicts.[57] Rationing continued for several years. From 1791, more regular supply ships arrived and farming gradually improved.[58] Governor Phillip calculated the required rations as follows,

> When the stores may permit the issuing of the established ration, the weekly expense [November 1791] according to our present numbers will be: Of flour 30,560 lb., of beef and pork 21,010 lb., of peas 179 bushels, of butter 1,432 lb., and our numbers will be increasing.[59]

Even at subsistence levels for a tiny population by today's numbers, the volume of food indicates the monumental task of supporting the colony.

By 1792, food stores had finally risen to surplus due to a range of imported sources and increasing local production. A dedicated marketplace in Parramatta sold grain, vegetables, livestock, poultry, cured meat and imported tea and sugar. Goods were also informally sold from boats at Sydney Cove near the government wharf.[60] By 1806, complaints ensued that "great inconvenience attends Boats which come loaded with Vegetables and other Articles for barter with the Inhabitants and others at Sydney".[61] Surplus was marred by a poor harvest in 1793 and Governor Phillip's return to England due to illness. Phillip was replaced by Lieutenant-Governor Francis Grose, who instigated military rule and was more concerned with his troops than welfare. Nevertheless, Grose's settlement of the Hawkesbury as the 'granary of the colony' and land grants issued to members of the Corps led to more farms. By 1794, the famine had been stemmed and the colony became self-sufficient in 1795. Albeit, surplus was not steady and subject to fluctuations from new influxes of convicts and settlers, as well as the impacts of unfamiliar climatic patterns.[62]

Social Failures

Accounts of the colony's near starvation tend to overemphasise the infertility of Sydney's inherent land capacity, rather than considering the social and cultural factors underlying agricultural failure.[63] Commentators have noted, for instance, that the "poor soils of the Sydney area militated against the 'vegetable production' envisaged by the Colonial Office in London",[64] and Sydney's "thin sandy clays begrudged the first settlers of even a subsistence crop".[65] This holds true if comparing Sydney's soils with ideal younger soils of Britain, Ireland and Europe[66] and in relation to Sydney Cove's immediate sandstone areas (Figs. 2.9 and 2.11). The infertility narrative, however, is frequently extended beyond Sydney's sandstone areas. While the wider region's soils include "red and yellow podsolics, high in acidity, low in phosphorus, resistant to nitrogen fixation, and deficient in important trace elements",[67] *Site and Topic*'s land and soil capability mapping shows considerable areas of agriculturally very capable lands (Figs. 2.9 and 2.12).[68]

Early Sydney's agricultural failures were the result of several key factors: initial overinvestment in lands containing sandstone geology (Fig. 2.11), a dearth of agriculturalists, skills, knowledge and infrastructures, as well as the imposition of imported practices and their regularly poor execution. Productive land capacity was hugely driven by colonial diet preferences for wheat, barley, beef and mutton, all of which are spatially extensive. Furthermore, Phillip's site selection, despite an improvement upon Banks' suggestion, offered naval rather than agricultural advantage. Derrubbin, or the Hawkesbury River, bypassed Port Hacking, Botany Bay and Sydney Cove. This deprived much of the area of significant catchments, with the Parramatta River lacking requisite freshwater significance (Figs. 2.9 and 2.11).[69] Accordingly, central and eastern Sydney did not have the necessary silt deposition to replenish alluvial flats suited to nutrient-intensive and extractive Western agriculture.[70] The Hawkesbury River's mouth was also considerable distance from Sydney Cove and Parramatta, challenging early movement.

The colony further lacked skilled agriculturalists and infrastructures. The effects of land enclosures prior to transportation had displaced rural populations and eroded traditional farming knowledge. This was greatly compounded by the onset of the Industrial Revolution, which dramatically reduced manual labour. Both led to increased crime. Consequently, most urban convicts did not have agricultural experience. Robert Hughes, for instance, argues that many convicts were "slum raised, utterly ignorant of farming" and would starve if left to their own devices, while others were too old, feeble, or sick to work the land.[71] Governor Phillip hoped to entice convicts and settlers to develop their farming skills through issuing land grants. Convicts were offered 30 acres, married men had 50 with additional acreage per child, discharged soldiers were allocated 80 and settlers given 130 acres.[72] It soon, however, became clear that many emancipists remained in the colony just long enough to pay for their passage back to England. Thus, neither "legal restrictions nor isolation made the 'settlers from convicts' better able to read the riddle of agriculture in a climate fruitful at times but fatally capricious".[73]

We can see by the list of convicts' trades that a "serious bureaucratic error was to ignore any need for a balance of skills", indeed, "three ivory turners could not be expected to be as useful as equal numbers of professional fishermen or farmers".[74] The colonial plan extended as far as earmarking convicts to provide the core of hard, unskilled labour but not to agricultural leadership. The questions of necessary expertise, skilled teams for implementation, and skills transfer were largely unaddressed. In the absence of a more robust plan, Governor Phillip's personal manservant, Henry Dodd, was installed as the colony's 'Head Gardener' based on his prior farmhand experience—a horticultural title more reminiscent of Banks' botanical interest rather than the agricultural production needed. Dodd was later promoted to Agricultural Superintendent.

Finally, poor execution of imported agricultural practices led to failures. Writing in 1826, James Atkinson recalls the colony's substandard farming techniques,

> If a foreigner who had travelled through England, were afterwards to visit New South Wales, he would scarcely be able to persuade himself that the inhabitants were derived from the same stock; he could hardly believe that the people, who, in the mother country, cultivate their lands with such persevering industry and intelligence, should here become so extremely slothful and negligent [..] many even neglect [...] (rotations) in good farming, but sow wheat on the same land, year after year, for a succession of seasons [...] The consequence of this miserable system is, that the land in a few years gets exhausted, and having very little tillage, is entirely covered with weeds.[75]

Consequently, Sydney's already delicate soils quickly degraded. This agricultural callousness is indicative of the wider colonial approach to land management geared towards extraction and private profit,[76] in stark contrast to First Nations conceptions of kinship and Country.

Adding to pressures, alcoholism beset the colony. In 1796, officer and pastoralist John Macarthur characterised Sydney's smallholder farmers as idle "drunkards who should be obliged to employ themselves in the services of an industrious and vigilant master".[77] Some hoarded meagre rations to trade for rum, which became an additional currency. Up to 28 kilograms of flour could be exchanged for just 1 L of the spirit. Convicts were allocated less than two kilograms of flour a week, thus "seven weeks' worth of food was equivalent to one bottle that would not last the night".[78] The New South Wales Corps, who had been sent to relieve the First Fleet officers, were recruited from military prisons or soldiers 'down on their luck' and were on half pay. The Corps became heavily involved in the alcohol trade, resulting in the moniker the 'Rum Corps'. The situation escalated to the 'Rum Rebellion' of 26 January 1808, in which Governor William Bligh's failure to curb trafficking resulted in his deposition in the only successful armed takeover of government in Australia's history.[79]

Marion Phillips delivered a scorching character assessment of Sydney's ruling class when reflecting on the colony's history in 1909,

> A great number of the 'gentlemen-settlers' were retired army and navy officers who applied with zeal the peculiar caste rules of the services. For the most part they were simple, commonplace men, physically courageous and intellectually vapid, men guided by a strange jumble of uncomprehended motives blind loyalty to the King, their regiment or ship, blind acceptance of the Church of England mingled with love of liquor, greed of gain and indifference to the usual tenets of morality. Few were men of striking ability or forceful character, for the colonial garrisons, which formed a back-water of the Service and the retired list, had little to show in those times of war in the way of brains or energy.[80]

The prevalence of drunkenness gave rise to speculators such as Edward Gibbon Wakefield seeking sober and industrious immigrants for his planned South Australian settlement scheme, partially to avoid the earlier troubles plaguing New South Wales.[81]

The Place Where the Eels Lie

How did the fledgling colony grow into current-day Sydney? Some agricultural prospects existed in the early years, although scant. After multiple failures in the Sydney Cove region,[82] attempts to feed the colony were directed westwards. Governor Phillip led an upstream expedition to locate more fertile agricultural land. In November 1788, he established a new settlement and Government Farm in Parramatta, called Rose Hill.[83] Here, the freshwater Burramattagal (Parramatta River) intermingled with

Port Jackson's brackish mangroves. Situated on what is now known as the Cumberland Plain,[84] the site's differing geology and soils were more receptive to the settlers' imported crop species. Convicts cleared and hoed land by hand to grow wheat, corn and barley, providing seed for further crops. In 1789, Phillip wrote to Britain to request more settlers with farming expertise. This proved timely, as Dodd died unexpectedly in 1791. This setback was compounded by a long and intense drought unknown in the English experience, evaporating hopes of self-sufficiency. Seed sown in 1790, for example, was barely recovered at harvest time.[85]

Rose Hill was Sydney's first agricultural lifeline. Depictions in 1791 (Fig. 3.2, bottom) and 1792 (Fig. 3.3, top, centre) distort the Australian landscape to convey British characteristics. The 1791 watercolour depicts an Arcadian vision of the Government Farm at Rose Hill. The moat is likely inaccurate and the image conceals the scale of production at the time, as well as excludes the convict labour facilitating it. By late 1791, over 200 acres (81 hectares) of land in the region were in cultivation, including wheat, corn and vines. Watkin Tench observed that 8,000 vines were planted there the same year, followed by 5,000 more by 1802.[86] Both Fig. 3.2 watercolour (bottom) and Fig. 3.3 1792 engraving (top, centre) show mature eucalyptus of the Cumberland Plain woodland having taken on leaf density and tones of European specimens. In the latter, a walled area containing rows of neat crops is visible, along with grazing cattle in a cleared pasture (centre). Figure 3.3 1798 depiction of Rose Hill (bottom) reveals a decade of nation building centring on Rose Hill's Governor's House, where gridded regularity and visual surveillance reflected colonial rule inscribed through planning and building.[87] Each convict cottage possessed land area for growing food, while the collective arrangement and streetscape frame the overseer Government House. A set of stocks is visible as a deterrent to law breakers.[88]

The Agri-production Epicentre

In 1784, Old Windsor Road was laid to connect Parramatta to Bulyayorang (Green Hills, now Windsor) and expansion continued outwards. By the end of 1792, 1,000 acres were under cultivation, mostly

Fig. 3.3 Engravings of Rose Hill in 1792 (top) and 1798 (bottom). Sources: State Library Victoria and State Library NSW

worked by convicts.[89] Although Parramatta was important, the Hawkesbury region was the most substantial district delivering food-supply salvation and compensating for Sydney Cove's agriculturally infertile sandstone areas. The early colony was in fact saved by the fertility of the Hawkesbury-Nepean River region, stemming from the township hub of Green Hills.[90] This became Australia's first significant food production area due to its richer flood-prone alluvial river flats and is still partially in use today. Green Hills, at 46 kilometres as the crow flies over challenging terrain, was at the outer limits to supply perishable food for the first two decades until roads and transportation improved.[91]

The Hawkesbury's fertile land came at a cost. The river has a propensity to flood to significant and destructive depths.[92] In less than two decades, floods occurred in 1799, 1800, 1801, 1806 (60 feet, or over 18 metres, above ground level) and in 1809.[93] Though flooding replenished soil fertility at a time when manure for fertiliser was scarce, it resulted in ruination for many farming ventures on its volatile plains.[94] In 1826, James Atkinson extolled the richness of the Hawkesbury Valley and chastised the first wave of farmers deficit in skills and suitable temperament,

> The alluvial lands in New South Wales are not surpassed in fertility by any in the world [...] the principal tracts are found along the banks of Hawkesbury, Nepean and the various branches of the Hunter; they consist of vegetable mould more or less mixed with sand of many feet in depth [...] The greater part of the alluvial lands upon the Hawkesbury and Nepean have been cleared and are under cultivation; and in the hands of any other than their present ignorant, indolent and improvident possessors, would produce the most plentiful and valuable crops.[95]

Helen Proudfoot contends that Atkinson's impressions should be countered by the limited alluvial lands, non-periodic rainfall and flood propensity and markets controlled by often "rapacious" Sydney merchants resulting in difficulties for many farmers and their families.[96] At this point, the alluvial lands were enough to sustain the population. While at times erratic, Sydney's rainfall was relatively consistent throughout the year, especially compared to the rest of the continent. Flooding was undoubtedly regularly ruinous but sociocultural factors were crucially at play in Sydney's early agricultural struggles. The failure to adequately attract and retain enough skilled and sober farmers had a far greater impact on colonial Sydney's agricultural challenges than inherent land capacity.

In 1799, the Hawkesbury region comprised 55 per cent of the colony's total cultivated area and 53 per cent of the colony's livestock, followed by Parramatta at 31 per cent and 33 per cent, with the Sydney region bolstering 14 per cent and 14 per cent respectively.[97] The Hawkesbury's small farms between 30 and 100 acres (12–40 hectares) became increasingly productive from 1810 to 1820, with the cropped area rising from 4,896 acres (1,958 hectares) to 14,927 acres (5,980 hectares), mostly from wheat and maize, with 17,339 sheep, 11,720 cattle, 16,855 hogs, 942 horses and 638 acres (254 hectares) of fruit and vegetables.[98] By 1815, more production districts had opened up (Table 3.1). We can further observe evidence of agricultural choices reflecting British diet preferences: cattle; sheep; goats; hogs; wheat; maize; barley; oats; fruit and vegetables, with horses facilitating agricultural production and transport (Table 3.1).

Recognising the district's importance to supplying fledgling Sydney, Governor Macquarie proclaimed five Hawkesbury River farming towns in 1810: Wilberforce, Pitt Town, Windsor, Richmond and Castlereagh. The initiative was essential, particularly given Macquarie's closure of Government Farms on less productive land. It proved to be the "most successful strategic food security intervention by the government in the new colony since the arrival of the First Fleet"[99] and it is doubtful whether it has been surpassed by any Sydney-based food security initiative since.[100]

Colonists quickly located and appropriated the most fertile agricultural lands.[101] By 1800, a settlement pattern emerged whereby smallholder cropping occupied more fertile pockets and riverbanks, while grazing took place on large landholdings and mostly uncleared land.[102] We can observe this makeup in a 1798/1817 map illuminating agricultural locations and favourable conditions (Fig. 1.1). The map indicates the extent of land assessed and occupied at the time and is suggestive of Sydney's lands most suited to Western agricultural production. Urban density was centred in Sydney Cove around today's central business district, east, south and inner west. Concurrent growth occurred in

Table 3.1 Greater Sydney's main production districts anc their crop and livestock elements in 1815[a]

	Sydney		Parramatta		Hawkesbury		Liverpool		Newcastle	
CROPS	Acres	%	Acres	%	Acres	%	Acres	%	Acres	%
Wheat and maize	2,210	13.1	2,591	15.4	9,344	55.6	2,541	15.3	116	0.6
Barley and oats	56	4.2	200	15.1	877	66.3	180	14.1	5	0.3
Fruit and vegetables	276	21.4	350	27.5	433	34.2	204	16.7	4	0.2
Total cropped	2,542	13.1	3,141	16.1	10,654	54.9	2,925	15.3	125	0.6
Total held	60,605	29.1	59,584	28.5	47,133	22.6	40,955	19.6	270	0.2
LIVESTOCK	No.	%	No.	%	No.	%	No.	%	No.	%
Horses	736	31.6	547	23.1	690	29.6	356	15.7	–	–
Cattle	6,361	27.6	5,748	24.9	4,960	21.5	5,931	25.9	13	0.1
Sheep	11,392	18.6	23,364	38.3	16,469	27.1	9,659	15.8	60	0.2
Goats	59	8.8	131	19.5	457	68.3	22	3.4	–	–
Hogs	1,190	11.7	1,558	15.4	6,275	62.1	1,000	10	83	0.8

[a] Adapted from Fletcher, 1976: 200

Parramatta as the primary agricultural hub. Windsor was the main focal point for more distant agriculture activity, with scattered hamlets and estates across many other districts.

Beyond the Basin

While the fertile Hawkesbury-Nepean floodplains initially overcame the problem of poor soils, it was not until the "formidable physical barriers" of the Hornsby, Blue Mountains and Woronora plateaus were passed in 1813 that the colony identified richer farmland.[103] The road over the mountains to Bathurst was completed in 1815 but steep grades in parts were unsuitable for horse-drawn vehicles. A new road from Mount Victoria to Hartley was built in 1832, led by Major Thomas Mitchell. It was now possible to travel safely by wagon or coach. This opened up the less humid regions west of the Blue Mountains that were better suited to Western pastoralism and arable crop production. Over time, this not only bolstered the colony's food supply but also income through exporting commodities.

Several decades in, the colony was finally gaining a more resilient foothold and approaching something closer to Banks' original promise. With new agricultural towns and farmland being established beyond the region's topographic basin, the tide had finally turned. A letter dated 30 June 1813, proclaimed,[104]

> The colony of New South Wales is in so flourishing a condition, that a memorial has been forwarded by the principal inhabitants, through the governor, to his Majesty's Ministers, praying, among other privileges, permission to distil from their surplus grain, and to export flour from thence to Great Britain; and pointing out that there is no farther necessity for any salt meat being sent thither, as the colony can furnish fresh beef, pork and mutton at a cheaper rate.[105]

Surpluses, however, are never assured in agriculture. The colony's delicate balance was significantly strained by double the number of convicts arriving between 1814 and 1820, placing new demands on food supplies.[106]

Growing Momentum

By the end of *Period One* in 1850, the British colony of New South Wales had grown from around 1,420 arrivals on the First Fleet in 1788 to approximately 50,000 settlers 62 years later. Its once-struggling colonial population jumped from around 30,000 to 50,000 between 1840 and 1850. On 22 May 1840, an Order-in-Council was issued to remove New South Wales from the list of places to which convicts could be sent,[107] which took a further decade to be implemented. The cessation of convict transportation to New South Wales marks a significant end point for our first period.[108] New, costlier labour arrangements needed to be forged, slowing the pace of large-scale agricultural developments.[109]

Beyond merely being an initial dumping ground for convicts, the colony was a haven for capitalistic speculation, including by those in government conducting deals in the manner of today's 'insider trading'. The Australian Agricultural Company (AACo), for instance, was established by British Parliament in 1824 and granted influential figures the 'right' to select 1 million acres (404,686 hectares) of land. Members included England's Attorney-General and Solicitor-General, 28 Members of Parliament, eight directors for the Bank of England, the Chairman, Deputy Chairman and five directors of the British East India Company as well as numerous English bankers and merchants. The AACo experimented in tobacco and flax production for export[110] and introduced sheep and cattle to Worimi Country (Port Stephens), which by 1829 had grown to 17,000. By then, the area had been taken over by 600 farms and an expanding road network.[111] This example alone highlights the degree to which power, wealth and speculative large-scale agriculture set the tone for Sydney, and indeed wider Australia's agricultural ethos, for the periods that followed.

The 11 Typologies: 1788–1850

The following section explores 11 key agricultural 'typologies'—*Agricultural Facilities and Education* (AF); *Mixed Farming* (MF); *Market Gardening* (MG); *Fruit Production* (FR); *Cropping* (CR); *Intensive Livestock* (IL); *Grazing* (GZ); *Vineyard Wineries* (VW); *Agricultural*

Infrastructure (IN); *Aquaculture* (AQ) and *Miscellaneous* (MC). These categories are used throughout the book and further outlined in *Site and Topic*. They are here examined over the course of 1788–1850. All figures in *Period One* are artistic depictions such as paintings, sketches, maps and engravings, and should therefore be viewed with an inquisitive and critical eye. Artworks can be misrepresentative or altogether false. It was not uncommon for works to be completed in Britain based on earlier site sketches (such as Fig. 3.9). Some are reproductions, duplications for records or alterations of earlier works. They have been shaped by bias, artistic licence or trends, and some have had fictitious elements added afterwards.[112] Colours were sometimes altered to suit British landscape palettes and tastes, with greens routinely deeper. Joseph Lycett's works, for instance, were based on memory and earlier sketches taken before leaving Australia in 1822 and critiqued for a "lack of originality" and a romanticised, Anglicised conception of landscape.[113] They are all visual indicators and interpretations of singular points in time in otherwise constantly changing landscapes. Nonetheless, the sources present an interesting insight into how the early colony's food landscapes were taking form.

Agricultural Facilities and Education (AF): Egalitarianism or Free Enterprise?

As mentioned, self-sufficiency was key to the venture's commercial viability, with the colony required to generate income through exports and free enterprise. In theory, self-sufficiency would principally result from establishing state-run production initiatives under convict labour. 'Government Farms' (and later 'Model Farms') were the agricultural centrepiece and sought to upskill convict workers and demonstrate food production techniques for emancipists and free settlers alike. Accordingly, *Period One* contains the highest number of entries for Agricultural Facilities and Education (AF) in the Map Database across all five time periods.

The first Government Farm was immediately established at Farm Cove in January 1788,

> His Excellency [Governor Phillip], seeing the state these poor objects were in, ordered a piece of ground to be enclosed, for the purpose of raising vegetables for them. The seeds were sown upon this occasion, on first appearing above ground, looked promising and well, but soon after withered away, which was not indeed extraordinary, as they were not sown at a proper season of the year.[114]

A similar fate befell most other crops and food gardens in the Sydney Cove region, despite portrayals of productive endeavours in early maps and paintings.[115] The scale of the crops depicted neatly within walled gardens or in rambling fields is also likely to have been exaggerated. By the end of 1788, "a miserly 20 acres of crops had been planted".[116] This placed unforeseen strain on government stores, requiring extra provisions to be imported from distant colonies and trading centres.

Initial experimentation into which species might flourish proved fruitless in the sandstone geology's challenging soils for British agriculture. Government Farms focused on establishing grain supplies for bread (predominantly wheat but also the less popular maize) and high-yield, hardy vegetables such as potatoes, turnips and cabbages. These provided the initial bulk of locally grown rations. Over time, private farming would increase, supplying the government stores as well as generating private markets and trade. Other Government Farms were established in Rose Hill (The Crescent, now Parramatta Park) in 1788; Toongabbie in 1791; Longbottom (Concord) in 1792; Petersham in 1793; Castle Hill in 1801 (Fig. 3.1 depicted in 1806)[117] and Emu Plains in 1819[118] (Figs. 8.2, 8.3, 8.4, 8.5, and 8.6), as well as Model Farms under Governor Macquarie.[119] Despite "limited success",[120] the farms nevertheless played a critical role in the early colony's survival. Toongabbie Government Farm proved particularly crucial during the 'hungry years' of 1791 to 1794. While all agricultural and gardening pursuits grappled with unfamiliar conditions and unknown insect pests, many private gardens were comparatively more successful than the public farms. These were also easier pickings for pilfering hungry convicts and settlers. Despite being provided with seeds and time to establish their own vegetable gardens, and the severe risks including possible death if caught, "convicts were proving more adept at stealing food rather than growing it".[121]

James Ruse was a notable rare beacon of initial agricultural success. Ruse's first successful venture in Parramatta (see the following section, *Mixed Farming*) led to Governor Phillip granting land, livestock, agricultural tools, seed and sometimes convict workers to emancipists and settlers willing to farm in the region. Initially, small land grants required the grantee to live on and work the land to support self, family, convict workers and possibly return a profit.[122] Land grants varied considerably during each Governor's reign in early Sydney. Governor Macquarie handed out official land grants comparatively sparingly during his long tenure from 1810 to 1821, granting fewer than 1,000 square miles (between 2,590 and 3,000 square kilometres depending on source).[123] This doubled under Governor Thomas Brisbane between 1821 and 1825, who also allowed settlers to purchase up to 4,000 acres (16 square kilometres).[124] By 1821, much of the land within a 320-kilometre radius of Sydney Town was unofficially occupied by graziers, farmers and squatters.[125]

Over time, agricultural production broadly transitioned from a welfare state to subsidised farming and then free enterprise, albeit, the latter was regularly enabled by the former. Government was central to Sydney's establishment,

> The first permanent building was a Government building, the first wharf and warehouse was a Government wharf and warehouse, the first food eaten was Government rations, and when the first wheat from the Government farm was ready for milling, it went to the Government mill [...] And so a tradition was established of great consequence for the future development of the County—the Government became a dominant influence in everyday life: it was often disliked and sometimes feared, but it was looked to in every emergency, in floods and droughts, famines and plagues—as indeed it still is.[126]

The colony's governance had two distinct approaches from its earliest years, apparent in the different approaches taken by its first two governors. The first 5 years under Governor Phillip were comparatively egalitarian compared to highly stratified eighteenth-century England. In the early New South Wales colony, Phillip distributed government rations equally between convicts and free settlers. After his departure at the end of 1792, however, the New South Wales Corps' incumbent relief of the

New South Wales Marine Corps under Lieutenant-Governor Grose scuttled Phillip's civic agenda, supplanted it with military rule and self-interest. Edward Shann noted that the "power of the British ruling class was as yet untempered by the constant responsibility to the opinion of the nation", with army and navy "prize-money" and the "spoils of war" whetting the appetite for further "enrichment by plunder or trade".[127]

Grace Karskens characterises the colony's lack of money, a treasury and plan for economic development under Phillip as "strange" and "naive".[128] It is perhaps then unsurprising that Grose promptly realigned this social anomaly that Phillip had painstakingly tried to establish. Grose instilled a military-controlled penal settlement that unashamedly favoured those in power, namely, the New South Wales Corps. Grose cut convict rations following the poor crops of 1793, discarded Phillip's social programmes, shunned collective farming, placed magistrates under military authority,[129] gave the Corps generous land grants and provided them with convict labour.[130] A "culture that allows commercial self-interest to become paramount creates the preconditions for exploitation and corruption", and New South Wales was no exception,[131]

> [Changes] Lieutenant-Governor Grose made as soon as 'the Governor's' back was turned seemed a deliberate plunge into anarchy. The civil court ceased to sit. Public agriculture was dropped. Phillip's admirable plans for the town of Sydney were ignored. The officers were granted land wherever they chose—land which soon included the former scenes of public agriculture. Public order collapsed and a general licentiousness raged unrebuked.[132]

Effectively, the "Corps began to control the food supply, selling goods to the government store".[133] When Grose returned to England in 1794 to continue his army career, the Corps' dominance over trade had been firmly established. The manipulated market was now engineered by the military and controlled by its 'Rum Corps', remaining tightly held until 1818.

Abuses of power were common. From 1841 to 1859, for example, Superintendent Charles Ormsby conducted unauthorised private farming at Biloela (Cockatoo Island), developing a reputation for treating the island as his personal fiefdom,

In 1849, he was reprimanded for running private enterprises from the island. This included keeping pigs, poultry and goats, fed from prisoner rations, and using convict labour to grow cabbages. In addition to pork and eggs, he had been selling up to 40,000 cabbages per year.[134]

Two other agricultural initiatives in *Period One* included shared access commons for livestock grazing, especially in the early decades (see GZ), and the first agricultural training facility. During his reign from 1821 to 1825, Governor Brisbane reputedly set up New South Wales' first agricultural training college[135], albeit details are scarce.

Mixed Farming (MF): Estate Ambitions, Convict Servitude

Mixed farming was a key colonial aim, providing grain, year-round vegetables, seasonal fruit and eggs.[136] Lofty agrarian intentions quickly dissipated following the long and arduous sea journey[137] and ideals soon gave way to the "eager materialism of this frontier society",[138] particularly after Phillip's departure.

Phillip's own farm and garden offered an initial model example of mixed-farming ideals with an array of species,

> Plants loaded onto the First Fleet at the Cape of Good Hope and Rio de Janeiro grew with some success. From Rio came plants of coffee, banana, oranges, lemon, Eugenia or pomme rose (*Syzygium jambos*), prickly pear, tamarind, jalap and ipecacuana, and seeds of coffee, cotton, cocoa, orange, lemon and guava. At the Cape, Phillip had loaded plants of fig, bamboo, Spanish reed, sugar cane, vines of various sorts, quince, apple, pear, strawberry, oak, myrtle, and grain seeds to feed the new colony. These, save the corn that was sown at the head of Farm Cove, were planted in the three gardens quickly established to the east of Government House.[139]

Government House at Sydney[140] and Parramatta[141] are both well illustrated with food plots, though their privileged access to resources and labour were not representative of the colony's wider mixed-farming endeavours.

The aforementioned James Ruse offers another example of mixed-farming success. After serving his sentence, Ruse was allowed to occupy an acre of cleared land near Parramatta from November 1789. Without animals, emancipated convicts were issued "meagre supplies" of a spade, shovel, hatchet, tomahawk and two hoes".[142] Ruse hoed the ground by hand,[143] sowing bearded wheat and maize. In the absence of manure, he fertilised the soil with potash obtained from burning timbers, using weeds and grasses as cover compost.[144] Ruse is also widely attributed to producing Australia's first successful wheat harvest.[145] By February 1791, he claimed that self-sufficiency was possible on the basis of supporting himself and his wife, Elizabeth Perry.[146] In April 1791,[147] he received the first land title grant, along with pigs and chickens.[148] His 30-acre (12 hectare) plot at Rose Hill was aptly named 'Experiment Farm'. Having potentially extracted the best yields that the land and circumstances would provide,[149] he moved to farm more fertile lands in the Hawkesbury region.[150] Since the mid-twentieth century, an agricultural High School in Carlingford has borne his name.

The grandest realisation of mixed farming in the period, however, was the so-called 'gentleman-settler' estates. They were usually built and operated by government-directed convict labour. Residents were often influential agricultural and horticultural figures, favouring pastoralism and viticulture.[151] These estates functioned as testbeds for speculative ventures to extend into the growing colony's rural hinterlands and beyond, eventually at greatly increased scales.[152] As centres of experimentation[153] and exploration,[154] estates helped to facilitate expansion from hobby, self-sufficiency, convict-labour-dependent and single-property scales to fully fledged industries generating commodities for export. Mamre house, farm and vineyard in Orchard Hills, for instance, was established by Reverend Samuel Marsden in 1804 from a land grant. As a model farm it included orchards, vineyards, exotic pasture and other crops, and it was particularly notable for helping establish the sheep and wool industry, including the Merino breed.[155]

Cobham Hall, the villa of emancipist entrepreneur Solomon Wiseman, included lemon hedges, orange trees, vegetable gardens and cattle and sheep runs on the alluvial Hawkesbury flats. Wiseman established Australia's first ferry service in 1827 and supplied provisions via lucrative contracts to convict road gangs in the area. These figures had widely varying philosophies to First Nations peoples, from enabling gatherings and corroborees on 'their' lands they had effectively stolen and establishing Christian missions, to leading calls for genocide and executing massacres.[156] The majority of early Sydney's records document influential figures' estates, which perhaps helps to explain a lack of prominent histories pertaining to more humble undertakings of smallholder farmers. John and Elizabeth Macarthur offer another example, subsequently discussed below (see *Grazing*).

Market Gardening (MG): From Convicts to Commerce

As opposed to mixed farming, a market garden is the small-scale production of vegetables, fruits and flowers. Market gardens are distinguishable from domestic and hobbyist pursuits by their commercial and professional nature, and from orchards by their intensive concentration of chiefly annual (as opposed to perennial) plants. As discussed above, some of the colony's initial vegetable and annual fruit production was for subsistence rather than commerce. Michael Christie claims that Australia's first market gardeners were convicts housed in Parramatta's huts as depicted in Fig. 3.3 (bottom)[157]—albeit as involuntary indentured labourers. Government Farms thus produced vegetables in a manner akin to market gardening (as already discussed in *Agricultural Facilities and Education*).[158] The aforementioned estates also produced vegetables and fruits as part of mixed-farming enterprises and regularly benefitted from convict labour. The success of private commercial market gardening ultimately required a functioning local economy outside of government purchasing, convict labour and the rum trade of the early decades.[159]

When examining market gardening in early Sydney, it is not always evident whether sources include the Hawkesbury, and sometimes the Parramatta, regions. This may help to explain why the National Trust date market gardens' appearance in Sydney from about 1835[160] when earlier examples existed across the County of Cumberland region. Successful plots resembling market gardens appeared on the Hawkesbury

River's alluvial banks and flats earlier, especially near Green Hills (Fig. 3.4). Land parcels hugged the fertile riverbanks that were well suited to intensive vegetable and crop production and were often smaller in size and oriented to maximise river frontage for the greatest number. Statistics from 1810 demonstrate that vegetable (and fruit) production was relatively balanced between three key agricultural production districts, Sydney Cove, Parramatta and the Hawkesbury.[161] Table 3.1 quantifies their makeup in 1815 and includes Liverpool as an emergent vegetable and fruit production district. By 1821, the Sydney region supported 442 acres of vegetable and fruit production, although the proportion of fruit versus vegetables is unclear.[162]

Early market gardeners in the Sydney Cove region were fragmented as they were restricted by small pockets of soil fertility and the lack of reliable fresh water sources for irrigation. Growers venturing further out to more favourable soils in Parramatta, Liverpool and especially Green Hills (Windsor), were in turn hampered by needing transportation for their perishable fresh produce. This created demand for increased and improved road access. A remnant market garden in La Perouse claims to be Australia's oldest continuing market garden and dates to the 1830s.[163] This is around the time that lectures in cultivation methods emerged from commercial nurseries.[164]

Despite a lack of specific details, there is a general consensus that Sydney's early market gardens were first cultivated by Cornish, Irish and German settlers.[165] Some posit that market gardens were initially an enterprise for the wealthy,[166] but if so, this dynamic shifted over time and was largely supplanted by Chinese immigration during the 1850s gold rush (see *Period Two*). Market gardening's profitability was significantly shaped by whether a gardener was a worker, lessee, owner or grant recipient. Market gardening's commercial viability increased as Sydney and Parramatta urbanised, constraining domestic production through diminishing plot and yard sizes. Accordingly, market gardens became more prevalent as the colony spatially concentrated throughout the 1830s, 1840s and into the 1850s. This pattern continued through *Period One* and into *Period Two* (1851–1900) with experimentations into suitable sites and regions that involved multiple failures along the way.

Figure 3.4 enlarges an 1813 engraving of Green Hills (Windsor) on the Hawkesbury River. The top image shows intensive farming resembling early market gardening. It is depicted from a similar vantage point to an 1809 painting in which flooding impacts are evident.[167] The bottom image shows a subdivision plan for 39 market garden allotments near the Sir Joseph Banks Hotel, dated 1840–1859. It reveals a process of existing larger market gardens being broken into smaller land parcels, and domination of British and European names—Saxby, Kellett, Phelps and Brown. The plan is suggestive of demand for market garden allotments in the area, later becoming known as 'Sydney's backyard vegetable garden'.

Fruit (FR): Hard Graft

Extensive trial and error were required to establish fruit production, let alone to create a commercial industry. Imported species and varieties needed to respond to unfamiliar macro and micro climatic patterns, dynamics and seasonality, soil capacity, ground and drainage conditions and unfamiliar pests and diseases. Experimentation within early estate orchards and orangeries greatly contributed to the colony's early fruit production. Governor Phillip's own garden successfully propagated species including mulberries, almonds and pomegranates. By the end of 1791 he was able to write to Banks,

> I had two or three bunches of grapes the year before last and last year several good bunches; at present the old vines in my garden are loaded with very fine fruit. One orange and two fig trees are also in good fruit. If I except the bananas, I do not know any fruit which has been tried and does not thrive as well here as it does in any part of the world.[168]

Phillip's horticultural success and passion, along with his resources, were not readily replicable by the average convict, emancipist or settler. An 1867 account reminds us,

> [W]e must recollect that [success] was not obtained without labour and difficulty, and the study of the most suitable soils and situations, when, all

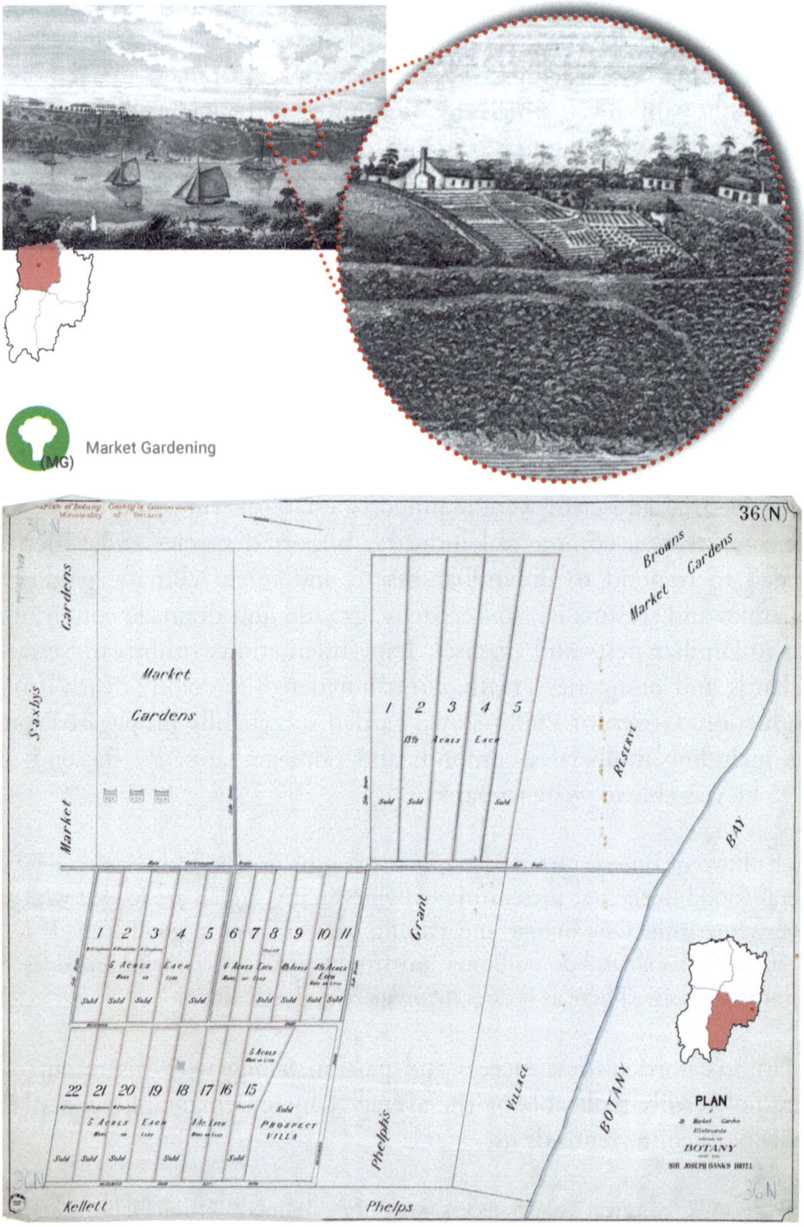

Fig. 3.4 An enlarged 1813 engraving of Green Hills (Windsor) on the Hawkesbury River (top), Botany subdivision Plan 1840–1859 (bottom). Sources: Philip Slaeger, Mitchell Library, State Library of NSW

> impediments having been overcome by judgement and perseverance, the experiments terminated in success, the trees became naturalised, were readily propagated, and after many years the result has been a source of great wealth to the colony, the produce finding a ready market, both for exportation as well as home consumption.[169]

Moderate- to large-scale fruit production therefore frequently took place in grand estates' orchards and orangeries. We can see fruit terraces as part of 'Mount Adelaide's' extensive garden design in Double Bay (Fig. 3.5, top). Terraces were laid out in the 1830s[170] alongside a large ornamental fishpond accompanying the garden and orchard, with vines likely producing table grapes rather than for viticulture.[171] When subdivided for sale in 1837, one 4-acre allotment included several hundred fruit trees and its vineyard was described as around 4,000 feet (1.2 km) long.[172]

In Annadale Estate, fruit production was part of a diverse range of mixed-farming enterprises as visible in 1843 (Fig. 3.5, bottom), with the enlargement emphasising its two orchard areas. Annandale contained a vineyard, orangery, cropping and granaries, dairy, livestock grazing and stables/breeding (cattle, sheep, pigs and horses), bakery, blacksmith forge and workshop, slaughterhouse/butchery, quarry, brick and tile yard and kiln and stores. The estate was the result of three land grants given to Captain George Johnston of the New South Wales Corps between 1793 and 1799. The holding extended more than three kilometres south from the harbour but the southern portion had the best soil and focused on farming activities. By his death in 1823, Johnston had acquired a considerable 3,689 hectares of land.

Outside of Sydney Cove's estates, the Hawkesbury River region recorded commercial fruit harvests alongside cropping and intensive farming successes from the early 1800s. By 1810, there were over 100 hectares of peach, plum and apricots and Kurrajong's elevated cooler climate supported apples and stone fruit (Fig. 5.9) following the 1820s.[173]

Three decades of relatively blissful production ensued before the emergence of the Queensland fruit fly.[174] By the end of *Period One*, fruit fly had become a serious pest, continuing to affect producers today.[175] Citrus proved less at risk and the Parramatta, Ryde and Hills Districts all became

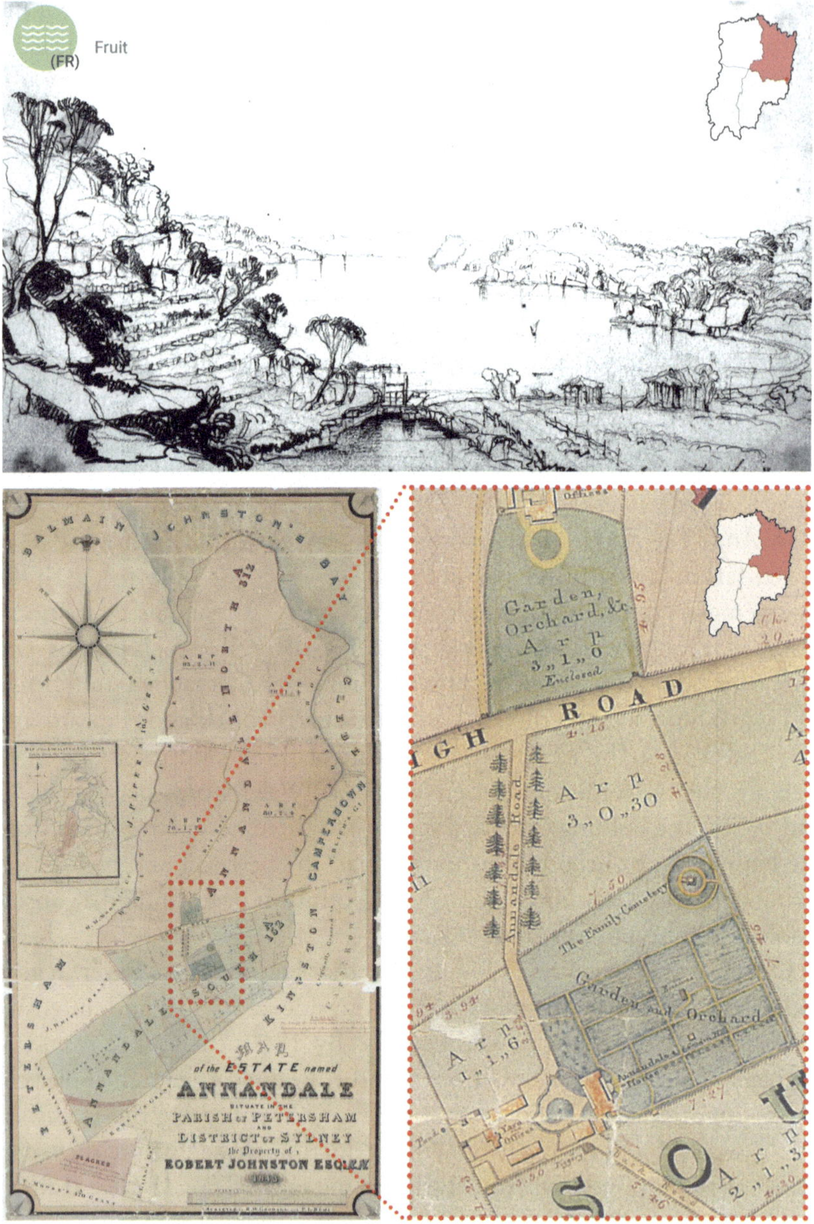

Fig. 3.5 Double Bay sketch showing terraces at Mount Adelaide (top), orchards at Annandale Estate 1843 (bottom). Sources: Georgiana Lowe; R.W. Goodall and P.L. Bemi, Mitchell Library, State Library of NSW

production regions (Figs. 4.7, 4.8, 4.9, 5.9, and 5.10). George Suttor claimed to be the first to send oranges to Sydney market in 1807,[176] though was impacted by a caterpillar plague around 1820.[177]

Cropping (CR): Sowing Rot

Period One was the most important and significant timeframe for Sydney's crop production. Faced with dwindling food stores and expensive additional imported supplies, self-sufficiency was essential. Crops such as barley, wheat and maize were key to feeding the settlers, given their dependence on non-indigenous foods. Maize seed, brought early to the colony, was more productive than wheat, which struggled in Sydney Cove's shallow soils. More regularly coming to be known as corn from the early nineteenth century, maize was cheaper and easier to produce, doubling as a crucial source of animal feed.[178] Lieutenant Watkin Tench commented in 1793 that,

> Of the vegetable productions, translated from other climes, maize flourishes beyond any grain. And as it affords a strong and nutritive article of food, its propagation will, I think, altogether supersede that of wheat and barley.[179]

Tench's prediction was not realised but not because his agricultural observations were incorrect. Rather, sociocultural influences drove diet preferences, in turn affecting landscape composition. Maize came to be maligned as the food of pigs and convicts, bearing the taint of poverty in a colony eager to establish social ascendency and soon disappeared from diets after convict transportation ceased.[180]

In the colony's early years, many sites near early Sydney's town settlement were cropped and a map of Farm Cove (now in the in the Botanic Gardens of Sydney) dated July 1788[181] includes nine acres of corn (Fig. 3.6). This was the first Government Farm and Phillip used a third of the colony's convicts, around 250, to cultivate the ground,[182]

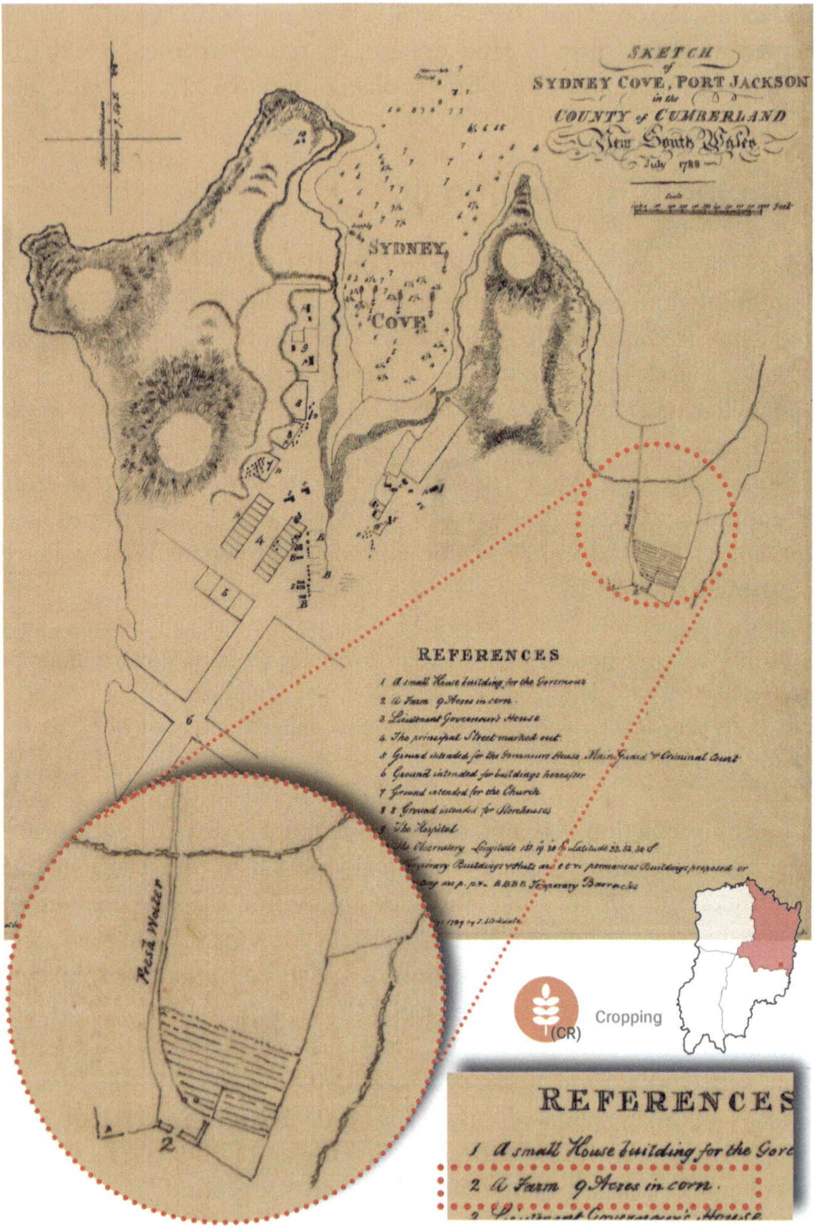

Fig. 3.6 Farm Cove corn cropping, July 1788. Source: Mitchell Library, State Library of NSW (Z/M2 811.16/1788/1A)

> A crop of wheat, barley and maize was sown. In Nov. less grain was harvested than originally sown. The crop was an absolute failure. Infertile soils formed on Hawkesbury Sandstone was the main factor contributing to crop failure. Trees blunted the axes, soil blunted the spades and picks, the men were not inclined to work hard. English wheat and barley rotted in the ground and the weevil destroyed the seed.[183]

Figure 3.6 shows a map drawn 6 months after British occupation in July 1788, revealing the concentration of settlement around Sydney Cove and the important Government Farm with its nine acres under corn production.[184] Farm Cove quickly became the area's namesake. It was abandoned by late 1791[185] and converted into a botanical garden in 1816 in the tradition of Joseph Banks' Kew Gardens in London.[186] Other Government Farm sites produced grain crops as well as inner-city regions such as Petersham Hill Reserve and Camperdown Park to the west of Grose Farm, now Sydney University. Convict and other workers were employed to clear 60 acres of government ground on the northern side of Paramatta Road, known as Orphan School Creek Valley, and sowed 20 acres with Indian corn.[187]

Cropping in Sydney Town was typically conducted destructively, in a 'slash-and-burn' manner. The rough-handed process involved clearing and burning trees, cropping the soil, leaving fields open to the elements causing nutrient mineralisation and organic matter accumulation, then repeating the process.[188] The lack of available animal manure and its high cost perpetuated slash-and-burn processes, with resultant wood ash assisting soils either deliberately or inadvertently. 1820, John Oxley feared that "without the assistance of manure, which is probably out of [the farmer's] power to procure, his lands will continue decreasing in produce, so that at the end of six years, the soil will be entirely exhausted".[189]

Oxley's prophecy came to pass for many early Australian agricultural districts, such as Rose Hill (Paramatta). The region's soils, whose enrichment was limited by the Paramatta River's small catchment basin, were soon exhausted by agricultural practices both on individual sites and across the district. Over time, the area transitioned to less nutrient intensive production such as livestock pasture and citrus orchards. As discussed earlier, the alluvial Hawkesbury siltation replenished soil fertility enabling ongoing cropping but was at the risk of the destructive flood events occurring at the bottleneck of the large Nepean and Hawkesbury Rivers. Maize remained undesirable and was grudgingly consumed when necessary. Once the hinterlands were opened up, cropping spread well beyond Sydney to arable districts more suited to wheat and barley production.

During the hungry years, canny farmers producing precious wheat supplies distilled their own spirits through small stills imported from England. This could return 250 per cent more from rum production than selling wheat directly to the Commissary.[190] The corrupt New South Wales Rum Corps shut down these farmers and proceeded to replicate this racket for themselves. The Hawkesbury district was renowned for its prevalence of stills, aided through its remoteness from the law and its ability to produce grain.

Notwithstanding rum production, grain became the basis of the colony's economy in its first few decades. The colony's largest grain producer, emancipated convict and later Windsor Magistrate Andrew Thompson, was the wealthiest settler by 1806 when Governor William Bligh arrived, though his success was short lasting as he died at 37 years old. [191] Grain's economic primacy was overtaken by wool by the middle of the period.[192] Crop fields are visible in both of the early engravings (Fig. 3.7), "The Brickfield-hill, or High road to Parramatta" 1796 (top, behind the large tree)[193] and "At Baker's farm, high land on the banks of the river" published in 1798 (bottom), where erosion and damage to the riverbanks are visible. Stumps, felled and burnt trees are evident in both depictions.

Intensive Livestock (IL): The Luxury of Livestock

The typology '*Intensive Livestock*' encompasses agricultural activities in which animals are confined in spatially concentrated conditions at higher stocking rates for meat, milk or egg production, including poultry, cows, sheep, pigs and goats. Standard practice holds that captive bred animals are provided with feed, as opposed to grazed on extensive pastures on rotational field systems (see *Grazing*).[194] Like grain, fruit and vegetable

Fig. 3.7 The Brickfield-hill, or High Road to Parramatta", 1796 (top), and "At Baker's Farm, high land on the banks of the river" published in 1798 (bottom). Sources: State Library of NSW (Edward Dayes, after Thomas Watling, artists. James Heath, engraver); William Lowry, State Library Victoria

production, animal agriculture also faced considerable barriers and setbacks. These were present from the early outset following the First Fleet's arrival,

> About twelve o'clock in the night [of 6 February 1788] one severe flash of lightning struck a very large tree in the centre of the camp under which some places were constructed to keep the sheep and hogs in. It split the tree from top to bottom, killed five sheep belonging to Major Ross and a pig of one of the lieutenants.[195]

The loss cannot be overstated. The value of the First Fleet livestock for breeding purposes was such that naval surgeon Arthur Bowes Smyth commented,

> [I]n England thieving poultry was not punished with death; but here, where a loss of that kind could not be supplied, it was of the utmost consequence to the settlement, as well as every other species of stock, as they were preserved for breeding. Therefore stealing the most trifling article of stock or provisions would be punished with death.[196]

Sydney's first execution (Thomas Barrett) took place in the Rocks on 27 February 1788, though for the theft of pork and split peas.[197]

One of *Intensive Livestock*'s key applications is dairying. Dairying was initially constrained by the lag time between breeding and establishing herds, identifying and clearing receptive lands, availability of fodder and pasture, climate (hot weather) and transportation challenges. Dairying suits moderately fertile and moisture-retentive soils experiencing longer rainy periods and cooler climates, hence its establishment often resulted from clearing the region's densely forested areas to create pastures. Dairying's spatial makeup related to the interplay of the proximity to markets, inherent land capacity and land values. Dairying spread south into the Illawarra region from the 1820s and north into the Hunter Valley,[198] albeit these areas faced numerous transportation challenges before refrigeration was developed (see *Period Two*).[199]

One of Sydney's earliest dairies was established on 700-acre Mona in Bungin (now Mona Vale), producing 'Foley's butter' from around 1825–1849. An early dairy site still exists in one of Australia's oldest buildings, 'Dairy Cottage' in Parramatta Park, part of a UNESCO World Heritage site. It began as the home of emancipist George Salter on a choice land grant of 30 acres given by Governor John Hunter in 1796. Salter established an operational farm, purchased back by Governor Lachlan Macquarie in 1813. Macquarie converted the cottage into an operational dairy that later became a part of a public park during the following period.

Prior to rail connections, early Sydney depended on urban production. Dairy cows were part of a menagerie of urban animals that wandered the streets, with various public orders put in place to attempt to curtail their nuisance. Governor Macquarie, for example, passed orders to keep livestock out of the Tank Stream and other important locations where their presence was a risk to public health. Prior to the development of wire fencing, containing livestock was additionally challenging and labour-intensive. As Sydney's population increased, so too did conflicts stemming from urban livestock—roosters' dawn cries, wafts of hog manure, and roaming cattle, goat and pig herds tested neighbourly relations, increasing the impetus to move intensive livestock further afield. Figure 3.8 includes a range of scenes from a book of sketches of Sydney from 1842 depicting livestock in the colony.

Grazing (GZ): Birthing a Nation's Economy

Pastoralism was a well-established tradition in Britain, accompanied by high social status.[200] Many of Sydney's aspirational colonists sought to establish sheep and cattle farms, becoming the driving economic force until the 1920s and catalysing a spatially unprecedented transformation of Australia's landscapes.[201] Yet pastoralism took some time to establish in the colony and challenges were considerable. Suitable animals and breeds needed to be selected and survive the considerable sea journey. Once on land, pastoralists and animals had to adjust to unknown conditions, plants and climates and time was required to breed and establish herds. Difficulties arose from shepherding, protecting and enclosing livestock. Suitable grazing areas needed to be identified and accessed beyond Sydney. Shifting dynamics around Crown land enclosure, grants, tenure,

Fig. 3.8 Views of Old Sydney 1842: South Head Road in Woolloomooloo Bay (top), sheep in Hyde Park (centre), and stables of Lyons Terrace (bottom). Source: Mitchell Library, State Library of NSW

leaseholds and squatting led to conflicts. Beyond the total enclosure of First Nations lands and destruction of their foodways and culture, violence regularly arose following spearing livestock.

Capital also created a challenge to herd establishment, thus an early pastoral stronghold by the Rum Corps was gradually infiltrated by powerful free settlers like Robert Campbell.[202] Between Campbell's arrival in 1800 and 1817, the colony's cattle grew from 1,000 to 45,000 and sheep from 6,000 to 170,000.[203] This signalled an opening of the continent's pastoral floodgates, in which non-native hard-hoofed animals and farming activities would significantly impact Australia's delicate biodiversity and its Indigenous custodians.[204] Locally, the Wianamatta shale soils of the Cumberland Plain's grassy woodlands provided Sydney's main grazing and agricultural lands until the opening of the western road to Bathurst in 1815.[205] This bolstered local meat supplies, reducing reliance on imported supplies such as Tahitian pork.[206]

The Cowpastures

The colony's precious initial cattle herd, purchased by Governor Phillip en route to South Africa on the First Fleet's voyage, suffered an early blow. During the King's Birthday celebrations on 4 June 1788, the seven surviving cattle (which included two bulls) escaped from their yards at Cattle Point (now the site of the Sydney Opera House). Matt Murphy attributes this to the drunkenness of their convict shepherd.[207] While the loss would eventually turn into unexpected agricultural serendipity for the colonists, it was a significant setback,

> Disastrously, [Phillip's] cattle strayed within five months and despite an extensive search were not found until seven years later, grazing contentedly on open pasture near the Nepean River and now numbering sixty one.[208] As it was presumed that the cattle had understood how to choose the best pasture for themselves, they were left to graze there and were protected by future governors. Not surprisingly, early colonists followed their example and set up their early pastoral holdings in the area.[209]

As Governor Phillip had already departed from Sydney when his now wild herd was located, their numbers grew significantly to 500–600 by 1801 and to several thousand by 1804. Over time this dissipated as they were opportunistically commandeered for private herds and slaughter.[210]

The loss of Phillip's cattle was incredibly serendipitous. It led to the establishment of south-west Sydney's 'Cowpastures' as the birthplace of New South Wales' and subsequently, Australia's pastoral industry, forming a foundation of the colony and a catalyst for the nation's prosperity.[211] Central figures in the region included the opportunistic and formidable John and Elizabeth Macarthur, who would also establish wool, wine and wheat industries.[212] The region's legend and the romanticism of its sprawling estates were still evident a century later in architect W. Hardy Wilson's *The Cow Pasture Road* (1920) (Fig. 5.12).

Sheep

The establishment of the sheep and wool industry traces several key controversial and intriguing figures characterised as 'heroes of the fleece'. These include George Johnstone (Annandale Estate—see *Fruit*) who arrested Governor William Bligh in the 1808 Rum Rebellion, Reverend Samuel Marsden, who imported and bred merino sheep at his 'Mamre Farm' and John and Elizabeth Macarthur. The Macarthurs had a profound impact on Australian agriculture. Their ability to realise the scale of their ambitions was unparalleled in the first decades of the new colony. John was a cunning, opportunistic, volatile and controversial figure who "as a disintegrating element" in New South Wales' administration "had no equal".[213] Jeanette Hoorn contends that large-scale land consolidation by successful entrepreneurs like the Macarthurs was a key factor in creating the necessary conditions for Sydney's pastoral industry to emerge.[214] Vast land agglomerations required for the commercial success of extensive grazing were dependent upon land acquisition through grants and purchases. In 1793, John Macarthur wrangled a 100-acre grant on some of New South Wales' best land,[215] naming it Elizabeth Farm after his wife (Figs. 3.9 and 8.7).[216]

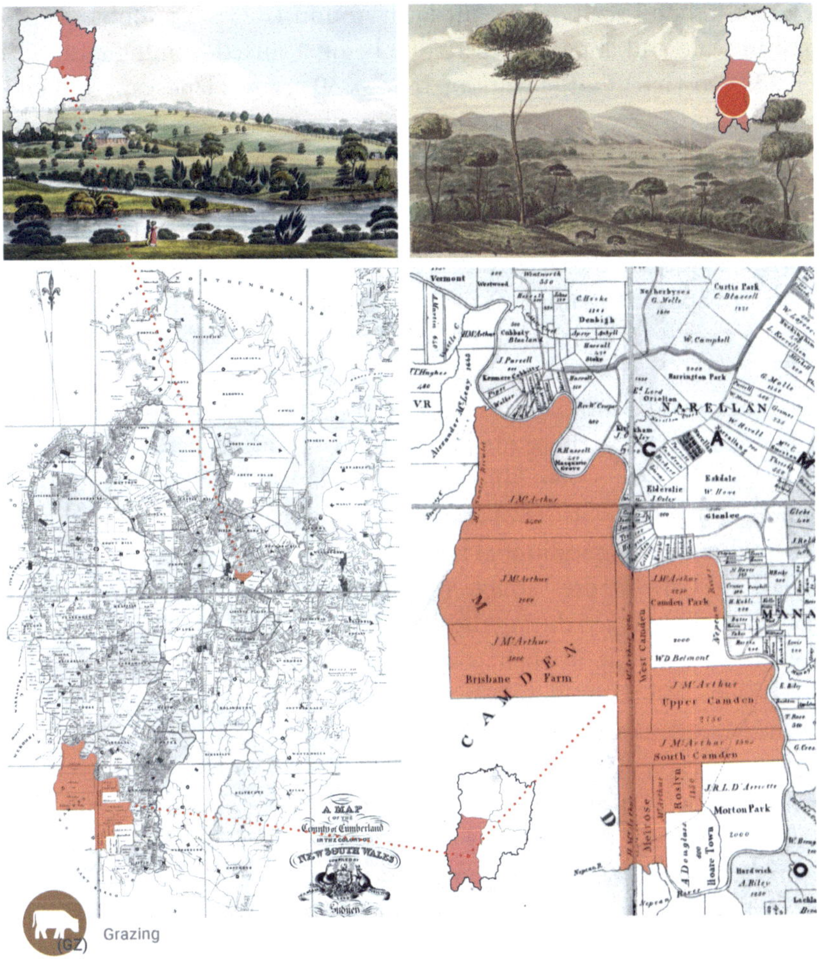

Fig. 3.9 The Macarthurs' Elizabeth Farm c. 1823 (top left), and a view of the Cowpastures district 1840–1846 (top right). The landholding maps (bottom) show Macarthur holdings c. 1840 at a County of Cumberland scale (left) and zoomed into the Cowpastures and Camden (right). Sources: Joseph Lycett and William Wells, State Library of NSW; Robert Marsh Westmacott, National Library Australia Bib ID: 3010283 and Joshua Zeunert (Adapted from Wells, 1840)

Artist Joseph Lycett depicts the farm around 1823 as undoubtedly pastoral in scale and appearance, despite being a mixed-farming enterprise (Fig. 3.9, top left).[217] The painting follows 30 years of landscape and built modifications. Yet it was comparably diminutive compared to the Macarthurs' unparalleled acquisitions at Camden Park Estate in the Cowpastures district (Fig. 3.9, bottom, red highlights). They became the largest landholder and recipient of the biggest single grant of 5,000 acres (2,024 hectares) in the coveted district.[218] At the time of John Macarthur's death in 1834, the family was one of the colony's wealthiest landowners, owning over 24,000 acres (9,500 hectares) of land with stock valued at £30,000".[219] This was over double the land area of the nearest rival.[220] The Macarthurs' land parcels in Camden (bottom right) are noteworthy in size compared to those on the north of the Nepean River. The map does not capture 'Seven Hills Farm' (Bella Vista Estate), which they owned from 1801 to 1821 but exchanged for Crown land in the Cowpastures,[221] nor additional holdings to the west[222] and further afield. Today, the Cowpastures' sizable parcels of land and the legacy of extensive land grants issued during the New South Wales Corps' self-interested reign remain visible in their continued lack of public open space and access (Fig. 3.9).

The Macarthurs' vision for the colony's wool industry was a mixture of feudal and capitalistic.[223] John sought to create a wool-exporting colony controlled by a colonial aristocracy with estates of at least 10,000 acres (4,047 hectares), maintained by convicts "in proper subjection" to guard against corrosion by "furious democrats".[224] Macarthur's preferred model for society was that of a "sort of vaguely despotic rural theme park" with an "agribusiness complex in colonial parliament".[225] Lucy Hughes Turnbull notes that free settlers in the region viewed Macarthur and his Rum Corps allies as avaricious and unrepresentative of their ambitions.[226] Governor William Bligh shared their view but ultimately to his downfall, as he was usurped in Australia's only military coup in 1808. The Macarthurs' experimentation and adaptation nevertheless pioneered primary industry translating to national wealth and associated comforts. Their success exporting wool from the 1820s was an early indication of an industry that would subsequently drive over a century of national prosperity 'ridden on the sheep's back'.[227]

Commons

The numerous 'commons' were a comparatively more egalitarian component in early Sydney's livestock grazing history. These lands were set aside as shared areas for free settlers[228] to graze their livestock herds.[229] One of the most prominent examples was Sydney Common, which would later become Moore Park, Centennial Park, Queens Park and Hyde Park (Fig. 3.8). The first official commons were the six Cumberland Commons established by Governor King in 1804. These comprised Nelson (5,650 acres/2,285 hectares and later Pitt Town Common), Richmond Hill (5,130 acres/2,076 hectares, later Ham Common) and Phillip Commons (6,150 acres/2,488 hectares, later Wilberforce Common), all in the Hawkesbury district. The Field of Mars and Eastern Farms Common (at 5,050 acres/2,044 hectares) ran along the southern side of the Lane Cove River from Hunters Hill to Pennant Hills. The Prospect Hill and Toongabbee Temporary Common was in Seven Hills; and the Baulkham Hills and Northern Districts Temporary Common was located in the north-western Cumberland Plain.

Although Sydney was the birthplace of Australia's pastoralism, graziers found that sheep fared more favourably in inland regions such as Bathurst, Yass, Goulburn and Orange. By the end of the period, focus had shifted inland to better climates and more expansive squats and landholdings. Like early cropping practices, pastoralism was accompanied by damaging agricultural practices. Major Thomas Mitchell noted that around the Wollondilly River on Dharug and Gundungurra Country, the "great number of sheep and cattle fed there" likely led to the area's "scantiness of grass".[230] Overstocking and overgrazing led to land degradation and negative succession of grazed ecosystems.

Vineyard Wineries (VW): A Civilising Drop?

Although not technically a food landscape, this book includes viticulture due to its significance within Sydney's and Australia's agricultural history. Wine has been a part of Australia's history since occupation in 1788. The first vines (and seeds) arrived as cargo on the First Fleet. Soon after disembarkation, they were planted at Farm Cove in what is now the Botanic Gardens of Sydney. Struggling to establish in the sandstone ground, the vines were transplanted to Parramatta where Governor Phillip had created a three-acre vineyard by 1791. German settler Philip Schaffer received 140 acres and 8,000 vines from Phillip and planted an acre of vines at 'The Vineyard' in Rydalmere. Given the colony was experiencing famine at the time, the investment in wine production indicates both the occupiers' penchant for alcohol and wine's status as a more 'gentrified' drink to that of harsh, concentrated rum.[231] In the following years, numerous other settlers attempted to establish viticultural ventures. Table grapes were sometimes grown alongside those destined for wine, although these were (and have remained) a distant secondary element.

Notable vineyards are recorded in the Map Database across each of the five time periods. *Period One* includes 20 entries with additional listings classified into (for example) 'mixed farming MF' pursuits that included vineyards in their agricultural activities. Several vie for the contested claim of being Australia's first, or first successful, vineyard. Notable vineyards within the period include: The Vineyard[232] (later known as 'Subiaco'); Mamre Vineyard in Orchard Hills (St Mary's); Marsden's Mill Dam Farm/The Vineyard (Betts and Blaxland's Redbank Mill Farm) in North Parramatta (which had over 5,000 vines in 1841);[233] The Macarthurs' at Camden Park; Varroville in Minto; Winbourne in Mulgoa; Wivenhoe vineyard in Narellan and Brush Farm Estate in Eastwood, amongst others. Free settler Gregory Blaxland was a successful wine producer across several properties. By 1822, he had shipped 136 litres of wine, primarily from his purchased Brush Farm Estate, to London where it won the Silver Medal of the Society for Encouragement of Arts, Manufactures and Commerce (now the Royal Society of Arts). A subsequent larger shipment of 1,800 litres won the Gold Ceres Medal in 1828.[234]

Before his death in 1834, John Macarthur petitioned for immigrant workers to be allowed into the colony under the new 'Bounty Scheme'.[235] This was initially capped at six men and their families, though Macarthur hoped to secure several hundred skilled workers from Germany, Switzerland and France.[236] Six experienced German vinedressers from Erbach on the Rhine were selected, and the Macarthurs' third vineyard was established in the Camden Park stronghold close to their winery in 1840. By 1849, Camden Park was producing 70,000 litres of red and white table wines and brandy annually.[237]

Viticulture increased momentum towards the end of the period following an 1830s "'fever' for growing vines".[238] 1843 statistics record 183.6 acres of grapes being grown in the Sydney Basin, producing 12,315 gallons of wine and 497 gallons of brandy, and expanding to 220 acres in 1845.[239] These official figures from the Colonial Secretary appear to vastly under capture the colony's scale of production as Camden Park Estate's in fact eclipsed that of the total colony's purported production. Another colossal viticultural undertaking was that of Sir John Jamison's 'Regentville' on the Nepean River near Penrith, whose 15-acre vineyard included extensive terraces and field vineyards with 30–40,000 vines and 200 varieties. It should be noted, though, that spatial scale and planting statistics do not necessarily equate with viticultural success.[240]

Agricultural Infrastructure (IN): Laying Foundations

The colony's agricultural infrastructure had an inauspicious start. All initial cultivation was done by hand, chiefly by convict labour using basic spades, hoes, tools and makeshift implements. Some ploughs became available from around 1800, [241] but overall, there was minimal assistance from animals and machinery into the 1820s.[242] Farm labour, transportation and exploration were constrained by a lack of horses until enough had been bred by around 1810.[243] Forced convict labour was essential to the colony's establishment in these difficult conditions. Sandstone was not easily hewn, for example, with early roads more accurately constituting tracks. The most important track ran for about a mile (1.6

kilometres) from Sydney towards Rose Hill, south of current Parramatta Road, and was created in the early 1790s.[244] This acted as Sydney's main terrestrial transport pathway and linked the two sister settlements of Sydney and Parramatta.[245] Parramatta Road was laid around 1797 and accessed by private coaches, although a regular service did not appear until 1823. With the challenges of road construction, the harbour and waterways remained central to transportation.

A lack of milling infrastructure caused further hurdles to self-sufficiency. The Scarborough (ship) delivered four millstones to Sydney Cove on 22 February 1788 but the colony lacked both a millwright and a reliable millstream, with the first watermill at Parramatta only functioning intermittently. Grain had to be ground day and night by 40 iron hand-mills that blunted within a year, forcing use of querns, mortars and pestles.[246] Sydney's first windmill was set up at Millers Point in 1797, and by 1809, there were six more. John Dickson's steam-driven mill opened in 1815 and could grind 10 bushels of wheat per hour by pumping water from a dam built at Cockle Bay.[247] Horse-drawn mills followed outside of the city (Fig. 5.13, top right). Technology remained rudimentary—the "grain was ground by a runner stone revolving on a fixed bedstone as it had been for centuries, and the flour was sifted through silk screens".[248] Mills continued to be constructed and by 1840, around 200 existed.

Flour was essential to the early colony's rations as perishable goods could not be stored. Potatoes, pumpkins, onions, long-lasting fruits and salted meat were transported from Tasmania or further afield but dairy and other fruits and vegetables could not withstand long journeys.[249] The first load of ice arrived in Sydney from Boston in January 1839, where it rivalled cotton as the second largest export commodity of the United States.[250] Frozen vegetables were processed as early as 1849[251] but it was not until the latter stages of *Period Two* that refrigeration became a viable industry. Growers in the County of Cumberland were therefore key to supplying the colony with fresh produce.

Markets were also crucial for food storage. The first markets were established near the temporary store houses east of the Tank Stream.[252] From 1792, regular marketplaces were set up on Sydney and Parramatta jetties. In 1806, Governor Bligh moved Sydney's market from Hospital Wharf to an open space above it in High Street (now George Street). In 1809, this was moved up the same street to the Old Parade Grounds.[253] Governor Macquarie prohibited the use of the hospital wharf for market activities in 1810, building a market wharf in Cockle Bay (Darling Harbour) and relocating the marketplace to the site of the current Queen Victoria Building. Macquarie also gazetted a market site in Parramatta in 1812. He utilised talented emancipated convict architect Francis Greenway to draw up an opulent vision for Sydney's central market building. To his and Greenaway's dismay, only a whittled down version was able to be completed in 1820 but its domed roof was still deemed ornate enough to be used as the Police Courts by the 1830s.[254]

High Street's central market and the original hay, corn and cattle market were well situated in relation to the main Sydney-Parramatta road and to city grocers and consumers.[255] This balance was disrupted when Governor Darling moved the cattle market to the base of Brickfield Hill in 1829 and Governor Burke transposed hay and grain to the cattle yards, rebuilding the sheds at the George Street Markets between 1831 and 1834.[256] In 1842, Sydney Council took control and moved the cattle market further out of town. These changes reflect the overall evolution of the city and its planning and lack thereof, particularly its increasing size, specialisation, conflicts and synergies between land uses and infrastructures.

Additionally, infrastructure needed to include processing facilities for crops, grain and livestock, including abattoirs. These gained momentum throughout the eighteenth century. In 1846, Sizar Elliot opened the first meat canning works utilising town gas supply.[257] Elliot was also a notable advocate for Australia's agricultural products, such as wine, lambskins and timber. Other canning plants soon followed and canned meat, initially beef and mutton, with rabbit from the 1870s, became a significant export to Britain.

The pace and magnitude of agricultural and infrastructure impacts saw the colony's first environmental laws passed in 1848 banning noxious trades within the city, including tanneries, woolwashers, rendering works, soap and glue makers and fellmongering. Just after the end of *Period One*, these were relocated outside of the city limits south of Cleveland Street to the water supplies of Waterloo and Botany Swamps.[258] Market gardens were included on the list of noxious trades and activities due to their use of fertilisers, including human manure.[259] Similar laws would continue to be passed throughout *Period Two*.

Aquaculture (AQ): Fishing, Sealing and Whaling

Despite dispossession and decimation, First Nations groups continued to live around the harbour and were forced to either find niches or live in government-constructed settlements. Governor Macquarie established a fishing settlement in Elizabeth Bay with nets and a boat, which was inhabited by 42 people. There are multiple accounts of Gadigal, Cammeraygal, Wangal and Burramattagal people trading fish or being 'employed' as fishermen.[260] As recorded in 1825,

> Sydney is most abundantly supplied with fish. They are caught with hooks and lines, chiefly toward the heads of the harbor, by the native blacks principally, and disposed of by them to the retailers, who hawk them about the town. The sounds of "Fish O!" chimes agreeably upon your ear early in the morning—agreeably, I say, because they recall to your remembrance, even in these unmusical strains, scenes one has so newly forsaken. Kingfish, mullet, mackerel, rockcod, whiting, schnappers, bream, flathead, are all found most abundantly about Sydney.

Port Jackson's minimal siltation also correlated with incredibly clear waters, supporting diverse aquatic life in crystalline conditions.[261] The ramshackle 'Fisherman's Village' known as 'Booralee Fishing Town' in Gwea (Botany) from the 1810s–1820s endured into the late 1970s until eviscerated by port development.[262] The Doyles, who settled in Kutti (Watson's Bay), built a hut from which they sold fish, later opening a small eating room that later developed into the iconic Doyles Restaurant.[263]

Sydney's waters were exploited for their seafood but to a much larger extent for their infrastructural role within a global trade network. Although occurring outside of Sydney, Whaling and sealing formed another lucrative marine venture farther from the coast, with animals slaughtered for trade of resources rather than food. Whaling and sealing were so significant that until the development of pastoralism, Australia rode on the 'whale's back' through the export of "whale oil, whale bone, sealskins, and seal oil that provided the first export staples of Australia".[264] In 1832, the whaling industry accounted for 52 per cent of Australian exports, falling to one per cent in 1855 as pastoralism expanded and petroleum replaced whale oil as fuel.[265]

British diets, however, favoured terrestrial foods for their primary intakes, with beef associated with the landed gentry and those seeking to emulate it. Even as a comparative dietary supplement, however, aquaculture took a significant toll on the environment. Sydney's marine environments were prioritised for transportation and trade, rather than for a sustainable aquaculture industry. Settlers' rapid pollution of Sydney's chief water source, the Tank Stream[266], would over time extend into marine environments. Natural oyster beds were overexploited, and as discussed in subsequent periods, aquatic malpractices continued to expand.

Miscellaneous (MC): Horticultural Emergence

Horticulture and associated expertise emerged in the 1820s and 1830s, as evident in several historic clues. The Royal Agricultural Society of New South Wales was established in 1822 under the patronage of Governor Brisbane.[267] In the same year, land in Parramatta was granted to the Parramatta Agricultural and Horticultural Society who established a nursery resulting in propagation and distribution of fruit trees and grapevines.[268] From 1827, Darling Nursery in Chippendale near Sydney University struggled to survive as the Sydney Botanic Gardens distributed plants for free,[269] while Camden Park's nursery had a catalogued inventory of 3,000 plants in 1840.[270]

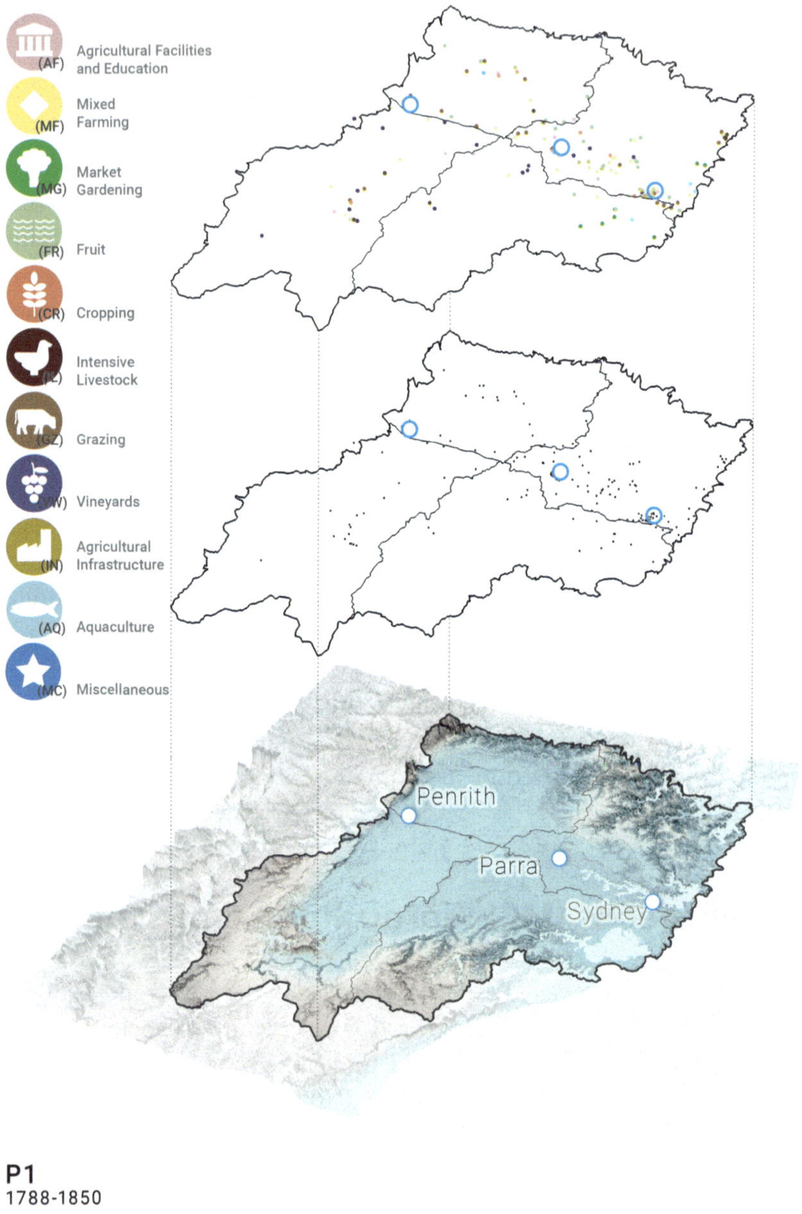

Fig. 3.10 *Period One*'s Food Landscape Map Database. Above the 3D base map, the total (middle, in black) and 11-typology-specific food-producing entries (top, coloured) are shown for *Period One*, 1788–1850. Three reference locations are shown as large blue dots: Penrith, Parramatta and Sydney (CBD). The entries are fanned across a wide region, reflecting early large landholdings and grants. Source: Joshua Zeunert, Josh Gowers and Stephanie Stankiewicz

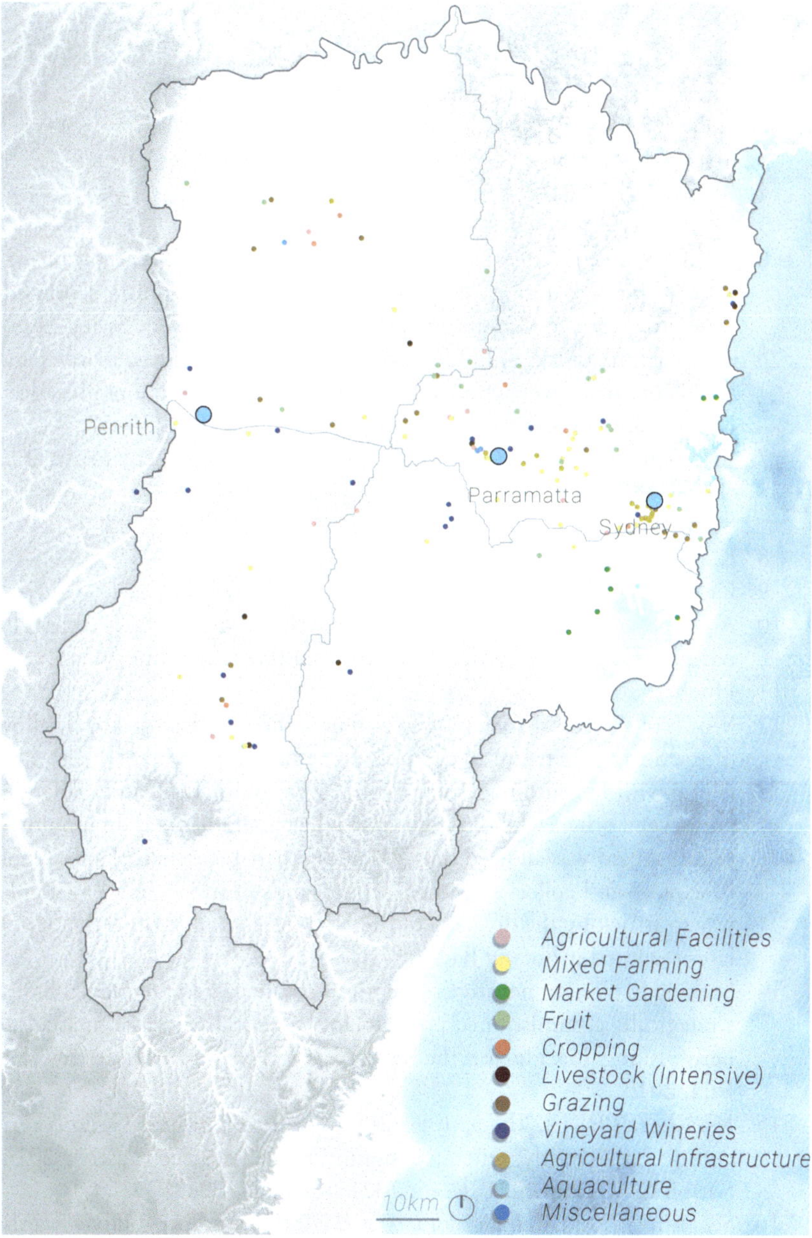

Fig. 3.11 *Period One*'s Food Landscape Map Database: plan view. Source: Joshua Zeunert, Josh Gowers and Stephanie Stankiewicz

Notes

1. Already established but renamed from prior 'Green Hills'.
2. Kohen, 2000: 82.
3. https://deadlystory.com/page/culture/articles/Standing_Strong
4. Blyton, 2015: 299.
5. Flannery, 1999: 3–4.
6. Lines (1991) challenges an oversimplified and popular notion of convicts as mere thieves, noting "lawyers, reformers, insurrectionists, land-and-tithe protestors, food rioters, conspiritists, radical weavers, Swing rioters and machine breakers, luddites, trade unionists, Chartists, martyrs and farm labourers…were effectively pushing back against the ramifications of land enclosures and mechanisation". See Lines, 1991: 55–57.
7. It is estimated that 52,000 convicted criminals were sent to America (Fitzgerald (1999: 13) claims 80,000) and 164,000 to Australia. See SLNSW, 2015.
8. Karskens, 2010: 64.
9. Crockett, 2017.
10. As opposed to agricultural preparation, the journey was successfully navigated with only 25 deaths, considered few at the time.
11. Hughes, 2003: 1.
12. Words of the Royal Society, who also note the Esq. was already a Fellow of the Society. Cited in O'Brien, 1997: 64.
13. Along with his entourage that included naturalist Dr Daniel Solander and several others who did not survive the trip. Banks was disappointed that his team was allowed only 7 days by then-Lieutenant James Cook to explore and collect specimens in the lands near Stingray Bay. Banks was so struck by the diversity of the flora in his explorations that Cook later renamed it Botany Bay.
14. Seeking to counter negative perceptions circulating in the media, Banks strategically commissioned portraits to shape his image as a "masculine hero risking his life for the sake of England and of science". See Fara, 2013.
15. Ma Rhea, 2017: 4. The nature and interpretation of First Nations agriculture has recently generated significant debate. See: Pascoe, 2014; Sutton & Walshe, 2021; Keen, 2021; Rintoul, 2021.
16. See https://joshuazeunert.com/2016/12/02/beyond-capability-towards-new-territory-in-british-landscape-architecture-and-education/.

17. Land unaffected by the influence of humans, harkening back to the legal term *terra nullius* or 'land belonging to no one'.
18. Gammage, 2011.
19. Joseph Banks' barrister father was a Member of the House of Commons from 1741 to 1747.
20. Hughes, 2003: 57.
21. Banks social status outranked that of James Cook and Banks is documented as exerting significant influence over Cook.
22. Along with James Matra (Magra) (to which the Sydney suburb of Matraville is named). Matra was a diplomat and sought to be Governor of Sydney.
23. Tink, 2020.
24. Ibid.
25. Hunwick, 2018: xv.
26. Flannery, 1999.
27. Karskens characterises Banks' and Cook's appraisals as "wrong, or irrelevant" (2010: 6).
28. Shann, 1948: 1.
29. Flynn & Sturgess, 2014.
30. Hughes, 2003: 92.
31. Flannery, 1999: 13–14.
32. Flannery, 1999: 81.
33. Flannery, 1999: 81.
34. Cited in Flannery, 1999: 81–82 and Hunwick, 2018: 2.
35. See https://aiatsis.gov.au/explore/map-indigenous-australia.
36. The First Fleet's inventory of animals was recorded as comprising: 18 Turkeys; 29 Geese; 35 Ducks; 122 Fowls; 87 Chickens; Kittens; Puppies; 4 Mares; 2 Stallions; 4 Cows; 1 Bull; 1 Bull Calf; 44 Sheep; 19 Goats; 32 Hogs; 5 Rabbits. A full inventory is listed here: https://firstfleetfellowship.org.au/library/first-fleetlist-livestock-provisions-plants-seeds/.
37. "And God said, Let us make man in our image, after our likeness: and let them have dominion over the fish of the sea, and over the fowl of the air, and over the cattle, and over all the earth, and over every creeping thing that creepeth upon the earth" (Genesis 1:26-31 King James Version). Also see more recent ecological interpretations of the Bible: https://www.vatican.va/content/francesco/en/encyclicals/documents/papa-francesco_20150524_enciclica-laudato-si.html.

38. Cicero, Nature of the Gods, Book 2.
39. Flannery, 1999: 13–14.
40. Morris, 2008: 22.
41. Ma Rhea, 2017: 3.
42. Pratley & Rowell, 2003: 22.
43. Pratley & Rowell, 2003: 22.
44. Ma Rhea, 2017: 161.
45. Hughes, 2003: 97.
46. Fitzgerald, 1999: 7.
47. Cited in Flannery, 1999: 80.
48. O'Connell, 2017; O'Connell, n.d., 1788.
49. Hunter, 1793: 70.
50. https://dictionaryofsydney.org/entry/phillips_table_food_in_the_early_sydney_settlement.
51. O'Connell, 2017: 7.
52. Farrer, 2005: 17.
53. Ma Rhea, 2017: 133.
54. Hunwick, 2018: xviii.
55. Ma Rhea, 2017: 11.
56. Stone, 2019: 3.
57. O'Connell, n.d., 1790.
58. Ibid.
59. Shann, 1948: 8.
60. Sydney Gazette August 24, 1806, cited in Hunwick, 2019.
61. Ibid.
62. https://australianfoodtimeline.com.au/tag/1780s/.
63. See Zeunert et al., 2025.
64. Farrer, 2005: 17.
65. Birmingham, 1999: 124.
66. Along with millions of years of relative geological stability equating to extensive natural weathering processes, tens of thousands of years of fire management reduced soil fertility for Western agricultural approaches (Pellegrini et al., 2017).
67. Short, 2000: 27.
68. Also see Zeunert et al., 2025.
69. Flannery, 1999: 10–11.
70. Western agriculture typically 'mines' soils for nutrients, which are transferred into foodstuffs. Excrement, human or animal, contains many of these nutrients and has thus traditionally been returned to the soil to create a loop or cycle of nutrients.

71. Hughes, 2003: 97.
72. Shann, 1948, 9.
73. Shann, 1948, 9.
74. Hunwick, 2018: xv.
75. Atkinson in Pratley & Rowell, 2003: Ch.1.
76. Lines, 1991.
77. Cited in Birmingham, 1999: 283.
78. Murphy, 2021: 50 (ebook pagination).
79. See: Shaw, 1966: http://adb.anu.edu.au/biography/bligh-william-1797.
80. Phillips, 1909: 10.
81. See Wakefield, 1829.
82. Further discussed in *Agricultural Facilities & Education*, *AF.*
83. The name was changed to Parramatta on the King's birthday in 1791. Parramatta (or Burramatta) loosely translates to "the place where the eels lie" in Darug.
84. See: https://www.environment.nsw.gov.au/-/media/OEH/Corporate-Site/Documents/Animals-and-plants/Biodiversity/bio-map-cumberland-subregion.pdf.
85. Shann, 1948: 8. Given time, however, farming in the area would be successful and the region's produce helped to support the colonists in its early years.
86. Read, 2015.
87. Karskens, 2010: 81.
88. See Fig. 8.5 for a vantage from the Government House site.
89. Christie, 1988.
90. Short, 2000: 27.
91. Clarke, 2017: 25.
92. An 1809 painting depicts an eroded 'Hawksburgh' (Hawkesbury)' riv-erbanks at Green Hills (Windsor) in 1809, a year it experienced its fifth flood in a decade. The foreground tree is visibly impacted by flood waters. See: https://digital.sl.nsw.gov.au File number: FL1152088.
93. Fletcher, 1976: 53–54. A disastrous Hawkesbury flood did not affect Governor Bligh's own 'model' farm, efficiently managed by Andrew Thompson but controversially utilising supplies from government stores and flocks.
94. As of mid-2022, the region has been subject to four floods in recent years.
95. Atkinson, 1826: 7–8, cited in Proudfoot, 2017: 42.

96. Proudfoot, 2017: 42–43.
97. Fletcher, 1976: 53. He outlines a range of statistical data outlining the state of agriculture in the colony from 1788 to 1821 including the proportion of government controlled crop acres and livestock (p. 28, 43), grain yield and its percentage provided by Government Farms (p. 40), the number of land grants and their acreage (p. 52) and the proportion of production by key regions as well as the crops and livestock these produced (p. 200).
98. Proudfoot, 2017: 42, citing Fletcher, 1976: 200.
99. Mason & Knowd, 2010: 63.
100. Thomas Mort's ability to apply refrigeration to the logistics of food transportation is noteworthy—see *Period Two*: *IN*.
101. Benson & Howell, 1990: 12.
102. Clarke, 2017: 25.
103. Short, 2000: 35.
104. Caledonian Mercury 7 March 1814, cited in O'Connell, 2017: 8.
105. Caledonian Mercury 7 March 1814, The Scots Magazine and Edinburgh Literary Miscellany, 1814, cited in O'Connell, 2017: 8.
106. O'Connell, n.d.: 1819.
107. Ibid.
108. Convict transportation to Australia ceased in 1868.
109. Morris, 2008: 52.
110. Stone, 2019: 18.
111. Stone, 2019:19.
112. E.g. Hughes Turnbull, 1999: 83.
113. Hoorn, 1990: 3.
114. By Surgeon John White, as cited in Flannery, 1999: 52.
115. See for example images 3, 4 and 6 at: https://mhnsw.au/stories/convict-sydney/convicts-colony/.
116. Stone, 2019: 18.
117. Figure 3.1's 1806 watercolour depicts the third Government Farm, 'Government Agricultural Establishment Castle Hill', around 5 years after it was established. The previously dense Blue Gum High Forest has been clear-felled and burnt by convict labourers, primarily to run livestock. Three areas for growing crops in rows are visible. After 1900, the site grew fruit in orchards in its south-west corner.
118. A successful venture until it ceased in 1832.

119. https://historyandheritage.cityofparramatta.nsw.gov.au/research-topics/suburbs/winston-hills-brief-history.
120. O'Connell, n.d., 1793.
121. Stone, 2019: 18.
122. https://www.sl.nsw.gov.au/stories/australian-agricultural-and-rural-life/first-farms.
123. Stone, 2019: 18.
124. La Croix, 1992: 4–5.
125. Stone, 2019: 18.
126. Winston 1957: 9.
127. Shann, 1948: Chapter II.
128. Karskens, 2010: 64.
129. Of Captain Joseph Foveaux. A parallel discussed in *Period Five* is authority for large planning approvals ultimately resting with the Planning Minister.
130. Fletcher, 1967.
131. Goodall, 2019: 46–47.
132. Shann, 1948, Chapter II.
133. O'Connell, n.d., 1793.
134. https://www.cockatooisland.gov.au/en/our-story/convict-era/
135. Heydon, 1966.
136. Christie, 1988: 23.
137. Shann, 1948: Chapter II.
138. Yarwood, 1967.
139. Morris, 2008: 22.
140. Governor's House at Sydney Port Jackson, 1791: https://sydneylivingmuseums.com.au/convict-sydney/convicts-colony#gallery-7 and 1809: http://digital.sl.nsw.gov.au/delivery/DeliveryManagerServlet?dps_pid=FL1152081&embedded=true&toolbar=false.
141. Government House, Parramatta, 1805: http://digital.sl.nsw.gov.au/delivery/DeliveryManagerServlet?dps_pid=FL1152082&embedded=true&toolbar=false and 1809: http://digital.sl.nsw.gov.au/delivery/DeliveryManagerServlet?dps_pid=FL1152083&embedded=true&toolbar=false.
142. https://collection.maas.museum/object/236809.
143. Ibid.
144. See Stone, 2019: 1, who uses Ruse as the introductory story to Australia's squatters.

145. Floyd (2015) states that Reverend Richard Johnson grew successful wheat crops before Ruse on his land near the Canterbury on river flats of the Cooks River.
146. Fletcher, 1967.
147. This date is contested and the date on the actual deed is mutilated.
148. See: http://archival.sl.nsw.gov.au/Details/archive/110363516.
149. Wheat and most field crops are heavy feeders, depleting phosphorus and nitrogen amongst other nutrients and minerals.
150. He later lived in Bankstown, Riverstone and Macquarie Fields.
151. Such as Gregory Blaxland at Brush Farm in Dundas Valley.
152. Such as by William Wentworth. See Persse, 1967.
153. Blaxland also established a saltworks, farm and factory at Newington Estate.
154. Such as William Wentworth in Vaucluse House (and Gregory Blaxland was also part of the expedition that successfully crossed the Blue Mountains).
155. Mamre House still exists and at the time of writing is run by CatholicCare.
156. See e.g. Raeburn et al., 2022.
157. Christie, 1988: 19.
158. The pseudo-definitive nature of the 11 typologies used in this book are not always helpful, as the nature of the typologies can change over the course of the timeframe of this book.
159. Whitaker, 2006: 23.
160. Multiple other sources quote this National Trust reference, who note, for example, the vegetable farms from 1830 in La Perouse cum market gardens that they Heritage listed in 2010–2011. See National Trust, 2010: 2.
161. Fletcher, 1976: 200. Parramatta's figures likely skew the fruit component of orchards which were emerging in the region (see *Fruit FR*).
162. Fletcher, 1976.
163. James, 2016: 113, citing Gardening Australia 2001 and Humphreys et al., 2001.
164. Multiple other sources quote this National Trust reference, who note, for example, the vegetable farms from 1830 in La Perouse cum market gardens that they Heritage listed in 2010–2011. See National Trust, 2010: 2.
165. Mulcair, 2007, cited: https://www.facebook.com/Igrewupinmortdale2223/posts/2902384879991880.

166. https://www.facebook.com/Igrewupinmortdale2223/posts/2902384879991880.
167. https://digital.sl.nsw.gov.au/delivery/DeliveryManagerServlet?dps_pid=FL1152088&embedded=true&toolbar=false.
168. Morris, 2008: 22.
169. The Orange in New South Wales (from the Intercolonial Exhibition Official Record), In The Inquirer and Commercial News, Perth WA, 14 August 1867: https://trove.nla.gov.au/newspaper/article/69386876.
170. Morris, 2008: 53.
171. Morris, 2008: 55.
172. Morris, 2008: 55.
173. http://hawkesburyheritage.blogspot.com/2015/.
174. *Bactrocera tryoni*, also known as Froggatt or 'Qfly'(Drew 1989, cited in Dominiak & Daniels, 2012: 106).
175. Ibid.
176. McClymont, 2013.
177. Parsons, 1967.
178. Owen et al., 2017: 34.
179. Tench 1793: 164–165, cited in Owen et al., 2017: 34.
180. Owen et al., 2017: 35.
181. Mitchell Library, State Library of NSW Z/M2 811.16/1788/1A.
182. Christie, 1988: 17.
183. Floyd, 2015.
184. Casey & Lowe, 2002: 37.
185. Although some experimental farm crops continued to be grown there.
186. Banks was Kew's first unofficial director following his South Sea voyage with Cook.
187. Campbell, 1930: 278–279, cited in Kerr, 2002: A15.
188. Pratley & Rowell, 2003: 8.
189. 1820 Bigge report, cited in King, 1948.
http://www.nswskn.com/wp-content/uploads/2019/05/Roy_Lawrie_The-soils-beneath-Australias-first-successful-wheat-crop-at-Parramatta.pdf.
190. Murphy, 2021.
191. Thompson further leveraged the generation of wealth through other enterprises such as a brewery and tannery. Byrnes (1967) writes: "Thompson's life and work in New South Wales and the administration of his estate illuminate many aspects of life in early Australian history. His swift rise to wealth illustrates the government attitude towards

emancipists from the beginning; his activities prove the importance of grain as a basis of the economy until it was replaced by wool; his enterprises show him to have been a powerful influence in the development of industry and commerce; his association with Bligh points to the exceptional prosperity of Bligh's farm as one important reason for Bligh's arrest; whilst his appointment to the magistracy became a lever for Macquarie's enemies in their attempts to have the governor removed".

192. Byrnes, 1967.
193. These could also be vegetable market gardens as it is hard to discern as a small background element.
194. Until the advent of feedlots from the 1950s.
195. Arthur Bowes Smyth, Cited in Flannery, 1999: 61.
196. Arthur Bowes Smyth, 6 February 1788, Cited in Flannery, 1999: 62–63.
197. Flannery, 1999: 69.
198. Wadham et al., 1964: 18.
199. Wadham et al.,, 1964: 17–18.
200. Crook, 2014: 6–7; Khara et al., 2021.
201. Zeunert, 2024.
202. Hughes Turnbull, 1999: 68.
203. Hainsworth, 1972: 206, cited in Hughes Turnbull, 1999: 68.
204. Flannery, 1999; Low, 1999, 2003; Ma Rhea, 2017.
205. Benson & Howell, 1990: 68.
206. Ibid.
207. No doubt significantly worsening the post-celebration hangovers. See Murphy, 2021: 43 (ebook pagination).
208. Some accounts say 40, see Stone, 2019: 3.
209. The specific details of the above incident vary according to the source. This account is from the Scenic Hills Association, see http://www.scenichills.org.au/history_6.html.
210. Stone, 2019: 3.
211. Robinson, 2008.
212. Robinson, 2008.
213. Ibid.
214. Hoorn, 2007: 63.
215. Steven, 1967.
216. Elizabeth has been extensively written about, including: Scott Tucker, 2018; Grenville, 2022; Atkinson, 2022.

217. Australia's oldest European olive (*Olea europaea*), for example, survives at Elizabeth Farm. See Morris, 2008: 46 and https://www.smh.com.au/lifestyle/a-hardy-evergreen-20120310-1uqld.html.
218. Steven, 1967, states that Camden Park incorporated over 60,000 acres (24,281 hectares) through grant and purchase. This does not correlate with the Macarthur's descendants who state that in "1837 The Macarthurs' Camden Estate reached its maximum size of 27,698 acres at the Cowpastures". https://www.belgennyfarm.com.au/history/site-history/the-cowpastures-and-its-wild-cattle.
219. https://sydneylivingmuseums.com.au/convict-sydney/john-macarthur.
220. McLoughlin, 1988: Fig. 3.6.
221. https://camdenhistorynotes.com/2017/04/16/bella-vista-farm/.
222. As shown in an c. 1840 map in: McLoughlin, 1988: Fig. 3.6. The map also depicts landholdings accumulated by other powers including King, Riley, Wentworth, Blaxland, Jamison and Lord.
223. Shann, 1948, 9.
224. Steven, 1967.
225. Birmingham, 1999: 318, 320.
226. Hughes Turnbull, 1999.
227. See e.g.: https://dl.nfsa.gov.au/module/1592/.
228. Burgin, 2010: 263.
229. Sydney's commons were established at a time when English commons were being enclosed and privatised. See: Baskerville, 2006.
230. Ma Rhea, 2017: 164.
231. McIntyre, 2009.
232. See e.g.: https://australia-explained.com.au/history/the-colonial-wine-industry-civilisation-class-and-alcohol#_ftn2.
233. https://historyandheritage.cityofparramatta.nsw.gov.au/blog/2015/08/20/betts-blaxland-house-1830-1948-north-parramatta.
234. Sydney Vignerons Association, 2011.
235. https://dictionaryofsydney.org/entry/german_vinedressers_in_camden.
236. https://dictionaryofsydney.org/entry/german_vinedressers_in_camden.
237. https://www.belgennyfarm.com.au/history/agriculture-at-camden-estate/horticulture-and-viticulture.
238. Morris, 2008: 53.
239. Sydney Vignerons Association, 2011.
240. Morris (2008: 51–52) notes that an Austrian visitor Baron von Hugel was critical of both the execution of the terraces, the site selection and its soils, as well as the end product which did not pass for wine. The vineyard was also described as ruins by 1846 (ibid).

241. Hughes, 2003: 97
242. Jeans, 1979, cited in Pratley & Rowell, 2003: 1.
243. Stone, 2019: 19.
244. Apparently along a First Nations track, see: http://nationalunitygovernment.org/content/every-symbolic-colonial-building-sydney-was-placed-upon-significant-first-nations-city-site.
245. Casey & Lowe, 2002: 37.
246. Farrer, 2005: 17.
247. Farrer, 2005: 18.
248. Farrer, 2005: 18.
249. Christie, 1988: 90.
250. O'Connell, n.d., 1839.
251. City Plan, 2014.
252. Christie, 1988: 17.
253. Christie, 1988: 152.
254. Christie, 1988: 48.
255. Christie, 1988: 83.
256. Illustrations are shown in Christie, 1988: 52–53.
257. Farrer, 2005: 31–33.
258. City Plan, 2014: 7, 14, 25.
259. City Plan, 2014: 25.
260. https://dictionaryofsydney.org/entry/aboriginal_life around_port_jackson_after_1822#ref-uuid=e369628f-ed6b-f2ee-be51-b12efe570e6b.
261. Flannery, 1999: 11.
262. https://dictionaryofsydney.org/entry/booralee_fishing_town.
263. https://www.doyles.com.au/our-history/.
264. https://slll.cass.anu.edu.au/centres/andc/meanings-origins/o See also: http://collections.Anmm.Gov.Au/collections;jsessionid=dd2a2470bd0f4651cf99de0aa85a241d.
265. O'Connell, n.d., 1805.
266. The Tank Stream was finally abandoned in 1826, although resembled an open sewer for at least a decade prior.
267. Heydon, 1966.
268. Sydney Vignerons Association, 2011.
269. Morris, 2008: 53.
270. https://www.belgennyfarm.com.au/history/agriculture-at-camden-estate/horticulture-and-viticulture.

References

Atkinson, J. (1826). *An account of the state of agriculture and grazing in New South Wales*. Facsimile Edition 1975. Sydney University Press.

Atkinson, A. (2022). *Elizabeth & John: The Macarthurs of Elizabeth Farm*. NewSouth.

Benson, D., & Howell, J. (1990). *Taken for Granted: The bushland of Sydney and its suburbs*. Kangaroo Press.

Australian Dictionary of Biography, 1972, 1974 (online 2006). https://adb.anu.edu.au/

Baskerville, B. (2006). The Hawkesbury Commons. *Journal of the Hawkesbury Historical Society, 1*, 57–66.

Birmingham, J. (1999). *Leviathan: The unauthorised Biography of Sydney*. Knopf.

Blyton, G. (2015). Hungry Times: Food as a source of conflict between Aboriginal people and British colonists in New South Wales 1804–1846. *AlterNative: An International Journal of Indigenous Peoples, 11*(3), 299–310. https://doi.org/10.1177/117718011501100307

Byrnes, J. V. (1967). Andrew Thompson (1773–1810): http://adb.anu.edu.au/biography/thompson-andrew-2728

Burgin, S. (2010). A natural history of the Ham Common, the lands sandwiched between Richmond and Windsor—two of Governor Macquarie's five towns. In D. Lunney, P. A. Hutchings, & D. Hochuli (Eds.), *The natural history of Sydney* (pp. 263–276). Royal Zoological Society of NSW. https://doi.org/10.7882/9780980327236

Casey & Lowe. (2002). *Archaeological Investigation Conservatorium Site Macquarie Street, Sydney, Volume 1: History & Archaeology*. For NSW Department Of Public Works & Services.

Christie, M. (1988). *The Sydney Markets 1788–1988*. Sydney Market Authority.

City Plan. (2014). *Report on city of Sydney industrial & warehouse buildings heritage study*, City Plan: October 2014: H-13070. Accessed 10 April 2020. <https://www.google.com/url?sa=t&rct=j&q=&esrc=s&source=web&cd=&ved=2ahUKEwianbmfqL7wAhVg73MBHft6Be4QFjAAegQICBAD&url=https%3A%2F%2Fwww.cityofsydney.nsw.gov.au%2F-%2Fmedia%2Fcorporate%2Ffiles%2F2020-07-migrated%2Ffiles_c-1%2Fcos-warehouses-industrial-buildings-heritage-study-report-final-2-13oct2014-accessible-version.pdf%3Fdownload%3Dtrue&usg=AOvVaw1h7o7uiw0KxM2GAvdoDJfy>

Clarke, M. (2017). *Values of the metropolitan rural area of the Greater Sydney Region*. AgEconPlus Consulting for the NSW Department of Planning and Environment and Greater Sydney Commission.

Crockett, G. (2017). *Convict Sydney*, Sydney living museums, https://sydneylivingmuseums.com.au/convict-sydney/convicts-colony

Crook, E. (2014). *Vegetarianism in Australia—a history*. Revised edition.

Dictionary of Sydney. https://dictionaryofsydney.org/

Dominiak, B. C., & Daniels, D. (2012). Review of the Past and Present Distribution of Mediterranean Fruit Fly (Ceratitis capitata Wiedemann) and Queensland Fruit Fly (Bactrocera tryoni Froggatt) in Australia. *Australian Journal of Entomology, 51*(2), 104–115.

Fara, P. (2013). Joseph Banks portraits of a placid elephant. *The Public Domain Review*. Accessed 14 February 2020. https://publicdomainreview.org/essay/joseph-banks-portraits-of-a-placid-elephant/

Farrer, K. (2005). *To feed a Nation: A history of Australian food science and technology*. CSIRO Publishing.

Fitzgerald, S. (1999). *Sydney: A story of a city*. City of Sydney.

Flannery, T. (1994). *The Future Eaters: An ecological history of the Australasian lands and people*. Reed New Holland.

Flannery, T. (1999). *The Birth of Sydney*. Text Publishing.

Fletcher, B. (1967). *Australian dictionary of biography* 1759–1837, Vol. 2, (MUP), 1967, Entry on Ruse, Accessed February 8, 2020. http://adb.anu.edu.au/biography/ruse-james-2616

Fletcher, B. (1976). *Landed Enterprise and Penal Society: A history of farming and grazing in New South Wales before 1821*. Sydney University Press.

Floyd, T. (2015). *Colonial Sydney, A case study*. Accessed March 6, 2020. https://tedfloyd.wordpress.com/soil-history/

Flynn, M., & Sturgess, G. (2014). New evidence on Arthur Phillip's first landing place—26 January 1788, history: Magazine of the Royal Australian Historical Society (122):3–5. http://www.rahs.org.au/wp-content/uploads/2015/12/New-Evidence-Online-Version1.pdf

Gammage, B. (2011). *The biggest estate on Earth: How Aborigines made Australia*. Allen and Unwin.

Goodall, J. (2019). *The politics of the common good: Dispossession in Australia*. UNSW Press.

Grenville, K. (2022). *Elizabeth Macarthur's letters*. Text Publishing.

Heydon, J.D. (1966). Sir Thomas Makdougall Brisbane (1773–1860). https://adb.anu.edu.au/biography/brisbane-sir-thomas-makdougall-1827#:~:text=Brisbane%20set%20up%20an%20agricultural,financed%20the%20importation%20of%20livestock

Hoorn, J. (1990). *The Lycett Album: Drawings of Aborigines and Australian scenery*. Melbourne.

Hoorn, J. (2007). *Australian Pastoral: The making of a white landscape*. Fremantle Press.

Hughes, R. (2003). *The Fatal Shore: A History of the transportation of convicts to Australia 1787–1868*. Vintage Books.

Hughes Turnbull, L. (1999). *Sydney: Biography of a city*. Random House Australia.

Hunwick, H. (2018). *The Food and Drink of Sydney: A history*. Rowman & Littlefield.

Hunwick, H. (2019). *To market, to market*...Sydney Living Museums. Accessed May 8, 2019. http://blogs.sydneylivingmuseums.com.au/cook/to-market-to-market/

James, S. (2016). *Farming on the Fringe: Peri-urban agriculture, cultural diversity and sustainability in Sydney*. Springer International Publishing.

Karskens, G. (2010). *The Colony: A history of early Sydney*. Allen & Unwin.

Keen, I. (2021). Foragers or Farmers: *Dark Emu* and the controversy over Aboriginal agriculture. *Anthropological Forum, 31*(6), 1–23.

Kerr, R. (2002). *University of Sydney grounds conservation plan: Appendix A: University of Sydney overview history*. Accessed January 21, 2021. https://www.sydney.edu.au/documents/about/heritage/gcp_appendix1.pdf

Khara, T., Riedy, C., & Ruby, M. (2021). The Evolution of Urban Australian Meat-Eating Practices. *Frontiers Sustainable Food Systems, 5*. https://doi.org/10.3389/fsufs.2021.624288

King, C. (1948). The first fifty years of agriculture in New South Wales. *Review of Marketing and Agric. Economics, 16*(11), 567–608.

Kohen, J. (2000). First and last people. In J. Connell (Ed.), *Sydney: The emergence of a World City* (pp. 76–95). Oxford University Press.

La Croix, S. (1992). *Sheep, squatters, and the evolution of land rights in Australia: 1787-1847*, conference paper presented at "Inequality and the Commons" 3rd annual conference of the International Association for the Study of Common Property, Washington DC, USA, 18–20 September.

Lines, W. (1991). *Taming the great south land: A history of the conquest of nature in Australia*. University of Georgia Press.

Low, T. (1999). *Feral Future*. Viking.

Low, T. (2003). *The New Nature*. Penguin.

Ma Rhea, Z. (2017). *Frontiers of Taste: Food sovereignty, sustainability and Indigenous-settler relations in Australia*. Springer.

Mason, D., & Knowd, I. (2010). The Emergence of Urban Agriculture: Sydney, Australia. *International Journal of Agricultural Sustainability, 8*(1–2), 62–71.

McClymont, J. (2013). *Pictorial history: The Hills District*. Kingsclear Books.
McIntyre, J. (2009). Not Rich and Not British: Philip Schaeffer, 'failed' colonial farmer [online]. *Journal of Australian Colonial History, 11*, 1–20. https://search.informit.com.au/documentSummary;dn=997044012583652;res=IELHSS
McLoughlin, L. (1988). Landed Peasantry or Landed Gentry: A geography of land grants. In G. Aplin (Ed.), *A difficult infant: Sydney before Macquarie* (pp. 120–147). UNSW Press.
Morris, C. (2008). *Lost Gardens of Sydney*. Historic Houses Trust.
Mulcair, J. (2007) The Leader, published 28/6/2007. Cited at https://www.facebook.com/Igrewupinmortdale2223/posts/2902384879991880
Murphy, M. (2021). *Rum: A Distilled history of colonial Australia*. HarperCollins Australia.
National Trust. (2010). *Classification report: La Perouse Chinese market gardens*. National Trust of Australia (NSW).
O'Brien, P. (1997). *Joseph Banks: A life*. University of Chicago Press.
O'Connell, J. (n.d.) *Australian food history timeline*. Accessed May 12, 2019. https://australianfoodtimeline.com.au/
O'Connell, J. (n.d.) 1788: First colonists on tight rations. https://australianfoodtimeline.com.au/first-colonists-food/
O'Connell, J. (n.d.) 1790: Food crisis in Port Jackson. https://australianfoodtimeline.com.au/1790-food-crisis/
O'Connell, J. (n.d.) 1793: Poor harvest causes hardship. https://australianfoodtimeline.com.au/poor-harvest/
O'Connell, J. (n.d.) 1805: Whaling industry begins in Australia. https://australianfoodtimeline.com.au/1805-whaling-industry-begins-in-australia/
O'Connell, J. (n.d.) 1819: Emu Plains Agricultural Farm: https://australianfoodtimeline.com.au/emu-plains-agricultural-farm/
O'Connell, J. (n.d.) 1839: First ice in Sydney: https://australianfoodtimeline.com.au/ice-trade-sydney/
O'Connell, J. (2017). *A timeline of Australian food: From mutton to MasterChef*. New South Publishing.
Owen, T., Casey, M., & Pitt, N. (2017). The Old Sydney Burial Ground: An inference of early colonial diet in Sydney and Britain. *Australasian Historical Archaeology, 35*, 34–42. https://www.jstor.org/stable/26503310
Parsons, V. (1967). George Suttor (1774–1859). https://adb.anu.edu.au/biography/suttor-george-1270
Pascoe, B. (2014). *Dark Emu*. Magabala Books.

Pellegrini, A., Ahlström, A., Hobbie, S., et al. (2018). Fire frequency drives decadal changes in soil carbon and nitrogen and ecosystem productivity. *Nature, 553*, 194–198. https://doi.org/10.1038/nature24668

Persse, M. (1967). Wentworth, William Charles (1790–1872*)*, Australian Dictionary of Biography, National Centre of Biography, Australian National University. https://adb.anu.edu.au/biography/wentworth-william-charles-2782/text3961

Phillips, M. (1909). *A colonial autocracy: New South Wales under governor Macquarie 1810–1821*. Hardpress Ltd.. http://etheses.lse.ac.uk/425/1/Phillips_A%20colonial%20autocracy.pdf

Pratley, J., & Rowell, L. (2003). Evolution of Australian Agriculture: From cultivation to no-till. In J. Pratley (Ed.), *Principles of field crop production* (pp. 1–25). Oxford University Press. Accessed February 23, 2020, https://www.csu.edu.au/__data/assets/pdf_file/0005/2805521/Chapter1_PratleyRowell.pdf

Proudfoot, H. (2017). *The Hawkesbury: A thematic history, Hawkesbury City Council* (2017 edition). Accessed August 21, 2020. https://www.hawkesbury.nsw.gov.au/__data/assets/pdf_file/0004/119911/The-Hawkesbury-A-Thematic-History-2017-February.pdf

Raeburn, T., Sale, K., Saunders, P., & Doyle, A. K. (2022). Aboriginal Australian mental health during the first 100 years of colonization, 1788–1888: A historical review of nineteenth-century documents. *History of Psychiatry*, 3–20. https://doi.org/10.1177/0957154X211053208

Read, S. (2015) *Early Vineyards and Viticulture in the Sydney basin*, Australian Garden History Society, 15 October, Adelaide.

Rintoul, S. (2021) *Debunking Dark Emu: Did the publishing phenomenon get it wrong?* Sydney Morning Herald, June 12, Accessed December 11, 2021. https://www.smh.com.au/national/debunking-dark-emu-did-the-publishing-phenomenon-get-it-wrong-20210507-p57pyl.html

Robinson, S. (2008). Grasmere. Camden Historical Society.

Scott Tucker, M. (2018). *Elizabeth Macarthur: A life at the edge of the world*. Text Publishing.

Shann, E. (1948). *An economic history of Australia*. Cambridge University Press. Available online http://gutenberg.net.au/ebooks13/1300861h.html

Shaw, A. (1966). *William Bligh*. Australian Dictionary of Biography. http://adb.anu.edu.au/biography/bligh-william-1797

Short, A. (2000). Sydney's dynamic landscape. In J. Connell (Ed.), *Sydney: The Emergence of a World City* (pp. 19–36). Oxford University Press.

SLNSW. (2015). Highlights from first fleet collections of the state library of NSW. https://sydneylivingmuseums.com.au/convict-sydney/convicts-colony

Steven, M. (1967). *John Macarthur (1767–1834)*. https://adb.anu.edu.au/biography/macarthur-john-2390

Stone, B. (2019). *The Squatters: The story of Australia's pastoral pioneers*. Allen & Unwin.

Sutton, P., & Walshe, K. (2021). *Farmers or hunter-gatherers?: The Dark Emu debate*. Melbourne University Press.

Sydney Vignerons Association. (2011). Historical Timeline: Wine and Grape History in the Sydney Region. Accessed April 13, 2020. https://www.hawkesbury.net.au/sydneywine/overview/history.html

Tink, A. (2020) Lord Sydney: The life and times of Tommy Townshend, Arden.

Wadham, S., Wilson, R. K., & Wood, J. (1964). *Land Utilization in Australia*. Melbourne University Press.

Wakefield, E. (1829). A letter from Sydney: The principal town of Australasia. Accessed February 11, 2020, from http://gutenberg.net.au/ebooks13/1306631h.html

Wells, W. H. (1840). *A Map of the County of Cumberland in the Colony of New South Wales*. Charles Turtle.

Whitaker, A. (2006). *Pictorial history Marrickville*. Kingsclear Books.

Winston, D. (1957). *Sydney's Great Experiment: The progress of the Cumberland County Plan*. Halstead Press.

Yarwood, A. T. (1967). Marsden, Samuel (1765–1838), Australian Dictionary of Biography, National Centre of Biography, Australian National University, https://adb.anu.edu.au/biography/marsden-samuel-2433/text3237

Zeunert, J. (2023a). *Soil Life(less)*, Chthonic: Inhabiting the Underworld, SASA Gallery, 9-22 October. https://doi.org/10.26190/150r-eg69

Zeunert, J. (2023b). FOOD | LANDSCAPES | AUSTRALIA, https://www.foodlandscapes.com.au/. https://doi.org/10.26190/s0kk-8x57

Zeunert, J. (2024). Bridging rural and urban disconnections: Spatial graphic explorations of Australia's livestock landscapes. In R. Pernice & B. Chen (Eds.), *Urban regeneration and rural revitalization in Australia and China* (pp. 110–143). Routledge.

Zeunert, J., Hawken, S., & Gowers, J. (2025). Visualising and valuing urban agriculture for land use planning: A critical GIS analysis of Sydney and neighbouring regions. *Land, 14*(4), 854. https://doi.org/10.3390/land14040854

4

Period Two: 1851–1900

Box 4.1 Population 1851 and 1900

Population 1851: c. 50,000
Population 1900: ~480,000[1]
Increase = nearly 10-fold

J. Zeunert, A. Daroy, *Sydney's Food Landscapes*,
https://doi.org/10.1007/978-981-96-0710-5_4

Fig. 4.1 Gelding's Victoria Nursery, 1883. Source: State Library of NSW

Box 4.2 Key Events 1851–1900

- Australian gold rushes from 1851;
- Changes in state and territory boundaries (Fig. 2.3);
- The NSW Land Acts' attempt to reduce squatting in favour of smaller (non-sheep) farmers from 1861;
- Railways expanded;
- Acclimatisation societies reshaped ecologies and landscapes and solidified British tastes;
- 1890s depression.

Box 4.3 Key Events in Sydney 1851–1900

- Passenger railways and tram infrastructure developed to facilitate trade, urban (and rural) development and passenger movement;
- Urban transition to mechanised transportation, particularly from the 1880s;
- Suburbanisation, especially along passenger railway and tram lines, expanded the city's urban form;
- Gas as a domestic cooking fuel from 1860s and 1870s;
- Bubonic Plague outbreak in 1900.

Box 4.4 Sydney's Metropolitan Agriculture Trends 1851–1900

- Liverpool became the colony's first irrigation district 1856;
- Market gardening increased, notably by Chinese-Australians following the gold rushes;
- Citrus and stone fruit orchards expanded;
- The feasibility of refrigerated transport increased making regional production outside of Sydney more viable, such as for dairy;
- Food processing and distribution increased through canning, drying, refrigerating and freezing, allowing some commodities and producers to shift from local to distributed models and export;
- Cropping and horticulture faced fungal disease rusts and pests such as caterpillars and fruit flies;
- Technologies and shifts to mechanisation aided production through ploughs, windmills, cultivators, seed and combine drills;
- Viticulture significantly expanded and was later devastated by *Phylloxera* from around 1888;
- Agricultural training institutes were established, such as Hawkesbury Agricultural College in 1891;
- Control and intervention in agricultural activities impacting populated areas increased, like the *Noxious Trades and Cattle Slaughtering Act*, 1894;
- Sydney's nursery industry was dominated by edible plant production;
- In 1892, 34 per cent of New South Wales' total population lived in Sydney, a lower proportion than in Victoria's and South Australia's capital cities.[2]

From 1851 to 1900, the previously comparatively struggling Sydney-region settlements ballooned from 50,000 people to 480,000—an almost 10-fold increase. Prospector Edward Hargraves' discovery of gold in Drunong Drung (Wiradjuri Country, now Ophir, Orange) triggered the search for gold across New South Wales. Finds throughout the colonies drew migrants from around the globe in one of the largest gold rushes the world has ever seen. The total colonial population quadrupled from 430,000 to 1.7 million in just two decades. New South Wales itself doubled its population in only 10 years between 1851 and 1861. Although few people realised their aspirations for quick riches, the manifold transformations across society and landscape necessitated increased food production.

Relevant Global and National Events: 1851–1900

In addition to increased food demand from population growth, the international phenomenon of the gold rushes significantly affected food systems' supply and import dynamics. Many emancipated convicts and free settler agricultural labourers deserted their farm jobs in the hope of striking gold. In the new colony of Victoria that was established in 1851, for instance, land cultivation was reduced by 40 per cent in 2 years following the onset of the gold rush (1851–1853).[3]

As we saw in *Period One*, international supply chains had been in place since the penal colony's inception.[4] Just as the gold rushes attracted immigrants from all over the world, a global food web helped to sustain the billowing population. Flour was imported from Chile and the United States of America, sugar from Mauritius and the West Indies, tea from China and beer, ale and port from England—often sold surreptitiously on the gold fields.[5] Britain sought ongoing colonial self-sufficiency and legislation was introduced to allow affordable land purchase for farming.

Victoria's *Nicholson Act* (1860) was the first of several statutes facilitating cheap land purchases relatively close to urban centres, with New South

Wales following with similar legislation in 1861. These acts not only sought to encourage more people to farm but also to diversify agricultural production, breaking up squatters' and pastoralists' sheep farming strongholds. Key acts such as the *Crown Lands Alienation Act* and *Crown Lands Occupation Act* (both in 1861) altered agricultural dynamics through helping would-be-farmers to select, purchase, or lease lands. Squatters, however, were frequently able to engage in 'peacocking', selecting the best lands for themselves.[6] Consequently, inexperienced farmers were frequently left with poorer quality lands, leading to failure and poverty. Preventative legislation was largely unsuccessful in New South Wales and Victoria but laid the foundations for a successful rural industry in South Australia and on Queensland's Darling Downs.[7]

Acclimatisation Societies

'Acclimatisation societies' were a curious phenomenon emerging in the period and highlighting an ongoing schism between Australian ecological systems and imposed degradative settler foodways.[8] Active from the 1860s, fluctuating from the 1870s but with some continuing until the 1950s, these societies sought to import and establish exotic plant and animal species for profit or recreation, including game hunting, livestock production and aesthetic scenery.[9] The Victoria Acclimatisation Society (VAS) was Australia's first and was established in 1861. The VAS were powerful, well connected and one of the wealthiest in the world.[10] Within 4 years, they had arranged for Cashmere goats, alpacas, deer, hare, pheasant, thrushes and sparrows to be imported and introduced into Australia's delicate and unique ecosystems.

In March 1861, the *Adelaide Advertiser* praised how "delightful" the effort would be in "peopling the woods, groves, forests, and rivers of Australia with the animals, birds, insects, and fishes of Europe, and especially of Britain".[11] Driven by nostalgic longing for predominantly English landscapes and bloodsports, the societies introduced species such as foxes and rabbits for hunting. Others were introduced merely for their entertainment value. Edward Wilson, a key figure in the movement, for example, lobbied the Government of Victoria to introduce monkeys "for the amusement of the wayfarer, whom their gambols would delight as he lay under some gum tree in the forest on a sultry day".[12] Acclimatisation societies were closely connected to the establishment of Zoological Societies (from 1879) whose animal captives were imprisoned within living 'cabinets of curiosities'.[13]

Acclimatisation societies were driven by more than quirks and homesickness, although these were key factors. There was an Imperial economic imperative and those eager to gain riches in implementing it. New South Wales and Victoria's acclimatisation societies were chaired by each state's Governors. The Sydney group was led by Dr. George Bennett from the Australian Museum and Charles Moore, then-Director of the Botanic Garden of Sydney.[14] Bennett praised the societies' work in bolstering economic development and future food supply, while simultaneously revealing his dismissal of First Nations' complex food pathways, foods and interconnected knowledge systems,

> From a soil producing only a few fruits barely edible, the aborigines [sic] merely subsisting on the precarious supply of food, obtained by hunting or fishing, we now obtain by Acclimatisation a large supply of food, luxuries, and important articles of commerce, affording subsistence for a large population of thousands of human beings.[15]

Bennett's sentiments not only reveal the commercial nature of the endeavour but are particularly offensive to First Nations Australians, especially considering Australia's Aboriginal population has been estimated in the range of 315,000–750,000 in 1787.[16] It took approximately 100–130 years post 1788 invasion—and many decades after Bennett's statement—for the settler population to support more than the original First Nations population.[17]

While Bennett's claim was certainly overstated at the time, it was prescient of Australia's capacity and trajectory. The national population of 26.8 million currently exceeds the estimated 1787 Aboriginal population by a factor of 35–83 times. This critical mass is sustained by almost exclusively non-indigenous diets and Western agricultural systems and underpinned by fossil-fuel-dependent and industrially dependent processes and systems.[18] The growing, continuance, adaptation and plant-breeding evolutions of many species imported by the Acclimatisation Societies and exotic species continue. This unsavoury reality can in part be traced to the associated devastation wrought upon Australia's First Nations Aboriginal peoples, and the country's ecological systems and environment, and is thus a thorny issue. The intergenerational sustainability of the imposed acclimatisation-style settler agribusiness is highly questionable and further discussed in the final chapter.[19]

The acclimatisation movements wrought devastation for native species, eventually turning most of the country into a wildscape of invasive flora and fauna. Rabbits, foxes, starlings, sparrows and blackberries are just a few species commonly classified as 'noxious' exotics within novel ecosystems. Acclimatisation societies, however, also played a central role in establishing many of Australia's agricultural commodities, systems and economy, along with subsequent intergenerational livelihoods for many farmers, and feeding the national and wider population through exports. The Queensland Acclimatisation Society alone, for example, experimented with many food crops that are now staples, including sugar cane, bananas, apples, pineapples, pasture grasses, maize/corn, olives, mangoes and commercialising the native macadamia nut.[20]

The societies' legacies are therefore fraught. On the one hand, they created irreversible cultural and ecological damage still being grappled with today. As early as 1868, farmers decried the sparrows destroying fruits and vineyards and holding the rabbit to be so destructive that the "income of Ballarat", as the epicentre of Australia's gold rushes, "would not save us from the consequences".[21] On the other, the societies established many agricultural commodities, systems, intergenerational farming livelihoods, and continue to support Australia's economy and the foods Australians choose to eat every day. The most impactful, and therefore destructive, influence on Australian species and landscapes has unquestionably been the imposition of Western agriculture based on (predominantly) British diet preferences.[22] Exotic species emphatically dominate and must therefore be either produced nationally or be imported through energy intensive supply chains possessing their own ecological ramifications in distant 'shadow places', further discussed in *Period Five* and *2021+*.[23]

National Parks

Like the acclimatisation societies' cultural drivers, national parks were initially established for recreation and pleasure rather than biodiversity preservation. Emerging in the period's latter decades, they were heralded by the establishment of the Royal National Park in southern Sydney in 1879. This was the second national park in the modern Western world after Yellowstone in the United States. Randwick and Marrickville were then Sydney's outer areas, with Alexandria and Waterloo in the peri urban fringes. Ku-ring-gai Chase National Park in Sydney's north followed in 1894. Over time, national parks shifted focus to become conservation areas, reflecting the environmental conservation movement's growing impact, especially during *Period Four.* As a general rule, protected areas tended to be established in regions less suited to agriculture and urban development, such as on sandstone geology. Ecological disturbance and species loss was consequently more widespread in areas better suited to Western agriculture—a continuing pattern today on Sydney's comparatively easily cleared, farmed and developed Cumberland Plain (Fig. 2.11: Vegetation). In other words, agriculture embodies significant ecological ramifications, especially in grasslands and vegetation communities where agriculture is most readily incised.

Booms, Boons and Busts

The 1880s were a time of considerable economic growth. Overseas capital poured into the colony, contributing to a land boom. Investment was particularly directed into pastoral industries. Unsurprisingly, these years were followed by a bust. Depression hit in the 1890s and Sydney was severely impacted from 1891. Various factors contributed—wool, which comprised around half of the colony's exports,[24] fell in price, overseas investment dried up, banks failed and unemployment soared. A severe drought hit in 1895, lasting until 1903. The national birth rate declined. The government realised that intervention was necessary. It fostered greater rural production diversity and shifted attention to the urban manufacturing sector, though this was not notably realised until the following period. This would in-turn exhibit significant negative impact on peri-urban food production.

Prior to the depression, some wins did emerge. Sydney was key to the development of refrigeration, transforming food supply and opening up export possibilities.[25] Entrepreneur Thomas Sutcliffe Mort established a cold store on the present site of the Chinese Garden of Friendship in Darling Harbour, creating refrigerated railway vans to transport meat and dairy and trialling a frozen meat shipment in 1868.[26] Additional plants soon appeared throughout Sydney. Exporting frozen meat proved more challenging, though a 40-tonne shipment from Sydney successfully arrived in England in 1879.[27] Refrigeration entered the domestic sphere more slowly and did not emerge until *Period Three*, only becoming commonplace in *Period Four*.

Nevertheless, experimentations with technology continued to advance agricultural development. Superphosphate fertiliser, for example, began to be applied to increase phosphate in Australia's ancient, weathered soils.[28] Although synthetic superphosphate fertiliser emerged in England around 1843, it was not applied until 1880 in South Australia and first nationally manufactured in Western Australia in 1910.[29] It became indispensable to Australian agriculture, though manure was still used in metropolitan agriculture where "home gardeners appear to have sought it as vigorously as commercial growers".[30]

Despite the development of agricultural processes, the lure of the gold fields diminished outputs. By 1886, agriculture and dairying comprised less than 10 per cent of New South Wales' domestic product. Even with the aforementioned land grants (see *Period One*), producers only increased by 21,000 from 1861 to 1884, meaning that by the "mid-eighties there was barely one acre of crop per head".[31] We can ascertain from Gibbs, Shallard and Co.'s *Illustrated Sydney guide* that horticultural commodities (in hierarchical order of acreage) included: wheat; maize[32]; oats; potatoes; barley; oranges; grapes (less than 20 per cent were for table fruit rather than viticulture); sugar cane and tobacco.[33] Wine, brandy and fruit production more generally also continued.[34] Primary industries, however, continued to dominate. Wool production dwarfed all agricultural commodities, followed by gold; tin; coal; copper; tallow; preserved meats; hides; leather and iron ore.

Relevant Sydney Events: 1851–1900

The gold rushes led to significant social upheaval due to creating haphazard and unpredictable spikes in wealth. The established social order and strata were initially thrown into disorder. Those who 'struck it lucky' frequently ruffled the feathers of the politically influential 'squattocracy' and their agrarian ideologies and political preeminence.[35] Urban Sydney was an unusual mix unto itself—geographically removed from agrarian life, where felons and emancipists mingled with speculators within a diversified economy. Drunkenness was still a considerable social issue, and related crime and vice.[36] Temperance movements sprang up, including the development of 'coffee palaces' in the late 1870s and 1880s as alternatives to the licentiousness of public hotels. Alcohol-free 'temperance hotels' increased in the 1880s, though the 1890s depression forced many to become licensed premises to survive, although some continued their sober trade into the 1920s.[37]

Inner-city grime became an increasingly serious problem leading to a smallpox outbreak in 1881 and the bubonic plague in 1900. These were only a taste of what was to come in the early decades of *Period Three.* Sydney's infrastructure did, however, advance amidst these challenges although unforeseen environmental problems would simultaneously ensue. The Board of Water Supply and Sewerage was established in 1888, following Adelaide as the first colonial capital to have an operational waterborne sewerage system in 1881. Reticulated water and sewerage systems improved water supply, drainage and hygiene, though the increasingly polluted river systems from noxious industries continued to worsen. Gas stoves, available from 1853 became more commonplace in the 1870s and relied on gas from coal-burning before a reliable reticulated gas supply became more readily available.[38]

Rail Infrastructure and Transportation

The Industrial Revolution was one of, and perhaps, the most significant influences on human and planetary history. *Period Two* witnessed the increasing emergence of machinery and technological innovations transforming agriculture. Sydney's industrial trajectory took flight, aided by substantial British investment. This included Sydney's domestic construction industry, transportation infrastructure, domestic and export food and drink industries, vehicle and carriage building and production of metals for urban and rural uses including machinery for farms and mining.

Australia's first steam-powered railway was privately constructed by the Australian Agricultural Company in their Newcastle pit coal mine in 1831. Sydney's first public rail was then opened in 1856, linking Sydney and Parramatta Junction (near today's Granville). The city enthusiastically adopted rail networks, initially for passenger travel moreover than freight. The Sydney-Parramatta line was linked to Liverpool a year later and extended to Campbelltown in 1858. Rail took a further 10 years to reach Picton but in 1869 had broken beyond Sydney's topographic basin to access Goulburn and beyond.[39]

Trams and coastal steamers (ferries) further bolstered the city's transport network, with the electrification of the tram network commencing in 1895. Until rail and ferries were established, most of Sydney market produce came from the County of Cumberland[40] but railways connected new regions and markets. Fresh food production and transportation expanded from local to state, and later national, supply chains and networks. Transportation altered produce as increased handling and distance required growing tougher varieties rather than those prioritised for flavour. Meanwhile, Sydney's trams, ferries and trains facilitated a nodal expansion of Sydney's urban form. This would be decentralised in the following period by the mass uptake of the private motor vehicle.

Urban Form

In the late 1870s and 1880s, Sydney's suburbs were expanding rapidly thanks to wealth flowing from the gold rushes and wool production, which led to a growing population and housing boom. Between 1871 and 1881, Sydney's population increased by 34 per cent. During this period, Sydney started to develop its "suburban character".[41] An 1879 quote by James Inglis[42] captures the frenzy of activity,

> The overflow of bricks and mortar has spread like a lava-flood, over the adjacent slopes, heights, and valleys, till the houses now lie, pile on pile,

> tier on tier, and succeed each other row after row, street after street, far into the surrounding country; and the eruption is still in active play, and everywhere the work of building and city extension proceeds at a rapid pace. The invasion of construction has bridged the harbour, and laid out streets innumerable on the North Shore: masonry crowns every island in the spacious basin—every projecting buttress of rock maintains a pedestal of wall and gable and roof. Verandas overrun the heights, and chimney-stacks peep out from the hollows. The sand drives are covered with cottages, the very marshes have a crop of dwellings, that are constantly springing up, like mushrooms;—often alas, like that, very fragile and brittle and little calculated to withstand a lengthened wear and tear, nurtured in corruption and redolent of putridity and decay. Land is so valuable, that open drains have been boxed in with timber, and weatherboard cottages have been in many cases erected on this fever propagating substructure. Handsome villa residences, match-box cottages, toy houses, and flimsy habitations stud the slopes in all directions round the city; and suburban extension is proceeding with wonderful speed. Everywhere the sound of the workmen's tools is heard, all through the busy day. Brick-yards are worked to their utmost capacity; iron foundries are taxed to their greatest powers, saw-mills and joinery establishments are in full activity, and at present the building trades are in constant and vigorous employment.[43]

An 1881 topographic map, likely prepared for an 1881 Royal Commission, charted Sydney's defence following British troops' withdrawal from the Australian colonies in 1870.[44] It offers a clear picture of the city's spatial makeup at the time (Fig. 4.2). Sydney's urban form and development tended to follow rail and tram transport corridors and station hubs. Railways also accelerated the Cumberland Plain's alteration through assisting the development of industries such as sawmills, tanneries, brick and cement making and sand and aggregate supply. Improved transportation systems further significantly accelerated an ongoing blight to metropolitan agricultural production through facilitating speculative land subdivisions. These new developments, along with the spread of built-up landscapes, routinely brought complaining residents with them. Grievances stemmed from their encroachment on pre-existing agricultural activities to a lack of adequate due diligence before purchase. This pattern of (sub)urban encroachment continues in present-day Sydney.

Fig. 4.2 This 1881 map illuminates the extent of continuous (red) and scattered urban form and roads, along with topography and vegetation character. Source: National Library Australia, Bib ID: 433368

Despite its notable expansion during the period, Sydney's built-up area was comparatively compact when compared to today's city (Figs. 2.7 and 4.2). This is primarily due to its much lower population, a lack of motorised vehicles to facilitate a more dispersed urban form, more condensed urban development and more residents per household.

Subsequently, metropolitan land for agricultural pursuits was readily available. Much of this land was not yet in intense competition with land development. Figure 4.3's engraving provides an incredible aerial perspective of the Parramatta River looking west in 1877.[45] Five Dock is at the bottom left, and Parramatta is the urban agglomeration at centre top. Note the orchards and farms carved out of the otherwise extensive tree canopy on both sides of the harbour. Creek and river flats and marshy, swampy or boggy areas were well suited for market gardening and vegetable production, as well as dairying and pigs. Lands on western Sydney's rural outskirts were suited for pasture production and grazing. Orchards could succeed outside of built-up areas, particularly in the north-west and west-north-west.

Significant urban growth was widely evident in the period, though intense competition between land development and agriculture would heighten in *Period Three* and then considerably again in *Periods Four* and *Five*. Urban growth and development nonetheless continued to encroach on Sydney's existing agricultural lands, particularly in proximity to Parramatta and Sydney and along rail corridors.

The 11 Typologies: 1851–1900

Although existing before the period's commencement, photography became more widespread in the latter decades of the nineteenth century, gradually supplanting illustrations and artistic depictions of landscapes. This transition would become normalised in the following period. Photography increases accuracy but also diminishes landscape's mystery and romanticism. *Period Two* therefore contains a mixture of illustrations and photographs. Photographs, although not objective by any means, did reduce some artistic licence in intentional distortion of scenes, as discussed in the previous chapter.

Fig. 4.3 Engraving offering an aerial perspective of the Parramatta River looking west in 1877. The hub of Parramatta is visible top centre. Source: State Library of NSW

Agricultural Facilities and Education (AF): Formalising Facilities

Hospitals, institutions, educational facilities and social and charitable organisations had become commonplace throughout Sydney by *Period Two*. Many, such as Carrington Centennial Hospital for Convalescents and Incurables (formerly Grasmere Estate, see *MF*), contained food-producing gardens, orchards and even kept livestock.[46] Parramatta Gaol's intensive vegetable farm offers a particularly productive example, producing 11 tonnes of vegetables in 1899 fertilised by the prison's night soil.[47] The prison farm continued until around 1970.[48]

Government agricultural activities also took on other incarnations. Cockatoo Island's (*Biloela*) illicit private farming through convict labour in *Period One*, for instance, transitioned to child labour by orphaned and neglected boys in *Period Two*. This time the practice was official, as part of the Public Industrial Schools and Reformatories for Children. The boys lived on two Nautical School Ships (*Vernon* and *Sobraon*) from 1867 to 1911. The captains' Annual Reports reveal that part of the boys' activities included farming the island, especially between 1869 and 1878.[49]

The early years of the 1890s depression also saw 930 hectares of Pitt Town Common at Scheyville set aside as a Government Labour Settlement Camp. The settlement offered selected families' breadwinners an alternative lifestyle, operating akin to a co-operative or commune. By 1894, over 500 residents were living onsite. The camp, however, was disbanded after only a few years of operation. In 1896, the site was converted to a Casual Labour Farm "to enable men who could not obtain employment through ill health [...] to put in a few weeks of comparatively light labour under wholesome conditions".[50] Such work consisted of 'light' manual duties around the farm, such as cutting of firewood.

More substantively, New South Wales' first agricultural college was established in Richmond in 1891 as Hawkesbury Agricultural College (Fig. 4.4).[51] It followed Roseworthy Agricultural College in South Australia (1883) and Dookie College in Victoria (1886). The College initially operated out of two historic residences in Richmond, 'Toxana' and 'Andrew Towns House', and offered a two-year full time residential course in general agriculture. The over-11,500-hectare campus was

Fig. 4.4 The 1892 photograph (top) shows drainage works for an orchard at Hawkesbury Agricultural College. The bottom montage of four images from 1899 shows (top to bottom): a stook of wheat; chaffing maize and filling a tub silo; cutting maize for the silo and harvesting wheat. Sources: Western Sydney University Archives Collection, P137, P474

previously part of Ham Common[52] and still contained dense pockets of woodland.[53] With relatively poor soils and a lack of permanent water supply, the land was useful in preparing agriculturalists for similar conditions in much of the continent's arable dryland farmland. The first intake comprised 26 male students and was operated by the State Department of Agriculture until 1976.

Mixed Farming (MF): Decline of Mixed-Farming Estates

Estates

Large private estates were centres of mixed farming at an individual or family scale. As in the previous period, a great proportion of mixed-farming and mixed-agricultural activities took place on estate landholdings, which experimented with species and varieties.[54] Estates' productivity relied heavily on convict labour and its cessation diminished their economic viability and/or productivity, especially for labour-intensive practices that weren't commercially lucrative. Although they frequently prized ornamental botany, the majority still contained noteworthy agricultural elements including: orchards; vineyards and wineries; kitchen gardens; orangeries and greenhouses; dairies, stables, studs, and occasional slaughterhouses; processing facilities and a bakery. Not all, however, were ideally situated for their agricultural endeavours in relation to inherent land capacity. In other words, the gravitation of grand residences to Sydney Cove's orbit, predominantly on sandstone geology, did not lend themselves to agricultural productivity.[55]

Influential estate owners were often agricultural and/or horticultural 'movers and shakers', including within viticulture[56] and pastoralism.[57] Estates therefore acted as testbeds and incubators for larger and more speculative ventures extending into Sydney's rural hinterlands and beyond. As centres for experimentation[58] and exploration,[59] private estates played an important role in enabling progression from subsistence or small-scale endeavours to fully fledged industries.

Joseph Cox's 21-acre North Ryde mixed farm was primarily designed for fruit production, for instance, with table grapes being the most lucrative element at the time (Fig. 4.5, centre). It was featured in a glowing article in *The Sydney Mail* in 1882, praising the "richness of the land",

> For miles back gardens and orchards, with soil of remarkable fertility, orange groves, vineyards, and orchards, each of which yielding handsome returns, are owned by prosperous yeomen, who toil industriously, and are well remunerated for such toil. Sydney affords a handy market for their fruits, and when the local market is glutted Queensland, Victoria, and New Zealand are available.[60]

Cox had seven acres under vines (with Black Hambro the principal variety), seven more grew oranges, apples, peaches, plums, loquats, and nectarines and another seven were utilised as pasture. A detailed overview of the composting and fertilising regime is outlined for the vines and oranges.

While some estates, like the Macarthurs' in Camden endured, many only lasted for the duration of their occupant's life. Some estates were passed down as bequests, with various degrees of success. 'Grasmere' Estate, for example, was a short-lived mixed-farming venture annexed from part of Camden Park Estate (north of Cawdor) in 1885. It was part of a 3600-acre (1457-hectare) subdivision of small farms of up to 130 acres (52.6 hectare). William Henry Paling bought several titles totalling approximately 450 acres (182 hectare), which he soon after bequeathed to the state in 1888, along with horses, pigs and 150 cattle. The estate, depicted in several 1888 illustrations (one in Fig. 4.5, bottom), not only included the donated land but also contained irrigation infrastructure and a lake/dam serving as a water reservoir, machinery, an operational dairy, an orchard and grapevines. In 1890, the estate re-opened as Carrington Centennial Hospital.

Intergenerational estate descendants navigated inheritance complexities, differing aspirations and work ethics and as such, estates and their agricultural activities overwhelmingly came to be halted, squandered, sold and subdivided. While their land sizes have considerably

Fig. 4.5 The top image is the earliest photograph in this book, showing Rouse Hill House c. 1859. Various fruit trees are visible in the garden and fenced crop fields can be seen beyond. Centre is Joseph Cox's 21-acre mixed farm in North Ryde, depicted in 1882, located 3.5 miles from Ryde Wharf. The bottom illustration shows 'Grasmere' Estate in 1885. Sources: Thomas Wingate, Caroline Simpson Library, Museums of History NSW; State Library of NSW

diminished, a surprising number of their imposing residences survived and were preserved, remaining publicly accessible today,[61] though others were not so fortunate and were demolished to make way for other land uses.

Market Gardening (MG): Chinese Immigration and Production Skills

In 1879, James Inglis[62] remarked on a lack of horticulture outside of Chinese immigrants' market gardens, which he diminished to "prosaic" and producing only "culinary roots and herbs". In reality, the Chinese market gardeners' horticultural prowess not only propelled but also sustained Sydney's fresh vegetable supply. Inglis' account is nonetheless useful for demonstrating a lack of widespread interest in domestic horticulture and commercial-scale urban food production, along with spatial-economic pressures,

> There are few cottage gardens about Sydney. Land is too valuable and too much cut up into fifty-feet sections, to admit of horticulture. What gardening there is, is in the hands of chinamen [sic], whose prosaic souls do not rise above the level of culinary roots and herbs. There are few flower-plots about the workmen's houses of Sydney. There are still fewer back gardens. The pot-herb patches, the mushroom beds, homely cabbage rows, of artisan dwellings in western Europe, are altogether wanting. The cottages are dependent entirely on purchased poultry and meat for the requirements of their cuisine [...] it is not considered worth the trouble to keep pigs or poultry in the back yard.[63]

Chinese-Australians became the central cultural group of market gardeners in Sydney in *Period Two*. This set a precedent in which market gardening came to be dominated by successive waves of immigrants beyond the early colony's dominant Anglo-Celtic descendants.[64] After convict transportation ended, indentured labourers were brought from China to work in rural estates. Over 1500 Chinese immigrants had arrived by 1852. Many leased or purchased Sydney's existing market gardens (such as La Perouse gardens during *Period One*),[65] while new sites and regions were also established. Some Chinese market gardeners had left the Victorian goldfields, with "post gold rush [Chinese immigrants doing] the jobs that Australian workers refused to do".[66] It appears that market gardening largely fitted this categorisation, with just 0.2 per cent of the European population engaged between 1891 and 1911.[67] By the end of the nineteenth century, "market gardens had become almost synonymous with the Chinese".[68] Chinese growers were already known as skilful, effective, and prudent in water use, as evident on the Victorian goldfields,

> Lease or sell half an acre of apparently worthless land to a small party of Chinamen [sic], and, if there is access to any kind of water or manure, they will transform it, by their system of intensive husbandry, into a most prolific garden.[69]

The statement highlights a crucial point. Despite the early British colonists' routine dismissal of Sydney's agricultural land as poor, marginal or agriculturally useless (and the persistence of this narrative to this day—see *Periods Five* and *2021+*), much of Sydney's land outside its sandstone geology was capable of respectable agricultural productivity. Chinese-Australian market gardeners demonstrated an ability to identify land suited to intensive production and cultivate it through willingness, ingenuity, and skill, even with basic resources at their disposal.

Eileen Price's local history reveals that "by 1858–59 in the Eastern Suburbs, excluding Double Bay, the Sands Directory lists 14 market gardeners", suggesting that Sydney's number was in fact far greater.[70] The 1885 Sands Directory catalogued 54 market gardens, almost half of which were in Alexandria, the other half in Botany (Fig. 4.6, top) and the remainder scattered throughout Canterbury, Marrickville, Rose Bay and Willoughby.[71] The 1886–1888 *Atlas of the Suburbs of Sydney* shows a large Chinese Market Garden on the site now occupied by Sydney Grammar Preparatory School in Paddington (Fig. 4.6, centre left).[72] Chinese lessees and some occupier names are evident in an 1889 Alexandria map (Fig. 4.6, centre right) The fluid character of the low-lying land contrasts markedly with the imposed geometric street grid. Swampy Sheas Creek

Fig. 4.6 Market gardens on Old Botany Road 1873 (top), in Paddington 1886–1888 (centre left and enlargement), Alexandria 1889 (centre right), and Kogarah c. 1890 (bottom). Sources: Samuel Elyard, State Library of NSW; National Library Australia, Bib ID: 4317208; Georges River Libraries

supported production and proved challenging to urban development, continuing to present flood risks following heavy downpours to this day. Additional extensive market gardens were present in Kogarah, as visible in a c. 1890 photograph (Fig. 4.6, bottom). The visible market garden adjacent to then-rural Kogarah Town Centre has since been considerably urbanised.[73]

Sydney's *Sands Directories* (residential and business listings, 1858–1933) did not capture all market gardens, nor reflect the impact of Chinese gardeners. As the Sydney Museum notes, "Chinese grew vegetables in over 100 densely cultivated gardens across the city" and by 1888, "Sydneysiders had come to rely on Chinese market gardens for their supply of fresh vegetables".[74] Despite being scattered across the city, clusters began to form. Greater Botany was a particular hub and continued to be affectionately known as 'Sydney's backyard vegetable garden.[75] Individual gardens could also house many gardeners, with one North Botany instance comprising 200 Chinese gardeners.[76]

Chinese market gardens' spatial coverage was collectively impressive, stretching well beyond what the above sources suggest. The great majority of Market Garden entries in *Period Two*'s Map Database[77] are associated with Chinese market gardeners. Gardens fan out from the city centre in most directions including regions in the North Shore that was yet to be connected by Sydney Harbour Bridge. Many locations are also surprising given that they are today located in inner urban or suburban areas not utilised for agriculture. These include sites across Alexandria; Waterloo; Mascot; La Perouse; Arncliffe; Kyeemagh; Monterey (Rockdale); Wolli Creek; Paddington; Rose Bay; Marrickville; West Ryde; Seven Hills; Winston Hills; Cabramatta; Fairfield; Auburn; Blakehurst; Wahroonga; Gordon; St Ives; Pennant Hills; Willoughby; Northbridge; Brookvale and Curl Curl. There were another 15 gardens along the Cooks River, which were devastated by an 1889 flood.

The majority of sites were selected in creek valleys or locations with some alluvial characteristics. They had naturally available freshwater access and possessed good drainage, albeit with flood risk. Gardeners therefore utilised the lands' productive capacity as well as was possible in light of Sydney's lack of alluvial deposition compared to the Nepean-Hawkesbury catchments. It was important for the gardens to remain close to town markets given the period's absence of refrigeration and inefficient transportation. In other words, gardeners traded off proximity to Sydney Cove against soil fertility and water availability, offset through their skill and ingenuity.

Fresh produce was sold by hawkers carrying produce in baskets from their shoulders selling door to door.[78] As the face of market gardening, Chinese immigrants "became part of everyday family life" through the exchange of vegetables such as cabbages and turnips.[79] Racism nonetheless persisted. Gardeners utilised millennia-old fertilisation practices of animal manure and 'night soil' (human excreta) to provide the intensive vegetable production process with vital nutrients. This prompted outrage at the "commercial collectors of night soil which was used untreated on the local market gardens, and the same carts often brought back the vegetables!"[80] Frequent complaints of the stench of fertiliser on market gardens led to their inclusion in the *Nuisances Prevention Act* of 1875.[81] Furthermore, Chinese gardeners worked in efficient cooperatives that were more profitable than those of Anglo-Australian growers. In a public meeting in Botany in 1878, attendants complained that "good white men had been crowded out by 'Orientals'.[82] As Michael Christie observes, Anglo-Australian settlers "had never been over-attracted to the back-breaking work that [market] gardening involved".[83] Cannier white settlers employed Chinese gardeners or leased their gardens to them instead of attempting to directly compete.

As Sydney rapidly developed in the 1880s and public transport lines and arterial transportation routes spread their tendrils across the city, market gardens were increasingly pushed from the urban edges.[84] This pattern not only continued, but significantly accelerated, in subsequent decades. Railway infrastructure, however, was not all negative for local food production. It benefitted growers in areas such as Liverpool and Fairfield, with rail reaching the latter in the 1880s. Some Chinese-Australian gardeners relocated to areas on rail lines such as Cabramatta, where cheaper land was available at less risk of subdivision.[85] By the end of *Period Two*, the impending *Immigration Restriction Act* (1901) significantly impacted the continuance of Chinese gardeners in a growing mood against non-British Isles minorities.

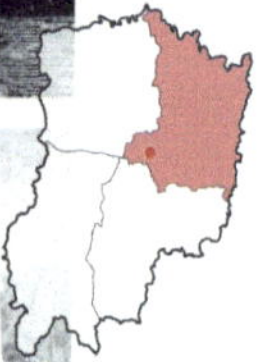

(FR) Fruit

Fig. 4.7 This figure depicts widespread citrus production in Parramatta. The first two images show Pye's orchards in 1878 and c. 1870–1880. The photograph directly evidences the illustration's scale of the immense trees. The bottom panorama was

Fruit (FR): A Thriving Industry

Stone, citrus and other fruit orchards expanded throughout the period and into *Period Three*, but not without significant challenges. Fruit production took place widely across what is now metropolitan Sydney. Commercial orchards require at least a moderately sized land area (relating to commercial economies of scale, at a given time). Fruit production was able to form a considerable spatial element in Sydney during *Period Two* due to the city's much lower population and comparatively compact urban footprint prior to decentralised suburbanisation. Accordingly, orchards were scattered across the city virtually in all directions, even in some unlikely locations.

An 1867 article gives some insight into the diversity of fruit cultivation in the Parramatta district alone,

> [To see] the naturalisation of choice fruit trees in New South Wales, [one] should visit the extensive gardens of orange and other fruit-trees near Parramatta, and in other districts of the colony. Oranges, lemons, apple, pear, loquat, apricot, peach, and other excellent fruits, together with extensive vineyards stocked with superior kinds of grapes, may be seen growing in the greatest luxuriance [...] the appearance of the gardens [...] filled with healthy trees in full bearing".[86]

Parramatta was such a prolific fruit production area that citrus scenes commonly appeared in promotional material for the district and a variety of orange was even named after it, as David Crichton noted (Fig. 4.7, bottom),

> *The Parramatta Orange.* This is an excellent and very useful variety, which is grown extensively in the Orangeries of New South Wales. It is one of the hardiest sorts, and yields regular and heavy crops of medium-sized, well flavoured quality. I can strongly recommend this variety for general cultiva-

Fig. 4.7 (continued) the feature image of a newspaper advertorial promoting Parramatta as a district—the full illustration encircled by depictions of its notable buildings. Sources: State Library of NSW, Museums of History NSW (State Archives Collection: NRS-15344 | 'Photographs of Sydney taken between 1860 and 1880'. NRS-15344-1-[6/15020]-6/15020–40 | Pye's Orangery—Parramatta), and National Library Australia, Bib ID:2929405 tion. Sometimes this sort is sent out by New South Wales nurserymen as the Pennant Hills or Kissing Point Orange, and under other local names.

The Acres, Pearce and Pye families were prominent citrus growers in the Parramatta and Hills Districts.[87] The Pye family, for instance, had numerous successful orchards following their establishment by John Pye in *Period One*. His son James became an authority on citrus. A seemingly fanciful illustration (Fig. 4.7, top) is validated by a photograph (Fig. 4.7, centre) evidencing the extensive scale of both individual trees and the landscape character of Pye's undulating Parramatta orchard on North Rocks Road, Northmead (then Governors Arms Road, Parramatta). Grafting and pruning had not yet become widespread and the orange and other citrus trees reached lofty heights well above those of today's orchards. The Estate was known as Rocky Hall and it situated on the south side of Darling Mills Creek, adjacent to Lake Parramatta.

Bold claims are made in an 1878 article in the *The Illustrated Sydney News*,

> It has been said, and with perfect truth, that in no part of the world can the orange be seen growing in greater perfection than at Parramatta, in the neighbourhood of Sydney. Visitors from foreign lands are lost in wonder at the amazing vigour and productiveness of the trees. The fruit has for many years formed a very important item of export to the neighbouring colonies; and with the extension of railways to the interior, a still larger demand will be created. But it must not be thought that the district suitable for the orange is at all circumscribed; there are thousands of acres in many districts suited for the fruit, which at all times commands a ready sale. The lemon, lime, citron, and every member of the citrus family, thrive with equal vigour. Nothing has as yet been done towards utilising the flowers for distillation, or the peel for preserving or expressing the oil, nor has the manufacture of citric acid, or the other products of the fruit, been attempted.[88]

By 1885, orchards and gardens had almost completely replaced native forests in Old Mans Valley, Hornsby (Fig. 4.9).[89] The region contains one of Sydney's rare instances of 'black earth' soils that formed over basalt, aiding productivity.[90] The Hawkesbury had also become a significant citrus producing region by the 1890s, particularly for oranges.[91] In 1892,

Bella Vista Farm in the Hills District was declared the largest orchard in New South Wales[92] and a major producer of citrus as well as wool.[93] To be so, it would have needed to exceed the Camden Park Estate orchard, which was expanded following flooding in 1889 to include an extra 2500 (mainly) peach trees and covered over 60 hectares containing various fruit trees.[94]

Pittwater offers an example of an unexpected orchard locale in a region now associated with bushland and sea, its regional food production associations likely linking with oyster farming histories on the Hawkesbury River. It might be surprising to see images of, for example, the undulating Elvina and Lovett peninsulas covered in orchards (Fig. 4.8).[95] Even more unexpectedly, Patrick Edward Quinn in 1889 described a fecund land "with deep, rich soil, in which the fig, the lemon, the orange, the olive, and the vine flourish to full perfection".[96] Further evidence of Pittwater's prior productivity is found in Roche's Bay View Orchard, Vineyard and Poultry Farm (Fig. 4.8). In 1895, the farm was described as having quality soil for production, 21 acres under citrus and stone fruits, and 600 poultry of various breeds.[97] The farm also contained a considerable extent of trellised vineyard. Roche's lemons and oranges were praised in an 1893 *Town and Country Journal*, which held them to "compare favorably with any fruit of the kind grown in this country or any other".[98] The Journal particularly prized the "lemons, which are of the Lisbon variety, and are very juicy, are simply superb, and grow close to the salt water in the greatest profusion" and "oranges, which are of the Siletta variety, are among the sweetest and best ever grown".[99]

The Inquirer and Commercial News reminds us that producing non-indigenous fruit in Australia was still highly challenging. In 1867 the newspaper reported that,

> [W]e must recollect that [the naturalisation fruit trees] was not obtained without labour and difficulty, and the study of the most suitable soils and situations, when, all impediments having been overcome by judgement and perseverance, the experiments terminated in success, the trees became naturalised, were readily propagated, and after many years the result has been a source of great wealth to the colony, the produce finding a ready market, both for exportation as well as home consumption.[100]

Fig. 4.8 Orchards on the Elvina and Lovett peninsulas in Pittwater c. 1887–1888 (top), lemons and 7 pound (3.1 kg) citrons from Sorrento orchard Blacktown 1893 (middle left), fruit trellises at Roche's Bay View Orchard and Poultry Farm 1893

Maria Smith, an orchardist in Ryde, offers a particular success story in fruit tree naturalisation. Smith sourced discarded French Crab apple remains from Tasmania and successfully grew a mutation or 'sport' apple seedling in 1868, which became known as the 'Granny Smith'. The apple requires fewer chill hours than most varieties and was able to succeed in Sydney's climate. Smith died in 1870 and another local orchardist, Edward Gallard, bought part of the Smith Farm and further developed the variety. Granny Smiths are now an Australian staple, also well suited to cooking due to holding their shape.[101] Grafting fruit trees onto ideal rootstocks was not yet the norm. Non-grafted (seedling) trees usually grow taller, as evidenced in two extraordinary photographs conveying the impressive heights achieved by some Sydney's orchardists. The first example is in Fig. 4.7, centre right. Harvesting fruit from such massive specimens was subsequently precarious, as also evident in Fig. 4.9. Orchardists also needed to overcome the challenges of nineteenth-century transportation. Once the fruit was at market, growers were largely at the mercy of agents who controlled the retail markets, with prices largely dependent on sales prowess rather than quality.[102]

Fruit Fly Challenges

One of the most insidious challenges to Sydney's fruit industry was fruit fly, which continued to escalate from *Period One* into *Period Two.* Thin-skinned fruits such as stone fruits were plagued by a range of fruit fly species, both native and exotic. The two key problematic species have been one native species, the Queensland fruit fly *Bactrocera tryoni* (herein *Qfly*) and one exotic, the Mediterranean fruit fly *Ceratitis capitata* (herein *Medfly*). The natural range of the Qfly in NSW is debated but evidence suggests that it may have originally extended as far south as Gosford.[103] There can be three or more generations of the QLD fruit fly per year and

Fig. 4.8 (continued) (middle right). Bottom shows fruit production in Pymble in 1887 looking east from the current train station with an orchard, nursery and fruit packing shed visible. Sources: Museum of Applied Arts and Sciences (85/1285–591 Glass negative, full plate, 'The Peninsular, Newport, N.S.W.' [sic], Henry King, Sydney, Australia, c. 1880–1900) and State Library of NSW

Fig. 4.9 Four members of the Harrington family casually grace a protracted and sturdy single pole ladder used to reach pears in their orchard in Old Mans Valley in Hornsby, c. 1885. Pears are slow growing, so this tree would be of a considerable age (p26, https://hornsbyshire.recollect.net.au/nodes/view/3053). Source: Hornsby Shire Recollects

numbers can increase rapidly over summer.[104] An 1819 report commented on the quality and quantity of stone fruits with no indication of the presence of pests.[105] By 1852, Qfly had become a serious problem[106] and by 1853, maggot-laden fruit was reported in fruit from Kiama, south of Sydney.[107] Fruit fly maggots were officially recorded in Sydney after 1886 but the process lacked species' identification.[108] In 1899 it was reported that Qfly had been transported from Queensland in infested bananas to Sydney.[109]

The Medfly would also pose a significant challenge to Sydney's fruit growers[110] but surfaced late in the period in 1898, thought to have arrived via Western Australia or directly from Mediterranean ports.[111] Medfly maggots were subsequently found extensively each year in Sydney's fruit.[112] Pests, and the need for efficient fertiliser, spurred the development of spraying technologies and techniques. The 1894 Universal Nursery catalogue, for example, includes a section on spraying for orchard management. It discusses 'Paris Green', an "arsenite of copper" and copper sprays are still used in orchards today. An 1894 illustration depicts a two or three-person operation plus horse(s), spraying from a pump loaded onto a horse cart, with one or two workers on hose attachments administering spray to the undersides of fruit tree foliage (Fig. 4.16, centre).

Another major fruit production challenge ironically stemmed from an Australian-first emerging in Liverpool in 1856, namely, dedicated irrigation districts. While initially attracting vegetable market gardeners, especially in locations with ready access to markets, towns, and rail transportation infrastructures, these later shifted the focus of Australian fruit growing to purpose-built new towns as regional irrigation production districts in *Period Three*. Mildura was established in 1886 in northern Victoria, followed by Renmark in South Australia and the Riverina in New South Wales. Development and eventual success took time. Initially, these far-flung regional areas were separated from capital cities, hindered by droughts and further affected by the 1890s economic depression. Rail connections and the economies of scale of agglomerated district irrigation incrementally shifted the bulk of fruit production away from metropolitan agricultural areas, especially for easily transportable fruits. This was concurrently assisted by late nineteenth-century developments in canning, drying, and refrigeration, expanding Australian fruit export capabilities.[113]

Fig. 4.10 Maize in Camden 1878 (top), maize in the Hawkesbury c. 1880–1900 (bottom). Sources: State Library of NSW and Henry King, Museum of Applied Arts and Sciences (85/1285–522 Glass plate negative, full plate, 'Farm, Hawkesbury River, N.S.W.', Henry King, Sydney, Australia, c. 1880–1900)

Cropping (CR): A Challenging Period

Native and exotic pests also proved to be increasingly troublesome for crop production. Imposed Western cropping practices were still grappling to adapt to soils, seasons, erratic climatic conditions, extreme heat and moisture and other dynamic and ever shifting factors. Plant breeding had not yet reached successful commercial scales suited to local soils and climatic conditions and most imported British crops preferred drier conditions to Sydney's wet and humid summers. Rather than change colonial diets to better-suited crops such as maize/corn (Fig. 4.10), an inflexible consumer market sought alternative locations suitable for wheat production. Drier areas were found west over the Great Divide beyond Sydney's humid basin.

The 1878 stylised engraving of maize in Camden (Fig. 4.10, top) offers a more idyllic scene than that of the c. 1880–1900 photograph of the same crop on the banks of the Hawkesbury River (bottom, mid-ground), despite its spectacular location. Grapes are also visible in the foreground. The engraving's article "convey[s] some idea of the growth of this plant in the fertile district of Camden. It thrives luxuriantly on all the rich alluvial lands adjacent to the rivers, growing to the height of ten or twelve feet, and is exceedingly productive".[114] The article also notes the chief use of maize as horse rather than human feed.

For much of the period, though, farmers grappled with a host of challenging conditions. As discussed in the previous chapter, early agricultural practices quickly stripped the soils of fertility, exacerbated by a lack of cover crops, soil conservation, and vegetation clearance. Wider consequences included soil erosion, rising water tables and soil salinity. A great extent of cropping practice continued to mirror slash-and-burn agriculture, necessitating moving to unfarmed lands to extract new yields. Outside of the Hawkesbury-Nepean River alluvial fans, this repeated pattern quickly exhausted soil fertility. Conversely, alluvial regions were subject to volatile and devastating floods that on the one hand, replenished nutrient-mined soils, but on the other, tested farmers' resilience to withstand devastation.

Monoculture farming can be highly productive but is prone to pest and fungal outbreaks.[115] In the 1860s, farmers removed large areas of ecosystems in regions such as Menangle and Greendale to yield successful wheat crops. Soon after, fungal disease rust struck, wiping wheat production from the County of Cumberland and Camden in favour of drier areas west of the Blue Mountains more climatically suited and also serviced by rail.[116] Troublesome introduced species both from the Acclimatisation movement and through global supply chain flows added further complexities for farmers to grapple with. One particularly noteworthy example is that of the rabbit. Though the First Fleet carried five rabbits, they were not released into the wild. On Christmas Day in 1859, 13 rabbits were released in Victoria. By 1880, they had vastly multiplied and crossed the Murray River into New South Wales and by 1886, they had reached as far as Queensland, with particularly devastating impacts for crops.

Solutions slowly emerged as the period continued. After years of cross-breeding, for example, William Farrer developed the wheat-rust resistant 'Federation' strain in 1900, proving essential to success in *Period Three.*[117] The emergence of superphosphate fertiliser in the latter part of the period transformed productivity, though accompanied by long-term soil and environmental impacts. Finally, cropping was significantly bolstered by emerging technologies and innovations that greatly increased efficiency and yield, including ploughs, mowers, reapers, harvesters, seed and combine drills and cultivators (see *Agricultural Infrastructure*).

Intensive Livestock (IL): Dirty Dairies

Pig farms, dairy herds, goats and poultry expanded throughout the period. These livestock were kept at commercial, hobby, and domestic scales and were part of everyday life within the urban fabric. Slaughter facilities are listed and detailed in *Agricultural Infrastructure* (IN), although domestic-scale and cottage-industry-scale slaughter was still commonplace in *Period Two.* Increased regulation of intensive livestock activities ramped up in the latter part of the period, culminating in *Period Three*'s multiple health crises. The photographs in Fig. 4.11 reveal

Fig. 4.11 Dairy livestock at the University of Sydney, 1860s (top) and a dairy cow and goats in Sydney c. 1885–1890 (centre, bottom). Sources: University of Sydney Archives (View of St John's College and Grounds with Cows in Foreground (1860s), [REF-00009926]. University of Sydney Archives, accessed 14/03/2024, https://archives-search.sydney.edu.au/nodes/view/101600) and Mitchell Library, State Library of NSW

intensive livestock as a part of daily life in urban Sydney. Dairy cattle grazed what was once Grose Farm[118] and subsequently Sydney University grounds in the 1860s (top). A dairy cow grazes a road verge adjacent to a Technological Museum (centre) and a herd of dairy goats enjoy grass opposite terrace houses (bottom).

Suburban, urban and peri-urban dairying were commonplace in Sydney throughout *Period Two*, along with dairy depots. During the early part of the period, a lack of refrigeration and pasteurisation necessitated localised self-sufficiency. Country dairies still struggled to supply Sydney with fresh milk, with a Wollongong to Sydney steamship service established in 1856 but failing after a few months when the unrefrigerated milk could not last the five-hour journey.[119] Thomas Mort (introduced in *Period One*) continued to advance the technical logistics of on-site and mobile refrigeration in train carriages and steamships from 1876. Convincing Southern Highlands dairy farmers to commit to an adequate minimum bulk supply for rail transport to be economically viable proved challenging. Pasteurisation and increased rail efficiency facilitated the process, initially from Bowral to Sydney. This marked a turning point for Sydney's fresh milk supply and incrementally shifted the emphasis of the supply chain from local to regional. By the end of the period, regional dairying had overtaken local production.[120]

Meanwhile, urban dairies did not help their own cause. Some dairy farmers did not realise that water must be boiled before its use for sterilisation. Milk adulteration was widespread,[121] with potentially infected water being added to milk to dilute it. Unsafe quantities of chemicals such as boric acid were used to kill bacteria and extend shelf life. Stored milk was regularly left uncovered by producers and sellers. Typhoid outbreaks in Leichhardt in 1886 and Randwick in 1890 were traced to contaminated water at local dairies.[122] Despite the *Dairies Supervision Act* 1886 being passed as a reaction,[123] milk supply was virtually unregulated and breaches were generally unprosecuted.[124] Poor practices continued and the industry was perpetually embroiled in numerous health and hygiene crises for half a century. Regional producers marketed their milk as clean and unadulterated and assisted their increasing market share. Regardless, urban dairying continued to boom. By 1891, Marrickville alone had about 40 registered dairies.[125] Although Sydney's expanding public transportation system forced many dairies further out of the growing city in the 1880s and 1890s, demand increased due to the fashion of substituting cow's milk for breastfeeding infants in the Victorian period and beyond into the 1960s.[126]

Regulation of Urban Livestock

Although dairying continued within the urban fabric, livestock production began to decline. A Royal Commission on Noxious and Offensive Trades in 1883 led to the *Noxious Trades and Cattle Slaughtering Act* of 1894. As an extension of British 'nuisance laws', the Act chiefly addressed complaints and symptoms of unpleasant odours rather than addressing the causes and their wider issues of generating environmental pollution.[127] The Act enabled the Board of Health to advise the government on suitable regulations of pollutants and activities arising from industries deemed noxious within the County of Cumberland, as well as registration and annual licensing. Trades included the "fat extractor and melter, bone boiler and grinder, blood boiler and drier, glue maker, pig keeper, poultry farmer, knacker, gut scraper, rag picker, rag dealer, flock maker, manure maker".[128] Accordingly, keeping, slaughtering and processing livestock were central features of the Act. Such regulations, along with an increasing ability to transport livestock and/or their processed end products through freezing and refrigeration, led to the very gradual elimination of livestock within Sydney's urban and suburban areas and its immediate fringes.

Grazing (GZ): Bunyip Aristocracy

Sydney's numerous Commons had predominantly been granted, divided or altogether dissipated by *Period Two.* Grazing and pastoralism continued but shifted from a prior prevalence in the Commons to increasingly privatised land divisions, though Pitt Town Common near the agricultural settlement of Windsor was an exception. Smaller, privatised landholdings than the extensive grazing properties beyond Sydney became widespread. Opportunistic grazing also took place on marginal, interstitial and transitional metropolitan sites. This regularly resulted in livestock wandering onto private property, with affected landowners withholding the offending animals until costs were paid, or selling on after an advertised period of time. Multiple disputes ensued. Sydney's centrality as a market, and increasing export hub, further added value to graziers. Some metropolitan landholders utilised their lands for value-adding services, such as the holding and/or fattening of herds before slaughter.

Grazing also continued on the site of the former Grose Farm (see *Period One*), now the University of Sydney since 1851. The farm had been primarily pastoral following attempted cropping in the 1790s in its heavy, adhesive clay soils.[129] It never excelled under ongoing tenure changes but housed various grazing animals in paddocks. Its ongoing use for grazing livestock is evident in an 1854 depiction following the University's establishment.[130] Grazing (and dairying) continued on University lands for some time, as shown in an 1860 photograph (Fig. 4.11, top) (noting Agricultural studies were not offered by the University until 1910). Pastoralism in rural contexts, however, was continuing to increase in scale, leveraging larger and cheaper lands, more sizeable herds, and receptive areas beyond metropolitan Sydney. By 1880, statistically speaking, most cattle were located beyond Sydney[131] though some of the city's largest landholdings contained sizeable herds. These centred on the Cowpastures district but extended across many regions of predominantly western Sydney.

Figure 4.12's engraved print from 1874 (top) shows a comparatively tamed Cowpastures when compared to the less cultivated 1840–1846 depiction (Fig. 3.9, top right). The landscape is nonetheless expansive and appears unfenced, but orderly cattle herds have passed through paddocks and over a bridge. An 1884 drawing titled 'The Scenic Hills from Denham Court', presented as part of the sale package of the estate, offers a colonial pastoral expression of landscape within the period (Fig. 4.12, bottom). The pencil sketch is a bucolic depiction of an Anglicised landscape. The distant farm cottage is framed by remnant trees "representing the heavily modified agricultural landscape with the vast expanse of indigenous vegetation beyond representing the picturesque wild landscape" that would have appealed to Victorian-era purchasers.[132] The view potentially looks over the Scenic Hills to the south-west adjacent to

Fig. 4.12 The Cow Pastures 1874 (top), The Scenic Hills from Denham Court 1884 (bottom). Sources: Robert Marsh Westmacott, National Library Australia, and Harold Brees, State Library of NSW

today's Campbelltown Road.[133] Surprisingly, the exact view in 2021, aside from significant roads now laid across the landscape, is reminiscent of the original.

As mentioned, pastoralism carried substantial social caché, carrying over from British landowner and class-based social hierarchies and certain members of the bunyip aristocracy lobbied for hereditary titles.[134] Landholdings carried social capital. Sprawling estates in Sydney's prime grazing regions were either gifted to influential men or purchased by wealthy families. Landholders commissioned leading artists of the day to immortalise their holdings within a Romantic pastoral idyll, as pioneered earlier by the Macarthurs' at Elizabeth Farm (Figs. 3.9 and 8.7). Three paintings of Bungarribee Homestead and Farm (two c. 1847–1852[135] and one 1858) are similarly aspirational. Bungarrabee/Bungarribee began as an extensive 2000-acre land grant in 1823. Its remnants are located in the now Western Sydney Parklands. Views from the three paintings feature its farm buildings and the subsequently demolished homestead (1957), along with its all-important paddocks. The fields were once a key thoroughbred horse stud for the East India Company, with long-horned cattle a secondary concern. The paintings present different foci, alternating between the horses and cattle presumably for differing clientele. The 1858 painting is more romantic but expressive of an American ranch than British gentleman's estate (see *Then and Now*, Fig. 8.8, top).[136]

Vineyard Wineries (VW): Devastation

Wine and brandy production continued to gather momentum from the previous period into *Period Two.* By 1855, 350 acres of vines were reportedly in production. Ten years later, conservative estimates hold the number as having grown to 511 acres, with 685 in 1876 and 940 in 1886, when production in the Hunter Valley overtook Sydney's.[137] Many of Sydney's large estates established vineyards, partly to fill pastoralism and grain production's void as they moved further afield, as well as due to popularity and prestige.

The Macarthurs' aforementioned Camden Park Estate continued to achieve success through the emigration of expert European viticulturists, such as Martin Thurn who established grapes on the site in 1852.[138] By 1853, Camden Park Estate had 33 listed grape varieties for sale.[139] Cawdor vineyard was also in the region and operated between around the 1850s–1930s, as was 'Elderslie', which was established in the 1880s and also produced table grapes. Tizzana winery in Sackville Reach commenced from 1882 by Italian Dr. Thomas Fiaschi[140] and continues to operate. A Mayor of Liverpool, William Smith had a vineyard, Arpafeeli, in Moorebank.[141]

The lithograph in Fig. 4.13 (top) depicts only one of Camden Park Estate's several vineyards in 1878, known as the 'third vineyard' and oriented to run across the contours below Camden Park House. The illustration was printed before *Phylloxera* devastated the property in the late 1880s. An agriculturally verdant 1892 photograph shows the Superintendent's vineyards at then Tarban Creek Asylum and property initially known as 'Longwood' (Fig. 4.13, bottom).[142] Dr. Frederick Norton Manning was superintendent of the Asylum from 1868 and established the vineyard in 1870. Mixed farming and grape production was undertaken at Longwood from *Period One* to *Period Four* following its establishment in 1836, adjacent to the now Gladesville Hospital.

Undoubtedly one of Sydney's most iconic viticultural establishments, 'Minchinbury', stemmed from Dr. James McKay's 1859 purchase of land previously granted to William Minchin in 1819. The winery was located south of now Rooty Hill train station, and north-west of Prospect Reservoir. McKay was a late beneficiary of convict labour that planted vines and built cellars.[143] In 1871, McKay's wines won awards at the Metropolitan Intercolonial Exhibition and 1877, he advertised the sale of 15,000 gallons of wine comprising varieties including Pineau, Riesling, Shiraz, Lambrascat, Burgundy and Hermitage, as well as Madeira and Colonial Port.[144] By 1881, Minchinbury had 50,000 vines.[145] The Cabramatta region had several wineries including Sunnybrook, Saint Johns Park, Orchardleigh, Mark Lodge and Sand Hill Farm (Sandal Farm) established in *Period One* and continuing into *Period Two.* There were also several viticultural ventures on the Northern Beaches in the

Fig. 4.13 Camden Park Estate's third vineyard 1878 (top), Superintendent's Vineyards at Tarban Asylum 1892 (bottom). Sources: State Library of NSW and Museums of History NSW (State Archives Collection: Government Printing Office; NRS 4481, Glass negatives. NRS-4481-3-[7/16175]-Sh346 I Government Printing Office 1—05466—View of grounds, Gladesville Asylum, Gladesville Hospital)

1880s. These included Rock Lily Hotel in Mona Vale, where vines were planted in 1886 (still signified in the enduring 'Vineyard Street'), and the Crawford Brothers' Ventnor and Frederick Chave's vines, both in Lovett Bay.

Some of Sydney's vineyards were established on pre-existing Estate ground, rather than lands best suited to producing wine varieties. In another case, Frere's Vineyard was established on purchased land 'sight unseen' adjoining the Georges River and Punchbowl Creek—the rugged terrain of what is now Heathcote National Park's Holsworthy Military Reserve. This was an attempt to avoid *Phylloxera* by growing grapes in sandy soils, albeit the prevalence of rock, lack of nutrients and poor overall soil quality led to its failure. A few miles to the north-east was the settlement of Eckersley,[146] where the more successful Grodno vineyard was forced to close when the Army took possession of the greater area for the military base in 1912–1913.

Viticulture and winemaking techniques and technologies had improved by the latter decades of the period. Dr. Thomas Fiaschi, president of the New South Wales Winegrowers Association for around 25 years, for instance, made important contributions through his experimental use of new grape varieties, aseptic conditions and new trellising methods.[147] James Angus purchased Minchinbury in 1895 and also began to introduce modern winemaking techniques, leading to its renowned sparkling wine, released from 1908.[148]

Despite advances, by the end of the period Sydney's viticulture was struck a colossal blow that changed the course of Australia's winemaking. A small, yellow, aphid-like root louse termed *Phylloxera* from eastern North America appeared in France in the late 1860s. *Phylloxera* feeds on resistant *Vitis labrusca* vines in its native context but the European *Vitis vinifera* vines were intolerant. The sap-sucking insect rapidly spread throughout Europe through infected rootstock, devastating up to 70 per cent of European vineyards by the end of the nineteenth century.[149] By 1877, it had arrived in Geelong, Victoria, and had promptly made its way up through New South Wales by 1884, reaching Sydney in 1888. South Australia, which had banned the movement of vine material in 1874, escaped infection and consequently holds the oldest Shiraz vineyards in the world today.[150] In New South Wales, however, the impacts were devastating for estates such as Eschol Park, which had produced 2000–3000 gallons of award-winning wine following its establishment in 1860 on a 15-acre vineyard. Responses meant burning existing vines and replanting them with *Phylloxera* resistant North American rootstocks, which was no small, quick or inexpensive undertaking, as further explored in *Period Three.*

Agricultural Infrastructure (IN): Innovation and Expansion

As expressed in the Map Database, agricultural infrastructure developments peaked in *Period Two*. There were more investments and greater expansion, particularly in relation to emerging mechanisation processes and technologies. These changes enabled British and European agricultural approaches to thrive in the challenging Australian conditions.

Significant morphological and infrastructural manipulations established irrigation districts, such as the previously mentioned Australian-first in Liverpool in 1856. This led to the large-scale development of dedicated irrigation districts across Australia from late in *Period Two* into the twenty-first century, most notably in the Murray-Darling basin. Large earthworks for field systems and dams were accompanied by drainage and diversion channels, pumping systems, pipelines and field irrigation systems (e.g. Figs. 4.4, top and 4.14, bottom right). Irrigation districts hence required significant investments of money, labour and embodied energy. As large infrastructures, they indicate the period's increasing impetus and ability to control the unpredictable vagaries of natural seasons and climatic patterns, as well as to adapt and direct natural systems. The micro and macro climatic and environmental impacts of such large-scale manipulations are ongoing.

The latter part of the period also witnessed significant innovations in agricultural machinery. Industrial developments arrived later in the British colonies than in their Imperial homeland, especially within processing and manufacturing facilities. These became increasingly prevalent in *Period Two*'s latter decades but even more commonplace in *Period*

Three. Perhaps unsurprisingly, given the colony's well-documented predilection for liquor since its inception,[151] alcoholic beverage production was an exception. Increasingly large-scale distilleries and brewing operations maintained a steady track of development, such as the industrial-scale hard liquor production occurring at Coopers Distillery on Black Wattle Swamp Creek in Chippendale in 1868 (Fig. 4.14, top).

Food and agricultural sectors were later adopters of mechanised and concentrated processing and production. This was in part due to their dispersed locations but also their distance from Britain's industrial hubs. Innovations frequently centred on agricultural field implements' invention and innovation, many of which generated significant transformations within farming. As mentioned above, this was especially the case in cropping that was becoming increasingly broadacre. Journals and catalogues featured sales advertisements for local and imported machinery, such as 1870 advertisements for Ritchie's ploughs and a Burgess and Key's Mower (imported from Britain) (Fig. 4.14, second row). Australian conditions, such as shallow topsoils and hardy stumps, often led to local innovations. Ploughs, mowers, reapers, harvesters, seed and combine drills and cultivators all assisted agricultural productivity, but the plough was particularly impactful. Ploughing was such a widespread practice that competitions drew large crowds (Fig. 4.14, bottom left). The first ploughing match was held in the Southern Highlands in 1828[152] but regular ploughing competitions became commonplace from the 1870s until the 1890s depression.

Despite its head start on the other Australian colonies, Sydney was less agriculturally advanced. Some machinery was engineered and manufactured locally, for example, Ritchie's steel working business founded in 1857 in Parramatta including agricultural machinery and ploughs built at his larger plant in nearby Auburn (Fig. 4.14, second row left). The Hudson Brothers also built ploughs and windmills, based on their own designs from 1884 in Granville, becoming Clyde Engineering from 1898.[153] Improved machinery not only assisted Sydney's production output, but manufacturing and distribution helped to service agricultural markets well beyond the city leading to bigger and more efficient farms (Fig. 4.14, third row). Accordingly, the latter decades of the nineteenth century witnessed a transition from domestic, family and 'cottage'

Fig. 4.14 Various agricultural infrastructure and machinery in *Period Two*. Coopers Distillery on Black Wattle Swamp Creek, Chippendale, 1868 (top). Advertisements in 1870 for machinery (second row). Hay and watermelon loading

industries into more uniform manufacturing operations. While the images in the bottom row are not located in Sydney, they are reminders of slow and laborious labour and earthworks processes prior to fossil fuels. Figure 4.14's largely agrarian scenes thus reveal the horse as the chief source of farm power in *Period Two*, while concurrently hinting agricultural industrialisation would soon triumph in *Period Three*.[154]

Refrigeration

Heat processing was of the period's major agricultural innovations, followed by refrigeration. James Harrison established the world's first mechanical ice-making machine at Rocky Point, Geelong, which was developed into a refrigerator in a Bendigo brewery in 1860.[155] Thomas Mort, who we met in *Period One*, continued to put Sydney on the refrigeration map through his Darling Harbour cold store which from 1872 received fresh milk via rail from Berrima and Moss Vale. 'Mort's' became the New South Wales Fresh Food and Ice Company and from 1878, Camden Park Estate sent milk, butter and cheese to Darling Harbour year round, with additional pick-ups during busy periods such as the Christmas rush.[156] Mort's facilities enabled milk and dairy products to be supplied from outside of the city, particularly the south coast and southern highlands from his Mittagong factory.[157] His refrigeration plant at Bowenfels in the Blue Mountains facilitated meat transportation to Darling Harbour's refrigerated warehouses.[158]

Refrigeration revolutionised the distribution of food products and led to the eventual demise of the city's previously ubiquitous local dairies.[159] Refrigerated railway cars in the 1890s made shipping frozen meat and

Fig. 4.14 (continued) at Windsor Railway Station c. 1885–1890 (third row). A ploughing match (bottom left). A team of horses working for months to complete dam earthworks c. 1890–1900 (bottom right) (https://collection.maas.museum/object/29201). Sources: Samuel Elyard, Dixson Galleries, State Library of NSW, Kingsclear books (Ted Books and HCCLS in Nichols, 2016) and the Museum of Applied Arts and Sciences (85/1284-70 Glass plate negative, full plate, 'Ploughing Match', Kerry and Co, Sydney, Australia, c. 1884-1917; and, 4. 85/1284-250 Glass negative, full plate, 'Tank Sinking', Kerry and Co, Sydney, Australia, c. 1884-1917) chilled dairy to major domestic ports more viable. After *Period One*'s challenges, the first successful 40 tonnes of frozen meat exports from Melbourne and Sydney reached London in 1879/1880 on the steamship *Strathleven*. It nevertheless took some time until there were enough ships fitted with refrigeration equipment to make carcase meat export a regular occurrence.[160] Canned meat therefore remained an export mainstay, including both farmed livestock such as mutton and beef and wild meat like rabbit, the latter of which was significant and still gaining momentum in *Period Two*.[161] Domestic refrigeration still remained decades away. The 'Coolgardie safe' was a short-lived precursor to the icebox and domestic refrigerator,[162] surfacing in the late 1890s in Western Australia but potentially an evolution of a South Australian water cooled safe produced in 1895.[163]

Processing

Abattoirs facilitated the production of both canned and fresh meat, becoming increasingly commonplace across Sydney in addition to roaming livestock[164] and carcasses hanging in butcher shops.[165] Sydney's abattoirs and slaughterhouses were part of the urban fabric despite their rank odours, noxious pollution and routine ability to attract rats on land or sharks in bays where wastes were deposited. Animal slaughter was thus far more apparent as a part of daily life than today where such activities are largely screened from consumer view.[166]

Glebe Island's abattoirs were a government affair and operated from approximately 1860 until 1915–1916, when they were superseded by an extensive facility in Homebush (see *Period Three*).[167] As Sydney expanded, complaints increased stemming from the abattoirs' stench and pollution. A noxious mix of sewers and blood and offal from the Glebe Island Government Abattoir were emptied into the harbour's waters.[168] The Glebe abattoirs featured prominently in the 1882 Royal Commission into Noxious and Offensive Trades, but nonetheless continued to operate for more than three decades. The Commission found that in 1882, 524,415 sheep, 69,991 cattle, 31,269 pigs and 8348 calves were slaughtered there. This immense number gives some insight into the extent of livestock kept and transported across greater Sydney. It also illuminates shifting dietary preferences and production outputs—sheep (lamb and mutton) dwarfed cattle, which has reversed over time to the present day.

The undulating topography at the Government Abattoir at Glebe Island offered the 1870s photographer a perspective into the back of the facility, defying the high stone wall (Fig. 4.15, top). This view reveals numerous pens, containing a horse and herd of sheep, but visually under-represents the degree of animal slaughter there. The four illustrations depict acts that would today be considered inhumane, along with extremely precarious working conditions in comparison to contemporary workplace health and safety requirements. The illustrations are barbaric, but somehow desensitised and more palatable than photographs.

Other common agricultural processing infrastructures included wool-washes, tallow works (of significant export value), tanneries and refineries such as Colonial Sugar Refinery (CSR), established in Pyrmont in 1879. Most of these urban features created unpleasant vistas, sounds and smells, explaining their presence on the Noxious Trades lists and their incremental segregation from the city through later industrial zonings in *Periods Four* and *Five.*

Markets

Period Two was tumultuous and a time of considerable flux for Sydney's fresh produce markets.[169] Markets included fruits, vegetables, hay, grain, livestock, fish, goods and entertainment. The George Street Markets were renovated and altered in 1858. New fruit and vegetable market buildings that became known as the Belmore Markets were opened on Campbell St on the prior cattle yards (now part of Belmore Park adjoining Central station). An open market opposite would evolve into what became known as Paddy's Market. The multiple decades of flux within Sydney's central markets continued until the end of the period. Increased wealth in the 1870s and 1880s led to more opulent and imposing city architecture. A consequent push ensued to eviscerate the central markets, which were frequently conceived as congested, tatty, tumble-down old sheds with urine-soaked odour and scattered horse manure. The markets were thus

Fig. 4.15 Government Abattoir at Glebe Island in the 1870s and 1880s. Charles Pickering and Australian Town and Country Journal, State Library of NSW

characterised as "the worm" in the heart of the Victorian "architectural rose".[170] This culminated in George Street Markets' demolition in 1891, with the Queen Victoria 'Market' Building (QVB) intended to act as a replacement. Architectural grandeur, however, stood in the way of practical produce needs and its unfeasibility for markets saw it transition to today's retail uses.

In the 1880s and 1890s the Belmore Markets were the main centre of fruit, vegetable and fish distribution but their efficiency was limited by becoming stranded at Darling Harbour's railhead.[171] A new fruit market was built on the Paddy's site in Campbell Street seeking to "woo back the growers" who set up their own co-operative fruit exchange on Bathurst Street.[172] An efficient central market required four acres in a central location with access to road, rail and port.[173] The decades of civic financial investment in redevelopments ultimately failed to meet these criteria. Despite multiple attempts of improvement, market facilities deteriorated in central Sydney from the 1870s.[174] Meanwhile, smaller village markets cropped up, such as Newtown's, visible in an 1886–1888 atlas map.[175] Small-scale street vendors and door-to-door vegetable hawkers also remained a common sight. The extension of rail services to fruit production areas such as along the Hornsby line supported large consignments, further creating links with Queensland's banana trades, Fijian produce and Tasmanian apple and pear imports (Fig. 4.14).[176]

Logistics

Albeit challenging and expensive, Sydney's fundamental infrastructure and logistics were treated with a laissez-faire and a haphazard approach. Consequently, rail and port facilities did not adequately link. An 1879 appraisal colourfully found "melancholy evidence of narrow vision and utter lack of prescience and common forethought, with the national rail terminus "cut off from the heart".[177] This resulted in the "accumulated wealth of natural produce of the continent" need to be "laboriously dragged [from Redfern] through the crowded city's streets [to port] by the primitive traction of oxen and horses".[178]

Aquaculture (AQ): Birth of an Industry

From the beginning of British occupation, naturally harvested oysters were sold cheaply without high status.[179] Sydney's natural oyster beds and shell middens were moreover sought out for lime for cement to aid the city's construction, with lime otherwise lacking in the local environment. Most natural oyster beds were destroyed by the mid-1860s following dredging, collecting, grinding down and burning to provide construction lime.[180]

Exhaustion of natural oyster stocks eventually led to successful attempts to farm Sydney Rock Oysters in the 1860s and 1870s by speculator and politician Thomas Holt. Holt procured and assimilated Weeney and Gwawley Bays in the Georges River into his substantial regional land accumulations of around 13,000 acres.[181] He developed a technique to successfully harvest oysters in the local environment, which included the significant undertaking of digging miles of claires (channels), enclosing the mouths of Weeney and Gwawley Bays with piles and constructing a bridge and floodgates to control water flow.[182] In doing so, Holt departed from earlier unsuccessful attempts to use the French method of canals that silted and harboured mud-worms that bored into the shells and destroyed the oysters. Instead, raised stones, beds or stakes were used to lift the oysters above the seabed into inter-tidal zones, where water flow prevented silting. Through Holt's connections to government and his subsequent lobbying, oyster leases were introduced in 1880.

Oyster parlours and saloons opened towards the late 1800s and were often operated by immigrant families, particularly Greek-Australians.[183] An 1889 article and engraving in *The Illustrated Sydney News* laments "restrictions now by which the law regulates oyster-gathering in our waters, and these regulations it is necessary to observe in order to avoid trouble", seemingly oblivious to widespread over-exploitation of Sydney's natural stocks.[184] Oysters were served as the opening course at the Federation banquet at Sydney Town Hall in 1901, revealing a rising social status that has continued.[185]

Fish Markets

The first official Sydney fish market was constructed on Forbes Street in Woolloomooloo by the City Council and opened in 1872. Prior to this, fishermen hawked their catch on the streets or sold directly from their boats in Sydney Cove or Woolloomooloo. The original market building, at 83 metres long by 22 metres wide, quickly became inadequate to serve growing Sydney's needs and was expanded in 1888 and 1893. A private fish market also opened in Redfern in 1891.

Miscellaneous (MC): A Golden Age for Nurseries?

Sydney's plant nurseries form a notable component of the city's food landscapes in *Period Two*. Numerous nurseries either specialised, in or incorporated, an array of edible plant species. Some of the most notable included Iliffe Nursery (later Milner and Smith/Milner and Co), Gelding's (Victoria Nursery), Universal Nursery, Darling Nursery and Wadd's Nursery.[186]

Unlike today, most nineteenth-century production nurseries grew the majority of their stock within their metropolitan Sydney sites, as opposed to acting as depots supplied by large external regional or interstate wholesalers. Growing locally reflects the period's transportation challenges, which could be substantial, even within the city itself. Nonetheless, some of *Period Two*'s nurseries had a similar model to that of today's and

Fig. 4.16 Universal Nursery Guide 1894 (top, middle), Milner & Co catalogue 1900 (bottom). Source: State Library of NSW

featured a production or research site beyond the urban fringes on cheaper lands. These could grow trees to accompany the urban site's more compact stock. Universal Nursery serves as an example, with their 1894 catalogue's illustrations of their sites at Wahroonga and Willoughby (Fig. 4.16, top, middle). Their Wahroonga nursery is carved out of the bushland, and with their Willoughby testing ground, together convey the scale of their operations. Despite roses gracing the cover, the majority of Milner & Co's 1900 Catalogue (bottom) is dedicated to edible species. Some large nurseries even had a (small) city-centre retail site, as in the case of Gelding (Victoria) Nursery in the old George Street Markets, and Milner & Co in Royal Arcade.

Gelding's retail outlet was more akin to a florist, focused on selling flowers and seeds. John and William Gelding's activities are worthy of closer inspection. The brothers commenced operations in 1861 in Rushcutters Bay. They soon moved to Double Bay, before relocating to the then urban fringe in Petersham in 1869. The journey to Petersham from Double Bay took them 13 hours by bullock train, roughly equivalent to a direct flight from Sydney to New Delhi in today's terms. Their new 12-acre site on Old Canterbury Road was infested with tens of thousands of rats, owing to the area's piggeries and slaughterhouses.[187] In spite of these challenges, they operated their Petersham nursery for over 30 years (today straddling Dulwich Hill and Summer Hill). An 1883 illustration offers an insight into their sizeable undertaking (Fig. 4.1). It was renamed 'Victoria Nursery' following the sale of half their land through subdivision to a booming housing market in the area. The sale funded renovations, including a number of modern glasshouses, and we can see an impressive and spatially significant operation despite effectively halving its former land area.

Iliffe's nursery is also worth noting as one of Rockdale's "success stories".[188] John Iliffe arrived in Sydney in 1850 from England and raised a wide variety of fruit trees, aided by his passionate and tidy manager John Ah Hee, originally from China and one of Rockdale's best-known personalities in the 1880s and 1890s.[189] Iliffe Nursery purchased an additional 31 acres in 1875[190] off Croydon Road and established a reserve nursery. By 1884, the *Illustrated Sydney News* declared Iliffe's the "most extensive nursery and plant establishment in the Colony",[191] best known for over 500 varieties of roses.[192] Despite this specialisation, their successor Milner & Co noted in their 1900 catalogue that the approximately 100,000 fruit trees were annually grown and the largest part of their business.[193]

An 1884 *The Illustrated Sydney News* article suggests common and aspirational landscape typologies of the day and their commercial appeal,

> Messrs. J. and W. Gelding issue a catalogue annually, which is well worthy a place in every gardener's library. It is prefaced by a calendar of monthly operations for the year, compiled with great care, and applies to every department of culture: the orchard, the flower garden, the shrubbery, the kitchen garden, and the conservatory, are all treated in a simple yet practical way, easily understandable by that large class of amateurs who are fond of a garden, but who often, from want of the simplest knowledge of plant life, find its management a wearisome and unprofitable task.[194]

The largest nurseries' dense annual catalogues further illuminate the array of available fruit varieties. In 1889, Gelding's Nursery alone listed 133 varieties of apples, far beyond any available in today's nurseries.[195] English and North American varieties dominated, with approximately 90 types of (mostly French and English) pears. Others included 68 plum varieties, 63 types of grape (mainly table) and 26 nectarine, 25 apricot, 24 fig and 18 gooseberry varieties.[196] The period's horticultural diversity is further evident in a collection of wax fruit and vegetables displayed in the Technological Museum in Sydney from the 1880s. In 1886,[197] for instance, the Museum's colourist produced wax models of 124 varieties of apples and 45 types of pears, as well as stone fruit, melons, tomatoes, carrots, cucumbers and potatoes.[198] This range highlights the agricultural biodiversity (agrobiodiversity) loss over the past century.[199]

Universal Nursery's catalogue further illuminates cultivars, as well as emerging mechanisation and technologies. The catalogue was as much a guide and repository of information for successful production as a means of hawking their wares. It contained a 'fruit clock' (annual calendar), statistics, techniques for pruning and training, machinery, pests and fertilisation, amongst other hints. Their 1894 guide details orcharding and horticultural practices (Fig. 4.16). While (organic) manuring was still then widespread (as evident in CSR's advertisement on the back page of the catalogue), fungicide use is illustrated by a horse-and-cart spraying setup (middle centre). Mechanical aids in harvesting and processing are also clearly emerging.[200] Milner & Co's 1900 catalogue is similarly rich (Fig. 4.16).[201] Both contain detailed horticultural and technical information in addition to their wide array of edible species for sale.

Other key events of the period included Arthur Yates' establishment of his seed business as a branch of Yates in New Zealand in 1887. In 1893 Yates launched his range of packet seeds for suburban home gardeners and in 1895, released the first issue of *Yates' Gardening Guide for Australia and New Zealand: Hints for Amateurs*. This was among the first guides written for home gardeners. Following this, Frank Finedon published *The Australian Kitchen Garden* in 1897.[202] These indicate burgeoning interest in domestic gardening and growing food. Agricultural shows were also established in *Period Two*, many of which continue to exist today. Sydney's first Easter Show was organised by the Agricultural Society in 1869, while the first Castle Hill District Show was held on Edwin Joseph Black's 30-acre property at Rogans Hill in 1886.

Colonial diets continued to centre on meat consumption, although a New South Wales Vegetarian Society was founded in 1891 following Victoria's 4 years earlier, though both arose decades later than England's in 1847. Dr. Philip Muskett despaired at the extent of meat consumption in his 1898 *Book of Diet*, recommending a more Mediterranean diet with fresh salads, vegetables, fish, wine and coffee.[203] As noted by Jan O'Connell, Muskett was "shocked by the amount of meat, tobacco and tea Australians consumed and was critical of their failure to consume vegetables and fruit other than cabbage and potatoes".[204] Muskett did not live to see his recommendations filtering into Australia following twentieth-century immigration from Mediterranean countries in *Period Three*, and more widely adopted late in *Period Four*. In *Period Two*, diets continued to depend on British foodways, along with the commodity-crop nature of production and importation of mutton, beef, dairy (milk, butter, cheese), sugar, tea, tobacco and a limited number of vegetables.

As *Period Two* drew to a close, the bubonic plague broke out in Sydney a year before Federation in 1901 and continued for several years. Its significance to Sydney's metropolitan agriculture is discussed in the following chapter.

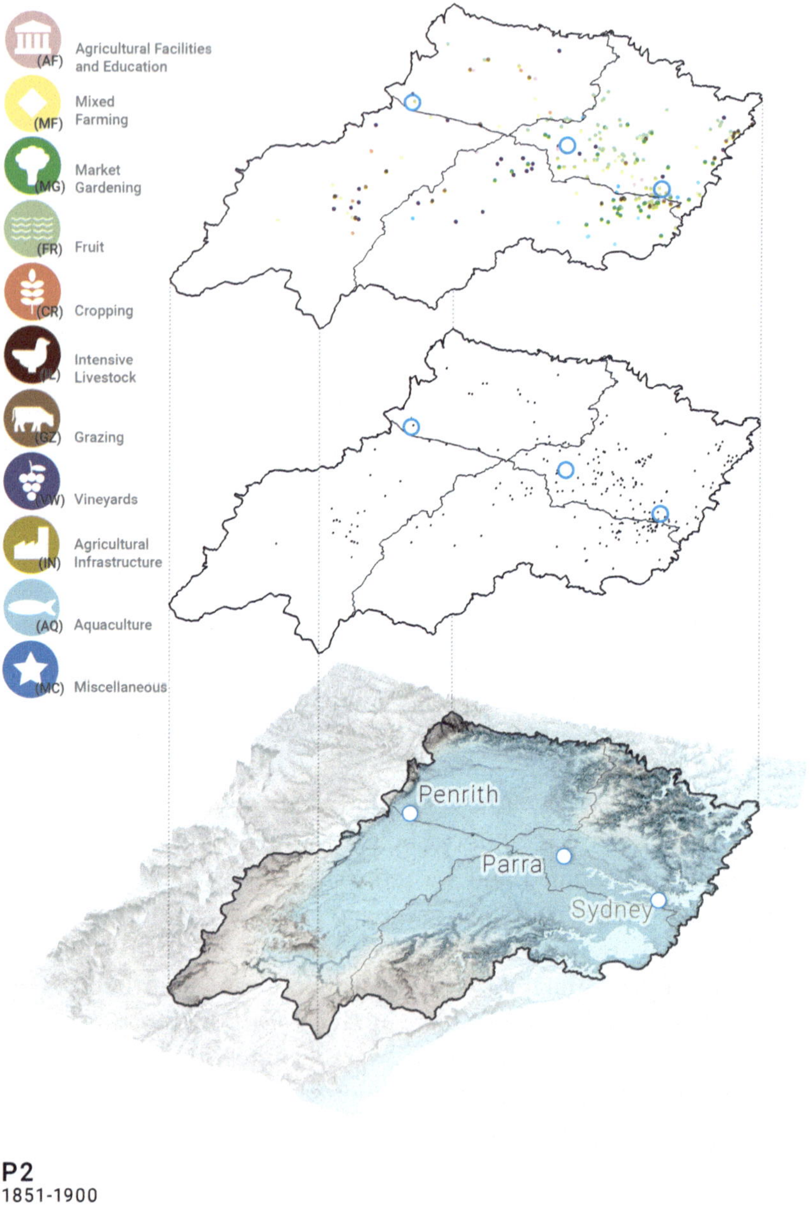

Fig. 4.17 *Period Two*'s Map Database: 3D. The number and diversity of entries have unsurprisingly increased from *Period One*, reflecting widespread agricultural expansion

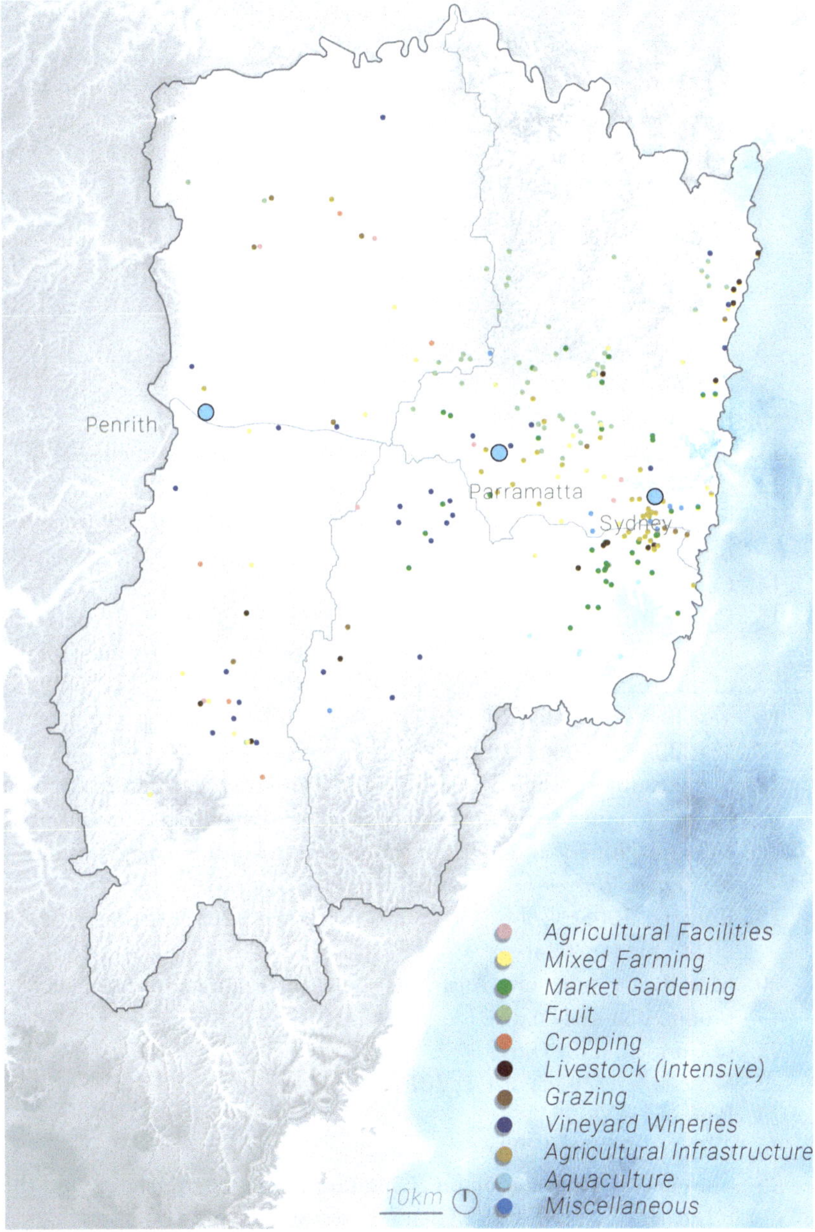

Fig. 4.18 *Period Two*'s Map Database: Plan

Notes

1. The 1901 Census listed 489,000, see: https://trove.nla.gov.au/newspaper/article/24730043.
2. Jeans, 1972: 295.
3. Mahar, 2015.
4. Most notably during the rationing period in early Sydney.
5. Mahar, 2015.
6. O'Connell, 2017, 14. Also see O'Connell, n.d., 1860.
7. O'Connell, n.d., 1860.
8. See Zeunert, 2013 for discussion of this in urban contexts.
9. de Courcy, 2008.
10. Ibid.
11. The Adelaide Advertiser in March 1861, cited in O'Connell, 2017.
12. O'Connell, n.d., 1861.
13. Only much later taking on a greater environmental and ecological role through breeding programmes.
14. O'Connell, n.d., 1861.
15. Quote attributed to Dr. George Bennett from Occasional Papers held in the Australian Museum archives. Bennett was Secretary and Curator of the Australian Museum from 1835 to 1841.
16. Year Book Australia, 2002.
17. Year Book Australia, 2002.
18. Significantly bolstered through the (misleadingly named) 'Green Revolution' in agriculture and its use of synthetic fertilisers and pesticides, machinery and logistics running on fossil-fuel energy sources—see *Periods Three—Five* and Zeunert, 2018.
19. Roberts, 2008; Cribb, 2010, 2017; Masters, 2019; Zeunert, 2018, 2023, 2024a, 2024b.
20. Macadamia nuts are Australia's only internationally commercially produced at-scale native food.
21. O'Connell, n.d., 1861.
22. See Zeunert, 2024a; 2018; 2013.
23. Plumwood, 2008.
24. O'Connell, n.d., 1893a.
25. The effect of refrigeration on civilisation cannot be overstated, and this is further discussed in this Chapter's sections *IL* and *IN*.
26. City Plan, 2014: 8 (68).

27. See O'Connell, n.d., 1879a.
28. Reuter, 2012.
29. According to company source here: https://www.csbp.com.au/about-us/history.
30. Gaynor, 2006: 26.
31. Christie, 1988: 90.
32. Chiefly for horse feed. See e.g.: *The Illustrated Sydney News*, March 23, 1878: https://trove.nla.gov.au/newspaper/article/63334806/5453000.
33. Gibbs et al., 1882: 136.
34. Gibbs et al., 1882: 136.
35. Cowey, n.d. Victoria was part of New South Wales until 1851, when it became its own colony. See Fig. 2.3.
36. See Murphy, 2021.
37. O'Connell, n.d., 1878.
38. O'Connell, n.d., 1853.
39. Yass Junction in 1876, Galong, Harden-Murrumburrah and Cootamundra in 1877, Junee and Bomen in 1878. The Murrumbidgee River Railway Bridge was completed in 1881 and the line was extended to Wagga Wagga, Uranquinty, The Rock, Henty and Albury in the same year.
40. Christie, 1988: 90.
41. Benson & Howell, 1990: 34.
42. ADB, Rutledge, 1972.
43. Inglis, 1880: 144–146.
44. Hoskins, 1997.
45. This was drawn decades before the first aeroplane flight and was most likely an estimated view based on ground studies. See Bell for a discussion on how such drawings were generated: https://www.academia.edu/36350956/Birds_eye_View_of_Adelaide_in_the_1870s.
46. See the Map Database.
47. Kerr, 1995: 30.
48. Kerr, 1995: 30.
49. Pers. Comm. Sarah Luke, 29 April 2020.
50. Nichols, 2015b.
51. Western Sydney University Archives Collection P137, P474.
52. Burgin, 2010: 264.
53. Bowd, 1986: 177, cited in LandArc, 2009: 36.
54. See, for example, Morris, 2008.

55. Such as vineyards at Robert Campbell's Mona Vale property c. 1819–; Dr. Alexander Berry's 'Crows Nest House' 1850–; Frederick Chave's and the Crawford Brothers surprising 1880s' wine and orchard ventures at Lovett and Elvina Bay's respectively and Patrick Fallon's 1887–c. 1910 'Mount Ramsey Estate' from Collaroy to North Narrabeen.
56. Such as Gregory Blaxland at Brush Farm in Dundas Valley.
57. Already well demonstrated by the Macarthur empire commencing in *Period One* and continuing to this day. See Fig. 3.9.
58. Blaxland also established a saltworks, farm and factory at Newington Estate. A 1893 map of the area (showing Ermington Orchard) is included in Benson & Howell, 1990: 48 and Dept. Infrastructure, Planning and Natural Resources, 2004: 12.
59. Such as William Wentworth in Vaucluse House (and Gregory Blaxland was also part of the expedition that successfully crossed the Blue Mountains).
60. The Sydney Mail and New South Wales Advertiser, 18820401: 497. see: https://trove.nla.gov.au/newspaper/page/16500548?#.
61. See *MF* in the Map Database, particularly in P1 and P2.
62. ADB, Rutledge, 1972.
63. Inglis, 1880: 147–148.
64. See the work of Sarah James.
65. James, 2016: 113–114.
66. Christie, 1988: 96.
67. Christie, 1988: 74–75, citing Yong's *The New Gold Mountain*.
68. Fitzgerald, 2008: 94.
69. From a booklet protesting against immigration laws on the Victoria Goldfields, cited in O'Connell, n.d.-a, n.d.-b, 1852.
70. Price, 2000: 11.
71. Fitzgerald, 2008: 94.
72. A photograph in Butel & Thompson, 1984 shows the area.
73. Bellevue Street runs left to right and the now Kogarah School of Arts is visible centre background.
74. Teffer, 2019.
75. Mulcair, 2007, citing the National Trust: https://www.facebook.com/Igrewupinmortdale2223/posts/2902384879991880.
76. Fitzgerald, 2008: 95.
77. These include both agglomeration districts as well as individual gardens.

78. See e.g.: https://digital.sl.nsw.gov.au/delivery/DeliveryManagerServlet?dps_pid=FL3313884&embedded=true&toolbar=false and https://www.flickr.com/photos/statelibraryofnsw/5775144516/.
79. Teffer, 2019.
80. Winston, 1957: 21.
81. Butlin, 1976: 11.
82. Christie, 1988: 75.
83. Christie, 1988: 75.
84. Christie, 1988: 75; Spearritt, 2000: 22.
85. Christie, 1988: 93.
86. The Orange in New South Wales (from the Intercolonial Exhibition Official Record), In The Inquirer and Commercial News, Perth WA, 14 August 1867: https://trove.nla.gov.au/newspaper/article/69386876
87. McClymont, 2013.
88. *The Illustrated Sydney News*, March 23, 1878: https://trove.nla.gov.au/newspaper/article/63334806/5453000.
89. Higgins, 2016: 25.
90. Kirkpatrick, 2022: 24.
91. Nichols, 2015a.
92. https://www.sekisuihouse.com.au/the-orchards/history.
93. https://www.thehills.nsw.gov.au/Library/Local-Studies-Family-History/A-Brief-History-of-the-Shire.
94. https://www.belgennyfarm.com.au/history/agriculture-at-camden-estate/horticulture-and-viticulture.
95. Also see images at: http://www.pittwateronlinenews.com/Pittwater-Summer-Houses-Rocky-Point-Elvina-Bay.php
96. Patrick Edward Quinn, Accessed 15 Nov 2019, http://www.pittwateronlinenews.com/Pittwater-Summer-Houses-Rocky-Point-Elvina-Bay.php. See here for further images.
97. https://trove.nla.gov.au/newspaper/article/162736592.
98. https://trove.nla.gov.au/newspaper/article/71191632.
99. https://trove.nla.gov.au/newspaper/article/71191632.
100. The Orange in New South Wales (from the Intercolonial Exhibition Official Record), In The Inquirer and Commercial News, Perth WA, 14 August 1867 https://trove.nla.gov.au/newspaper/article/69386876.
101. O'Connell, n.d., 1868 and https://arthurtuckwell.wordpress.com/george-thomas-tuckwell/granny-smith-connection/.

102. Metzke, 2004: 34.
103. Dominiak & Daniels, 2012: 106.
104. Meats, 1989.
105. Bellas, (1995–96). https://www.aph.gov.au/About_Parliament/Parliamentary_Departments/Parliamentary_Library/pubs/rp/RP9596/96rp29.
106. Dominiak & Daniels, 2012: 106 citing Froggatt, 1909.
107. Bellas, (1995–96). https://www.aph.gov.au/About_Parliament/Parliamentary_Departments/Parliamentary_Library/pubs/rp/RP9596/96rp29.
108. Dominiak & Daniels, 2012: 106, citing Froggatt 1899.
109. Dominiak & Daniels, 2012: 106, citing Lea (1899).
110. See Meats, 1989, Bonizzoni et al., 2004: 3846, Dominiak & Daniels, 2012.
111. Dominiak & Daniels, 2012: 106.
112. Dominiak & Daniels, 2012: 106.
113. https://www.sl.nsw.gov.au/stories/australian-agricultural-and-rural-life/fruit-industry.
114. *The Illustrated Sydney News*, March 23, 1878: https://trove.nla.gov.au/newspaper/article/63334806/5453000.
115. While it can also deplete soil fertility, as already observed in *Period One*. See also: Zeunert, 2018.
116. Clarke, 2017: 25.
117. O'Connell n.d., 1903.
118. See 1854 depiction: https://dictionaryofsydney.org/media/1451.
119. http://www.marrickvilleunearthed.com/209-livingstone-rd.html.
120. Lewis, 1980: 196.
121. Lewis, 1980: 196–198.
122. Gaynor, 2006: 26.
123. http://www.marrickvilleunearthed.com/209-livingstone-rd.html.
124. http://www.marrickvilleunearthed.com/209-livingstone-rd.html.
125. See http://www.marrickvilleunearthed.com/209-livingstone-rd.html and https://cityhubsydney.com.au/2019/08/dairy-beloved-building/142188/.
126. http://www.marrickvilleunearthed.com/209-livingstone-rd.html.
127. Butlin, 1976: 15.
128. Butlin, 1976: 291.
129. Kerr, 2002: A16.

130. https://dictionaryofsydney.org/media/1451.
131. Clarke, 2017: 25.
132. Davies & Britton, 2011: 15.
133. Davies & Britton, 2011: 15.
134. Deniehy, 1853.
135. https://archival.sl.nsw.gov.au/Details/archive/110339020.
136. See https://commons.m.wikimedia.org/wiki/File:H.E.I.Cos_Depot._Bungarrabbee_(i.e._Bungarribee)_N.S.W.,_ca._1847-1852_T._Rider_FL542914.jpg.
137. Sydney Vignerons Association, 2011.
138. https://www.belgennyfarm.com.au/history/agriculture-at-camden-estate/horticulture-and-viticulture and https://dictionaryofsydney.org/entry/spring_farm.
139. Sydney Vignerons Association, 2011.
140. Nichols, 2016: 93.
141. Liston, 2009: 61.
142. There are conflicting historical accounts about the exact names and dates of the property. Three sources present different dates of owners/operators and associated namings: https://huntershilltrust.org.au/archive/the-priory/; https://drnorrie.info/html/article_austasylum-vineyards.html; and Sydney Vignerons Association, 2011.
143. https://australianfoodtimeline.com.au/minchinbury-champagne/.
144. https://trove.nla.gov.au/newspaper/article/13401570.
145. http://thedistillerywoodfire.com.au/about-us/our-story/.
146. McGill, 2017: 37.
147. Sydney Vignerons Association, 2011.
148. Sydney Vignerons Association, 2011.
149. https://australianfoodtimeline.com.au/phylloxera/.
150. Langmeil winery, Barossa Valley. Western Australia and Tasmania also evaded infection.
151. See for example Murphy, 2021.
152. https://collection.maas.museum/object/30123.
153. https://collection.maas.museum/object/367773.
154. Pratley & Rowell, 2003: 1.
155. http://www.austehc.unimelb.edu.au/tia/089.html.
156. https://www.belgennyfarm.com.au/history/agriculture-at-camden-estate/dairying/milk-for-sydney.

157. See Lewis for various statistics outlining the shifting power dynamics in milk supply and distribution in late P2 and early P3 (Lewis, 1980: 196).
158. Technology in Australia 1788-1988, 1988: 91 - 92, Sourced via Online Edition 2000, https://www.austehc.unimelb.edu.au/tia/089.html. See also O'Connell, n.d.: https://australianfoodtimeline.com.au/fresh-food-ice-co/.
159. City Plan, 2014: 37–38.
160. https://australianfoodtimeline.com.au/frozen-meat-exports/, citing https://www.austehc.unimelb.edu.au/tia/089.html.
161. https://labourhistorymelbourne.org/the-rabbit-industry/.
162. In rural areas it persisted for much longer. See O'Connell, n.d.-a, n.d.-b.
163. See https://trove.nla.gov.au/newspaper/article/38868891.
164. Hughes Turnbull (1999: 106) notes roaming goat herds in Sussex Street in the late 1850s.
165. See https://trove.nla.gov.au/newspaper/article/13045034, https://dictionaryofsydney.org/place/glebe_island_abattoir and https://pyrmonthistory.net.au/abattoirs.
166. Zeunert, 2024a, b.
167. The Dictionary of Sydney states 1830, but this is hard to substantiate given the following sources says, "In 1848 the government began to discuss building an abattoir on the island, to remove animal slaughter from the centre of Sydney. The island was considered 'admirably suited for the purposes.' Architect Edmund Blacket designed the buildings, which were erected over a period in the 1850s". https://www.dailytelegraph.com.au/news/today-in-history/colonial-vision-would-have-turned-glebe-island-wasteland-into-a-harbour-gem/news-story/090307f6b7f24585807e1915e80ed2b1, https://dictionaryofsydney.org/place/glebe_island_abattoir.
168. Winston, 1957: 21.
169. See Christie, 1988.
170. Christie, 1988: 76–77.
171. Christie, 1988: 84.
172. Christie, 1988: 152.
173. Christie, 1988.
174. Ibid.
175. Ashton & Waterson, 2000.
176. Christie, 1988: 83.
177. Inglis, 1880: 200–201.
178. Inglis, 1880: 200–201.

179. O'Connell, n.d., 1872.
180. There are two main indigenous oyster species—the Sydney rock oyster, which prefers warmer waters; and the Angasi or flat oyster, found in colder temperatures. See O'Connell, n.d., 1872.
181. Sutherland Shire Environment Centre, 2008.
182. Sutherland Shire Environment Centre, 2008
183. O'Connell, n.d., 1872, 1879a, 1879b.
184. https://trove.nla.gov.au/newspaper/article/63621626/5789340.
185. To the 1000+ seated male-only guests. See O'Connell, n.d.-a, n.d.-b, 1901: https://australianfoodtimeline.com.au/federation-banquet/.
186. In Waterloo, see https://nla.gov.au/nla.obj-231131950/view.
187. See quote at: https://reflectivegardener.blogspot.com/2012/08/geldings-victoria-nursery.html.
188. Rathbone, 1986: 11–12, see: https://catalogue.nla.gov.au/Record/6704750.
189. Rathbone, 1986: 11–12, see: https://catalogue.nla.gov.au/Record/6704750.
190. Rathbone (1986: 11) records this as "44 acres in present day West Bexley" but the *The Illustrated Sydney News* on 7 June 1884 reported it in 1884 as 44 acres of land "at Hurstville in the centre of Gannons Forest".
191. *The Illustrated Sydney News*, 7th June 1884, as cited in Lawrence, 2011: 40.
192. Rathbone, 2006: 39.
193. See https://digital.sl.nsw.gov.au/delivery/DeliveryManagerServlet?embedded=true&toolbar=false&dps_pid=IE3795333.
194. *The Illustrated Sydney News*, 1 Sep 1883, Accessed 16 April 2020: https://trove.nla.gov.au/newspaper/article/64034792.
195. The largest variety of apples the author has found in Australia is through 'Heritage Fruit Trees' in Victoria, see: https://www.heritagefruittrees.com.au/.
196. https://reflectivegardener.blogspot.com/2012/08/geldings-victoria-nursery.html.
197. https://collection.maas.museum/object/221333.
198. Mrs. A.G. Goodman was the colourist at the time. No photos of this exhibition are available at the time of writing, but a similar model (1899) is here: https://collection.maas.museum/object/223801.
199. See Zeunert, 2018.

200. See page extracts here: http://digital.sl.nsw.gov.au/delivery/DeliveryManagerServlet?embedded=true&toolbar=false&dps_pid=IE3702649.
201. https://digital.sl.nsw.gov.au/delivery/DeliveryManagerServlet?embedded=true&toolbar=false&dps_pid=IE3795333.
202. https://sydneylivingmuseums.com.au/exhibitions/eat-your-history-library#item-details-5.
203. https://australianfoodtimeline.com.au/1892-mediterranean-diet-recommended/.
204. O'Connell, n.d., 1893b.

References

Ashton, P., & Waterson, D. (2000). *Sydney takes shape: A history in maps*. Brisbane.

Benson, D., & Howell. (1990). *Taken for granted: The bushland of Sydney and its suburbs*. Kangaroo Press.

Bonizzoni, M., Guglielmino, C. R., Smallridge, C. J., Gomulski, M., Malacrida, A. R., & Gasperi, G. (2004). On the origins of medfly invasion and expansion in Australia. *Molecular Ecology, 13*, 3845–3855.

Burgin, S. (2010). A natural history of the Ham Common, the lands sandwiched between Richmond and Windsor—two of Governor Macquarie's five towns. In D. Lunney, P. A. Hutchings, & D. Hochuli (Eds.), *The natural history of Sydney* (pp. 263–276). Royal Zoological Society of NSW. https://doi.org/10.7882/9780980327236

Butel, E., & Thompson, T. (1984). *Kings cross album: Pictorial memories of kings Cross Darlinghurst Woolloomooloo & Rushcutters Bay*. Atrand.

Butlin, N. G. (Ed.). (1976). *Sydney's environmental amenity 1970–1975*. ANU Press.

Christie, M. (1988). *The Sydney markets 1788–1988*. Sydney Market Authority.

City Plan. (2014). *Report on City of Sydney industrial & warehouse buildings heritage study*, City Plan: October 2014: H-13070, Accessed April 10, 2020. https://www.google.com/url?sa=t&rct=j&q=&esrc=s&source=web&cd=&ved=2ahUKEwianbmfqL7wAhVg73MBHft6Be4QFjAAegQICBAD&url=https%3A%2F%2Fwww.cityofsydney.nsw.gov.au%2F-%2Fmedia%2Fcorporate%2Ffiles%2F2020-07-migrated%2Ffiles_c-1%2Fcos-warehouses-industrial-buildings-heritage-study-report-final-2-13oct2014-accessible-version.pdf%3Fdownload%3Dtrue&usg=AOvVaw1h7o7uiw0KxM2GAvdoDJfy

Clarke, M. (2017). *Values of the metropolitan rural area of the Greater Sydney Region*. AgEconPlus Consulting. Accessed December 10, 2018, from https://gsc-public-1.s3.amazonaws.com/s3fs-public/Values_of_the_Metropolitan_Rural_Area_of_the_Greater_Sydney_Region_(Ag_Econ_Plus).pdf

Cribb, J. (2010). *The coming famine: The global food crisis and what we can do to avoid it*. University of California Press.

Cribb, J. (2017). *Surviving the 21st century: Humanity's ten great challenges and how we can overcome them*. Springer.

Davies, P., & Britton, G. (2011). *Visual analysis of Campbelltown's Scenic Hills and east edge scenic protection lands*. Prepared for Campbelltown City Council.

de Courcy, C. (2008). *Acclimatisation society of Victoria*, School of Historical & Philosophical Studies, The University of Melbourne. Accessed April 1, 2020. http://www.emelbourne.net.au/biogs/EM00033b.htm

Deniehy, D. (1853). The constitution of New South Wales. Speech documented in The Sydney Morning Herald, New South Wales, Australia, 16 August 1853, p. 4. Accessed February 2, 2021, from https://trove.nla.gov.au/newspaper/article/12948089

Dominiak, B. C., & Daniels, D. (2012). Review of the past and present distribution of Mediterranean Fruit Fly (Ceratitis capitata Wiedemann) and Queensland Fruit Fly (Bactrocera tryoni Froggatt) in Australia. *Australian Journal of Entomology, 51*(2), 104–115.

Fitzgerald, S. (2008). *Red tape gold scissors: The story of Sydney's Chinese*. Halstead Press.

Gaynor, A. (2006). *Harvest of the suburbs: An environmental history of growing food in Australian cities*. UWA Press.

Gibbs, Shallard and Co. (1882). *Illustrated guide to Sydney and its suburbs and to favourite places of resort*. Sydney, Angus and Robertson.

Higgins, L. (2016) The family of old Mans Valley, https://hornsbyshire.recollect.net.au/nodes/view/3053

Hoskins, I. (1997). Topographic map of the country around Sydney. *Powerhouse Museum*.

Inglis, J. (1880). *Our Australian cousins: 'Maori'*. Macmillan & Co.

James, S. (2016). *Farming on the fringe: Peri-urban agriculture, cultural diversity and sustainability in Sydney*. Springer.

Jeans, D. (1972). *An historical geography of New South Wales to 1901*. Reed Education.

Kerr, J. (1995). Parramatta correctional Centre: Its past development and future care, Commissioned by NSW Public Works for the Department of Corrective Services, Department of Corrective Services, Sydney. https://csa.intersearch.com.au/csajspui/bitstream/10627/851/1/Parramatta%20CC.pdf

Kerr, R. (2002). *University of Sydney grounds conservation plan: Appendix a: University of Sydney Overview history*. Accessed January 21, 2021. https://www.sydney.edu.au/documents/about/heritage/gcp_appendix1.pdf

Kirkpatrick, A. (2022). Laying the groundwork for soil biodiversity. *Landscape Architecture Australia, 173*, 22–26.

LandArc. (2009). Ham common plan of management. Accessed July 16, 2020. https://www.hawkesbury.nsw.gov.au/__data/assets/pdf_file/0006/6297/HamCommonAdoptedPOM.pdf

Lawrence, J. (2011). *St George pictorial memories*. Kingsclear Books.

Lewis, M. (1980). Milk, mothers, and infant welfare. In J. Roe (Ed.), *Twentieth century Sydney: Studies in urban & social history* (pp. 193–207). Southwood Press.

Liston, C. (2009). *Pictorial history: Liverpool & district*. Kingsclear Books.

Mahar, C. (2015). *Food and provisions*, eGold. Cultural Heritage Unit, The University of Melbourne. Accessed March 30, 2020. http://www.egold.net.au/biogs/EG00116b.htm

Masters, N. (2019). *For the love of soil: Strategies to regenerate our food production systems*. Printable Reality.

McClymont, J. (2013). *Pictorial history: The Hills District*. Kingsclear Books.

McGill, J. (2017). *Pictorial history Campbelltown & District*. Kingsclear Books.

Meats, A. (1989). Bioclimatic potential in Fruit flies: Their biology, natural enemies and control. In A. Robinson & G. Hooper (Eds.), *World crop pests* (Vol. 3B, pp. 241–252). Elsevier.

Metzke, M. (2004). *Pictorial history Hornsby Shire*. Kingsclear Books.

Morris, C. (2008). *Lost gardens of Sydney*. Historic Houses Trust.

Mulcair, J. (2007) The Leader, published 28/6/2007. Cited at https://www.facebook.com/Igrewupinmortdale2223/posts/2902384879991880

Murphy, M. (2021). *Rum: A distilled history of colonial Australia*. HarperCollins Australia.

Nichols, M. (2015a). Fruit growing in the Hawkesbury. http://hawkesburyheritage.blogspot.com/2015/

Nichols, M. (2015b). Government labour settlement at Pitt Town. http://hawkesburyheritage.blogspot.com/search/label/Scheyville

Nichols, M. (2016). *Pictorial history: Hawkesbury*. Kingsclear Books.

O'Connell, J. (n.d.-a) *Australian Food History Timeline,* Accessed May 12, 2019 https://australianfoodtimeline.com.au/

O'Connell, J. (n.d.-b) Late 1890s Coolgardie safe invented. https://australianfoodtimeline.com.au/coolgardie-safe-invented/

O'Connell, J. (n.d.) 1852: Vegetable garden on Victorian goldfields. https://australianfoodtimeline.com.au/1852-vegetable-garden-on-victorian-goldfields/#more-9742

O'Connell, J. (n.d.) 1853: First gas stoves in Australia: https://australianfoodtimeline.com.au/first-gas-stoves/#more-3471

O'Connell, J. (n.d.) 1860: First of the Land Acts. https://australianfoodtimeline.com.au/first-land-acts/

O'Connell, J. (n.d.) 1861: First Acclimatisation Society. https://australianfoodtimeline.com.au/1861-first-acclimatisation-society/

O'Connell, J. (n.d.) 1868: Granny Smith apple appears: https://australianfoodtimeline.com.au/1868-granny-smith-apple-appears

O'Connell, J. (n.d.) 1872: Oyster farming begins in Sydney: https://australianfoodtimeline.com.au/oyster-farming-begins/

O'Connell, J. (n.d.) 1878: Coffee Palace Hotel Company formed: https://australianfoodtimeline.com.au/coffee-palace-hotel-company/

O'Connell, J. (n.d.) 1879a: First frozen meat exports: https://australianfoodtimeline.com.au/frozen-meat-exports/

O'Connell, J. (n.d.) 1879b: First Greek fish and chip shop. https://australianfoodtimeline.com.au/first-greek-fish-and-chip-shop/

O'Connell, J. (n.d.) 1893a: Height of 1890s depression. https://australianfoodtimeline.com.au/1890s-depression/

O'Connell, J. (n.d.) 1893b: Mediterranean diet recommended. https://australianfoodtimeline.com.au/1892-mediterranean-diet-recommended/

O'Connell, J. (n.d.) 1901: Federation banquet. https://australianfoodtimeline.com.au/federation-banquet/

O'Connell, J. (n.d.) 1903: William Farrer's Federation wheat. https://australianfoodtimeline.com.au/william-farrer-federation-wheat/

O'Connell, J. (2017). *A timeline of Australian food: From mutton to MasterChef.* New South Publishing.

Pers. Comm. Sarah Luke. April 29, 2020.

Plumwood, V. (2008). Shadow places and the politics of dwelling. *Australian Humanities Review, 44.*

Price, E. (2000). *Some market gardeners in Randwick and Coogee 1840s–1880s.* Randwick and District Historical Society.

Rathbone, R. (1986). *From white gum flat to the suburb of Rockdale*. Rockdale City Council.

Reuter, D. (2012). The South Australian Superphosphate Story—part I. *Agricultural Science* (24–2) Aug 23–33. Accessed February 23, 2020. https://search.informit.com.au/documentSummary;dn=809294082396749;res=IELHSS

Roberts, P. (2008). *The end of food*. Houghton Mifflin.

Spearritt, P. (2000). *Sydney's century: A history*. UNSW Press.

Sutherland Shire Environment Centre. (2008). *Occupation: Thomas Holt*. Accessed April 20, 2020. https://www.ssec.org.au/our_environment/our_bioregion/kurnell/history/occupation/tomholt.htm

Sydney Vignerons Association. (2011). Historical Timeline: Wine and grape history in the Sydney region. Accessed April 13, 2020. https://www.hawkesbury.net.au/sydneywine/overview/history.html

Teffer, N. (2019). Celestial City: Sydney's Chinese story. *Museums of History NSW*.

Winston, D. (1957). *Sydney's great experiment: The progress of the Cumberland County plan*. Halstead Press.

Year Book Australia. (2002). Commonwealth of Australia: Australian Bureau of Statistics, Accessed July 27, 2011. http://www.abs.gov.au/ausstats/abs@.nsf/94713ad445ff1425ca25682000192af2/bfc28642d31c215cca256b350010b3f4

Zeunert, J. (2013). Challenging assumptions in urban restoration ecology. *Landscape Journal, 32*(2), 231–242.

Zeunert, J. (2018). Challenges in agricultural sustainability and resilience: Towards regenerative practice. In J. Zeunert & T. Waterman (Eds.), *Routledge handbook of landscape and food* (pp. 231–252). Routledge.

Zeunert, J. (2023). *Food | landscapes Australia*. https://www.foodlandscapes.com.au/. Doi: https://doi.org/10.26190/s0kk-8x57

Zeunert, J. (2024a). Bridging rural and urban disconnections: Spatial graphic explorations of Australia's livestock landscapes. In R. Pernice & B. Chen (Eds.), *Urban regeneration and rural revitalization in Australia and China* (pp. 110–143). Routledge.

Zeunert, J. (2024b). States of Australia's Agri-Environment: Visual extractions from degraded landscapes. *Visual Studies, 39*(4), 609–621. https://doi.org/10.1080/1472586X.2023.2294004

5

Period Three: 1901–1945

Box 5.1 Population 1901 and 1945

Population 1901: 488,000
Population 1945: 1.5 million
Increase = 3-fold (over 1 million)

J. Zeunert, A. Daroy, *Sydney's Food Landscapes*,
https://doi.org/10.1007/978-981-96-0710-5_5

Fig. 5.1 Mascot, 1928. Source: National Library Australia, Bib ID: 3930726

Box 5.2 Key Events 1901–1945

- Federation in 1901 unified the previously six self-governing British colonies still subject to British Parliament with individual laws. Colonies became states and territories of the Commonwealth of Australia;
- Formalisation of the *Immigration Restriction Act* in 1901, also known as the 'White Australia Policy', particularly to restrict Chinese and Pacific Islander immigration;
- (Non-First Nations) women gain the vote in New South Wales through the passing of the *Women's Franchise Act* in 1902 (following South Australia as early as 1861 for local elections);
- Increasing employment for women, especially within manufacturing;
- Solidification of state and territory boundaries after 1911, with New South Wales taking today's form (Fig. 2.3);
- World War 1 erupts in 1914, ending in 1918;
- The Great Depression hits in 1929 until the late 1930s;
- World War II takes place from 1939 to 1945;
- Increasing use of concentrated fossil fuel energy sources, chiefly crude oil, gas and coal, incrementally diminishing local supply chains;
- Technological and production advancements, particularly increasing mass production scales and uniformity through assembly lines, the internal combustion engine and private motor vehicles;
- Emergence of synthetic chemicals, including industries and agricultural applications, accompanied by a shift from organic to synthetic agricultural production approaches.

Period Three (P3) spans a turbulent time of immense change including industrialisation, war, disease, depression and population growth. At Federation in 1901, Sydney had a population of 481,830.[1] This had more than tripled by 1945, reaching at least 1.5 million and changing the city's fabric and agricultural practices.[2] The tumultuous 45-year timeframe encapsulated multiple significant events, from unification of previously disparate colonies, a decade riddled with the bubonic plague[3] followed by the even more devastating Spanish influenza, two world wars and the Great Depression. These events impacted agriculture—as Jeff McGill notes: "Local farmers felt they were the backbone of the nation during the wartime emergency".[4] In spite of, and in some cases because of, these challenges, Sydney entered a new era of accelerated technological development. This was one of the most transformational periods in human history, during which fossil fuel-driven industry and mass-produced

Box 5.3 Key Events in Sydney 1901–1945

- Bubonic plague breaks out in 1900;
- State government investment in infrastructure, with electricity (from 1904), gas, potable water, sewage, tram, rail (including underground), roads and bridges;
- Sydney's tram network significantly expands, with tracks reaching 181 miles (291 kilometres) in 1923 and peaking at 405 million journeys in 1945;
- *The Meat Industry Act* passed in 1915 and the Metropolitan Meat Industry Board established as a statutory body to control and maintain abattoirs, cattle sale yards, meat markets and slaughterhouses;
- Spanish influenza outbreak in 1919;
- The emergence of manufacturing as a significant industry and employer;
- The *Local Government Act* passed in 1919, allowing for the declaration of land use zoning;
- Sydney Harbour Bridge opens 19 March 1932, greatly supporting development in Sydney's north;
- Formation of the County of Cumberland Council in 1945 to prepare Sydney's first metropolitan-scale plan;
- The ongoing establishment of National Parks and Nature Reserves.

Box 5.4 Sydney's Metropolitan Agriculture Trends 1901–1945

- A shift from prior agrarian agriculture (chiefly supported by horse and human labour and manure fertiliser) to increasingly mechanised, industrial and synthetic approaches;
- Decreasing proportions of the population employed in agriculture;
- Continued advancements in mechanisation, such as machine graders and tractors and their attachments;
- Increased employment in food processing factories facilitated through, for example, the State Abattoir's relocation from Glebe to large facility in Homebush;
- Increasing agricultural training opportunities, formal education and lengths of study;
- The emergence of private motor vehicles facilitating supply chains' decentralisation;
- An overall decline in urban dairies and stone fruit orcharding;
- A steady decline in citrus orcharding;
- Increase in poultry-keeping and poultry-stocking densities;
- The loss of market gardens in areas of urban expansion;
- Shifting immigration patterns stemming from the *Immigration Restriction Act*, or 'White Australia Policy' (1901) and world wars affecting agricultural typologies such as market gardening.

transport accelerated the time-space continuum. This increased the pace of life and compressed distances and the time taken to cover them, making the world smaller and much faster to access.

The central transformative agents were crude oil and its refined fuels and by-products in the form of petroleum (petrol, gasoline), distillate (diesel) and heavy fuel oil in shipping, deployed in internal combustion engines. By midway through *Period Three*, the mass production of engines, vehicles and an array of motorised-powered transportation and availability of their fuel sources facilitated significant local, national and global changes. This included urban expansion away from heavy rail and tram infrastructures, beginning a widespread and mass process of spatial reconfiguration of cities and significant decentralisation and suburbanisation. Not altogether dissimilarly, petrol or diesel fuelled tractors and harvesters incrementally supplanted millennia of human and animal-based farm labour over the course of the period. This significantly reduced the intensity of labour required and transformed farming's spatial characteristics through opening up new forms of broad-scale agriculture and increasing production scales. This was especially the case in cropping but also in horticulture, amidst others. It also removed agriculture's chief organic supply of nutrients, namely horse manure, which would be replaced with synthetic alternatives and fossil fuel derivatives.

By the latter part of the period, increasingly larger, mechanised production signalled a decline within Sydney's metropolitan context as a focal point for many forms of agricultural production, especially dairy, fruit and, to a lesser degree, vegetables. At the same time, the urban form that previously clustered around transport nodes was becoming widely decentralised and supplanting agricultural production lands of all typologies. Suburbanisation had commenced in earnest. Given the often subsistence nature of metropolitan farming, increased land values enabled farmers to capitalise on new urban land development opportunities that were considerably more lucrative than most forms of metropolitan food production.

Sydney's streets, for instance, had previously been extensively used to move livestock. A 1929 photograph presents an anachronistic capture of urban livestock and a horse-driven wagon still present alongside a truck mustering 24 pigs (Fig. 5.2, top). Private vehicles' mass uptake led to increasing congestion and incompatibility with unruly livestock roaming

Fig. 5.2 Pigs being herded by car on Day Street, Sydney 1929 (top). Spraying and ploughing at Hawkesbury Agricultural College c. 1900–1910 (bottom). Sources: Sam Hood, Mitchell Library, State Library of New South Wales; Western Sydney University Archives Collection P1496

the city streets. Growing production scales also rendered smaller urban operations less competitive and livestock began to diminish in urban areas, shifting to the peri-urban fringe and beyond. Improved transportation systems enabled city livestock markets to be relocated to Homebush, 20-kilometres from Sydney city. Outside the city, horses remained in service to plough fields, pull machinery and carts (Fig. 5.2, bottom).

The first decade of the twentieth–century also marked the onset of chemical and synthetic agriculture, which was particularly relevant for crop production. The c. 1900–1910 photograph of Hawkesbury Agricultural College (bottom), for example, shows spraying or 'dusting' of vegetable crops, here for cauliflower (Fig. 5.2).[5] Manual application would later be upscaled through mechanised processes such as tractor and in some instances, aerial chemical spraying.

The growing normality of synthetic agricultural production represented another seismic historical shift of the period. Previous agricultural approaches utilised animal wastes and organic amendments for soil nutrient replenishment. Synthetic agricultural production introduced, or significantly expanded, chemicals, fossil fuels (especially natural gas)[6] and mined rock phosphate. These were used to manufacture, for example, granular and liquid ammonium phosphate fertilisers like superphosphate[7] and animal feed supplements, while industrial wastes and by-products were adapted for fertilisers and pesticides.[8] Metropolitan Sydney's dwindling animal numbers, especially horses, reduced the availability of manure, while industrial manufacturing created abundant supplies of synthetic products. Synthetic agriculture was therefore symbiotic with the rise of vehicles, with chemical fertilisers becoming the central replacement for manure. Synthetic chemicals were transformative for killing agricultural pests, facilitating, alongside the large machinery applying them, *Period Four*'s 'Green Revolution' and its greatly increased crop yields.

Fossil fuels therefore became embedded in day-to-day life, food production, processing and distribution, facilitating population growth.[9] Accordingly, these dramatic industrial, technological and sociocultural developments were accompanied by increasing anthropogenic environmental impacts, from pollution and the release of greenhouse gases, to accelerated consumption of natural resources and the destruction of ecological systems. Awareness of these impacts, however, did not widely surface until *Period Four*.[10]

Relevant Global and National Events: 1901–1945

At the beginning of the period in 1901, the Australian Population Census recorded 14 per cent of the settler population as working in agriculture and pastoralism.[11] This number would decline over time. Sydney housed a third of New South Wales' population, 488,000 of 1.365 million.[12] Wool production initially continued to dominate land use and the economy at a national scale, even with the setbacks of drought, World War 1 and the economic depression. Its significance diminished by the end of the period to be overtaken by manufacturing industries. Rhetoric around the Australian national identity still greatly privileged agrarian industry, despite the growing urban-rural disconnect of a national population clustered in capital cities and urban agglomerations (Fig. 5.3).

Although national in scope and not necessarily indicating Sydney's specific agricultural makeup, the Australian Bureau of Statistics offer a clear picture of Australia's commodity focus of the time,

> About 284,000 tonnes of beef and 400,000 tonnes of mutton and lamb were exported in the first decade of the century [...] In the 1901–02 season, wheat accounted for 2 million hectares of the area under crop, followed by oats (187,000 hectares), maize[13] (119,000 hectares), and barley (30,000 hectares). The areas planted to orchards and fruits, sugar cane and potatoes amounted to 59,000 hectares, 54,000 hectares and 44,400 hectares respectively. Most cropping was done in the eastern States [...] Other broadacre crops, such as grain sorghum, canola, field peas and safflower, had not yet made an appearance [...] Attempts had been made to grow cotton and coffee, but without much success.[14]

Agriculture continued to reflect British, Irish and European culinary traditions as reflected in the makeup of Australia's production of commodities.[15] At the turn of the century, Australian diets largely consisted of beef, mutton, potatoes, oats, sugar and tea, supplemented by barley and cabbage. Fruit and varied vegetables featured minimally, despite the many species present in *Period Two*'s nurseries (see *Miscellaneous*).[16] Culinary diversity was comparatively narrow compared to Sydney's today.[17]

Fig. 5.3 Million Farms Campaign Illustrations, 1921. Source: National Archives Australia (A457, 1400/5 PART 2, Immigration Encouragement. "Million Farms" Campaign. Sir Joseph Carruthers Scheme, 1921–1922: 154, 169)

Australia's productive and consumptive patterns reflect Hoorn's notion of the "making of a white landscape".[18] The *Immigration Restriction Act*, or 'White Australia Policy', was one of the newly formed federal Parliament's first pieces of legislation to be passed in 1901. The openly racist policy not only reflected ongoing exclusion of Australia's First Nations Aboriginal peoples, but sought to further socially engineer the nation's fabric through stipulating who could emigrate.[19] The Act particularly impacted further Chinese immigration, with run-on effects for Sydney's existing Chinese-Australian community and their significant influence in market gardening (see below and *Market Gardening*). Like the cessation of convict labour in *Period Two*, stemming immigration created agricultural challenges. Labour-intensive pursuits, such as fledgling tea plantations, were particularly dependent on Pacific Islander and Asian immigrant labour.[20] An increased focus was placed on labour-saving innovations and mechanisation that were already gaining momentum. Immigrants beyond the British Isles nevertheless continued to shape Sydney's food landscapes, particularly following World War II when the Act was relaxed to admit certain refugees deemed an asset to the policy's exclusionary white standards and post-war labour needs. As noted by Peter Spearritt, subsequent "low-grade resentment" and friction nevertheless arose as these 'new Australians' "rubbed up against the old British monoculture", with immigrants often facing hostility, discrimination and overt racism.[21]

Market gardening hence suffered considerably in the process of city remaking. Chinese-Australian dominated market gardens were partially targeted due to their traditional use of human and animal excrement as fertiliser but also due to racism legitimised by the White Australia Policy. This contributed to urban agriculture's lack of agency, its marginalisation and its subsequent erasure from the modernising city. This was especially the case when taking place within close proximity to existing and growing urban areas. At the same time, many forms of metropolitan agriculture on the once urban-fringe were being increasingly and significantly eradicated under Sydney's expanding spatial footprint. Lands once on the city's edges were subsumed and developed into factories and new suburbs (Fig. 5.4). This urban pattern continues in Sydney to the present day.[22]

Fig. 5.4 Botany's Davis Gelatine Factory in Spring Street, 1937 (top) and Kellogg's Factory on Stephen Road, 1938 (middle), and nearby Mascot in the 1940s (bottom). Sources: Royal Australian Historical Society (Set 72157640448188933, ID 29834398536, Original title: Davis Gelatine Factory - 15th June 1937; Set 72157640448188933, ID 29705226650, Original title: Kelloggs Factory—1938), and Milton Kent, Mitchell Library, State Library of New South Wales

Despite Australia largely realising a white food landscape and immigration policy, ongoing attempts to develop an agrarian society ultimately failed to eventuate, at least beyond the realm of narrative and imaginary. While an agrarian national landscape was spatially achieved through pastoralism's land use dominance,[23] agrarian culture remained largely a myth. Australia's social fabric has never taken to the land to the extent that would qualify it to be a truly agrarian society,[24] instead clustering in urban agglomerations. The period's onset of mass production and manufacturing further concentrated employment within factories, dooming any notion of a fully realised agrarian ideal. This, however, was not necessarily apparent at the time—or certainly not to former New South Wales Premier Sir Joseph Carruthers who proposed and lobbied for a national agrarian vision of behemoth proportions. Carruthers' 'Million Farms Campaign' (1919–25) envisioned a 'Big Australia'[25] and aimed to settle 1 million farms with 1 million farmers (Fig. 5.3).[26] Given Australia's 1.5 million population at the time, the proposal was hugely ambitious, if not altogether unrealistic.

It is perhaps naïve, though, to overstate the Million Farms Campaign's agrarian ideals given the scheme's political and financial motives, which sought to guarantee investment for British and local financiers.[27] As Gary Lewis observes, the scheme was a "sterile [...] cuckoo egg" but an unqualified success with respect to Carruthers' "cunning, energy and rhetorical power of a shrewd and highly gifted conservative tactician".[28]

Meanwhile, as agrarianism faded, manufacturing accelerated. By the 1947 census, Sydney housed 31 per cent of the country's manufacturing jobs and its industries employed more people than farming and mining nationally.[29] By the end of the period, food production faced further disruption due to World War II's petrol and food rationing, with official restrictions beginning in 1942 for certain foodstuffs, principally tea, sugar, butter and meat.[30]

Relevant Sydney Events: 1901–1945

There are multiple key intertwined themes that signalled the incredible levels of transformation to Sydney's makeup that took place in *Period Three*. Technological developments facilitated Sydney's expansion through city-defining infrastructure projects. The manufacturing sector's emergence reshaped infrastructure, while employing a growing number of Sydney's workers. Motorised transport decentralised Sydney's form, signalling metropolitan agriculture's decline as former productive lands were increasingly eviscerated by an expanding and suburbanising city footprint. Early in the period, bubonic plague outbreaks and the influenza pandemic generated significant health and hygiene responses that increased regulation and more formalised planning processes, leading to huge changes in Sydney's built fabric. Working class families moved out of the city centre into new fringe suburbs serviced by comparatively improved infrastructure and more sanitary living conditions. Buses and cars following World War I then enabled continual suburban spread from these areas. Consequently, *Period Three* witnessed the loss of much of Sydney's productive agricultural land.

Shifting Labour Markets

Period Three witnessed a colossal shift from a land-centric economy to the creation of a 'modern' nation—a movement from "sheep's back" to "industry stack".[31] Sydney's remaking as an industrial city not only impacted the national economy, but also played a major role in developing Australia's self-sufficiency, growth, urbanisation, society and contribution to World War II.[32] The large tracts of land required for pastoralism meant that the city only facilitated supplementary production, rather functioning as a processing and distribution hub and commercial centre. The severity of the 1890s' depression had made the need for government economic intervention apparent and accordingly, local manufacturing industries received support following the start of the century and even more so after World War I. Australia's manufacturing industries were heavily bolstered through government subsidies, high tariffs and industries concentrated in urban contexts near labour bases.

Industrialisation and mass production upheaved established mechanisms of generating wealth, previously based on land-based processes controlled and manipulated by the Crown and colonial government. Manufacturing disrupted this order. Between 1817 and 1939, for example, New South Wales' 50 wealthiest citizens were graziers and agricultural landowners.[33] The 1920s and 1930s witnessed increasing wealth generated by factory production and consumer manufacturing, while wool exports continued to decline from World War 1.[34] This signalled a power shift from the prior agrarian and supposedly genteel pastoral squattocracy and Australian 'gentry' as the urban, *nouveau riche* industrialist entered into Sydney's social foray.

For Sydney's working class confined within gritty factories and manufacturing plants, the hardships of unsanitary life in urban slums at the turn of the century were supplanted with hazardous working conditions. Architectural veils of secrecy, often fenestration-denied boxes with sawtooth-roofs, obscured dangerous working environments with a concentration of workers housed inside. Beyond the sheer size of some factories, a billowing smokestack was often the only external symbol indicating the intense industrial hearth at its production epicentre. Conditions were particularly bad for women who largely joined the workforce following World War I's onset, ruffling the feathers of gendered labour dynamics. The New South Wales Board of Trade decreed in 1918 that women were to be paid half the wage of male workers. By that time, women comprised 72 per cent of Sydney's clothing and textile workforce, also working alongside minors in the lowest paid industries such as food, drink and tobacco production.[35]

Any lingering agrarian romanticism within Sydney's labour markets was increasingly imaginary. Regimented, imposed industrial rhythms permeated the city's sizeable manufacturing sector. Factory work became highly significant for Sydneysiders, through the "sheer size and importance of Sydney's role, both in the warehouse of goods and as a manufacturing city in its own right".[36] By the start of *Period Four*, 47 per cent of Sydney's population were employed in the industrial sector.[37] This was especially apparent in southern city areas such as Redfern, Eveleigh, Waterloo, Alexandria, Zetland and Rosebery, where factories were "vying with each other to be considered the most 'industrial'", with three-quarters of the County of Cumberland's factory employees living within a "three and a half mile radius of Redfern station".[38]

Urban manufacturing and industry directly impacted Sydney's metropolitan agriculture, as evident in Alexandria's, Botany's and Mascot's market gardens that were supplanted by factories and warehouses. With the added impacts of the White Australia Policy on Chinese-Australian market gardeners and the vogue of inorganic chemical use, Sydney's former 'backyard vegetable garden' was enveloped by industry and its seeping airborne and aquatic plumes of noxious wastes. Aerial images dating to 1937 reveal the encroachment of industry in Botany's once extensive and verdant market gardens, for instance. Facilities such as the Davis Gelatine Factory and Kellogg's Factory are fitting emblems of city food production transitioning from garden to factory, and from whole foods to processed (Fig. 5.4).[39]

By *Period Four*, further industrialisation, Port Botany's growth and Kingsford Smith Airport's expansion in the 1950s and 1970s, all but displaced the area's market gardens and only two remained in Matraville and La Perouse. A similar pattern of urban development emerged across the wider St George region previously characterised as Sydney's 'salad bowl'.[40]

Remaking the City: Decentralising City Form

Metropolitan Sydney had no formal planning machinery to guide its significant population and spatial growth in the first half of the twentieth–century. While advocates had for several decades promoted the efficiency, economic growth and social dividends that would come from greater development controls and coordination over market-driven development, formal state planning legislation was not enacted until 1945. Accordingly, the planning discipline did not formally emerge in Sydney until *Period Four*. Sydney's development largely followed its train and tram routes until the onset of motor vehicles, which effectively enabled the city's spread.

Two transportation elements were key to *Period Three*'s unparalleled urban expansion. Firstly, the continued development of heavy passenger train and light rail (tram) networks necessitated new corridors, lines and routes. Significant rail events include the opening of Central Railway Station (1906), an underground rail loop through the city (1920s–1950s), and the construction of the Sydney Harbour Bridge (1926–1932). These concurrently enabled the urban footprint to be further concentrated and expanded. The intergenerationally momentous infrastructure and city building projects employed numerous workers. Sydney's electrification and public transport required large powerhouses to operate, leading to further ancillary industries and manufacturing processes and structures.[41] Secondly and most significantly for its agricultural impacts, cars, commercial vehicles and buses decentralised urban development, with the 1920s heralding a "transport revolution"[42] as palpably evident in Fig. 5.1.[43]

Prior to World War I, the two primary modes of transportation were horse-drawn vehicles or trams. The Automobile Club of Australia (A.C.A.) was formed as early as 1903 but operated mainly as a gentleman's social and sports club.[44] Sydney ordered its first truck in 1908 and became Australia's first city to have garbage trucks in 1910.[45] There were 104 registered commercial motor vehicles in 1910 and 816 in 1920.[46] More widely, 3978 vehicles were recorded in New South Wales in 1911, swelling to 14,973 in 1916, 33,214 in 1921 and a colossal 127,160 in 1926.[47] The National Roads and Motorists Association (N.R.M.A.) was established in 1924 and became a strong pressure group.[48] A survey of all vehicles passing 61 city checkpoints between 7:00 AM and 6:00 PM on 2 July 1923 logged 40,253 motor wagons and trucks, 50,333 motor cars and 58,436 horse-drawn vehicles.[49] As cars became cheaper, their number increased steadily alongside garages, petrol stations, showrooms and mechanical industries, threatening state-owned tram and train services and signalling the eventual extinction of blacksmiths, stables and carriage works.[50] Nevertheless, horse-drawn vehicles clung on throughout the period and remained a part of the city's infrastructure and soundscape, with bread carters and milkmen still using horses after World War II.[51]

Hygiene, Health and Planning

Multiple interrelated planning outcomes cascaded into the years and decades that followed the plague outbreak in Sydney's inner harbour areas in 1900. In the first 8 months of 1900, 103 people had died but a more significant toll was avoided thanks to swift responses by the authorities, including city cleansing, rat extermination and quarantine. Having navigated the plague that continued to erupt seasonally over the next 10 years, a pneumonic influenza pandemic more commonly known as the 'Spanish flu' broke out in 1919. Between January and September 1919, 6387 people died in New South Wales. Infection rates in metropolitan Sydney were immense, with 290,000, or almost one-in-three, contracting the virus at one time.[52] Densely populated areas and industrial workers suffered the highest death rates, while lower-socioeconomic demographics were notably affected due to lower levels of health and unsanitary living conditions.[53] To further rub salt into Sydney's wounds, another small outbreak of plague in 1921 killed 10 residents.[54]

In 1905, the state government granted the City Council the power to remove what they considered to be slums. These areas of urban decay harboured unhygienic conditions and disease, evidenced through being flashpoints during the period's health crises. The city government enthusiastically razed such areas, exorcising "powers which might astonish present day planners".[55] Another concerted campaign in the 1930s continued to seek out slum areas for demolition with successful clearance.[56] Sydney's slums were replaced by commercial and 'modern' redevelopments. This also led to the intentional relocation of urban residential land use to new suburbs outside of the city centre. The suburbs were serviced by new transportation networks (rail, tram and road) and centralised infrastructures, such as gas, electricity, potable water and sewerage systems. The significance of these centralised infrastructure systems cannot be overstated. They positively transformed health and wellbeing and improved living standards and efficiency. Instating them, though, came at the cost of both built heritage and agricultural land.

Disease outbreaks hence decentralised the city's urban form but so did hygiene concerns directly stemming from food production and processing. As noted by Andrea Gaynor, metropolitan food production is a two-way process: "just as the safety of suburban food production could be jeopardised by urban wastes, food production itself could be a source of pollution".[57] Following a Royal Commission in 1908–1909, the *Quarantine Act* of 1908 heralded a national approach to control, prevent and remove imported plant, animal and human diseases. A Government Bureau of Microbiology was established in 1909 to conduct investigative research. This resulted in outcomes such as the appointment of a New South Wales Meat Inspector in 1908. Prior to this, many butchers' shops had open fronts with hanging carcasses. By 1913, all but five had enclosed their premises. Removing livestock also cleansed the city of cattle and suburban swine herds polluting and degrading waterways. The city's expanding edges butted up against meat processing plants, pushing them further afield, as evident in Sydney's main abattoir relocating from Glebe to Homebush.

Urban agriculture was further affected by aesthetic trends. Following Federation, British and North American concepts such as 'City Beautiful' became increasingly fashionable in Australian urban planning and civic design.[58] Despite the movement's concern for 'beautility', or well-executed functionality,[59] agriculture was thought to be better suited to distant 'countryside' regions, where it could be pleasantly viewed from a safe and waft-free distance. The economic downturn of the 1930s Depression diminished the movement but metropolitan agriculture continued to be maligned. Market gardening's classification as a 'noxious trade' (see *Period Two*) further affirmed its undesirable status. Despite some recognition throughout the twentieth–century within town and country planning,[60] agriculture near concentrated settlements tended to be considered unclean rather than recognised as an integral component of life and city planning.

Other British and American initiatives were also influential,[61] as evident in the construction of Sydney's underground that opened in 1926 and not only mitigated congestion but sought to beautiful the city. A.J. Brown's vision of the city surrounded by a 'green girdle' was a noteworthy local planning precedent published before the onset of World War II.[62] Unlike City Beautiful, Brown understood the importance for interconnectedness between life sustaining landscapes and the city, holding that,

> As social and economic forces respect no arbitrary bounds, the planner must also ignore them and, acknowledging the interdependence of town and surrounding country, provide for ample play of those forces along well-ordered lines. The bustling city cannot be considered as separate from its rural background, which is essential to its existence, and the converse is also true.[63]

Period Four examines whether Brown's vision would come to pass. *Period Three* therefore witnessed the emergence of a more formalised and orderly approach to town and urban planning. Advocates argued that formal planning would improve hygiene, establish order and address prior approaches to city development susceptible to "selfish individualism and laissez faire capitalism".[64] *Periods Five* and *2021+* critiques the extent to which these have been achieved.

A Changing City

At the 1911 census, one-third of Sydney's population lived in the city centre and inner ring. Just two decades later, this had halved to one-sixth.

As noted by John Birmingham in 1920, most urban clusters within 13 miles (21 kilometres) of the city "still had the appearance of villages".[65] These dotted scatterings of satellite and agrarian towns were encircled by 'countryside' buffers. Agricultural land occupied the in-between spaces of city and village but was increasingly being pushed further outward to a more distant urban fringe by growing suburbanisation. Fields, market gardens and farmland, often already fallow through anticipating subdivision, were subsumed by the onslaught of detached housing. This is visible in Milton Kent's 1930s and 1940s aerial images that reveal industrial and residential land use patterns and processes eviscerating interstitial rural lands and metropolitan agriculture.[66] New roads dissected lands as suburbs and factories enveloped the peri urban countryside and conjoined the urban footprint into a large mass resembling an "octopus like suburban conurbation".[67] Sydney city sat at its 'head' of this cephalopod as the central hub of activity. This urban spatial pattern continued to dominate to the present day (see *Period Five* and *2021+*).

Although a long time coming, by 1921 Parramatta was regarded as part of Sydney.[68] The metropolis' inexorable growth between the wars erased much prior agricultural activity separating the two original settlements. Before World War I, for example, southern suburbs on the Illawarra railway line included many small farms and vegetable gardens that disappeared with residential expansion. Sydney's built-up area "expanded enormously, engulfing many erstwhile agricultural areas".[69] The urban-rural fringe was on the move, and during the same period, new farms were appearing further out of the city in places like Prospect and Castle Hill.[70] By the early 1940s, half of NSW's population lived in Sydney. World War II lifted Sydney out of the lingering effects of the Great Depression by stimulating economic activity and growth,[71] also significantly reinvigorating agriculture. The armed forces became the priority channel for fresh, local foods.[72]

Finally, the period saw several noteworthy Sydney occurrences relating to biodynamic and organic philosophies and practices relevant to agriculture. As discussed by John Paull, biodynamic agriculture founder Rudolph Steiner's influences constituted the first of four waves of Australian organic agriculture.[73] Initial dissemination can be traced to Sydney, with Bob William's inaugural biodynamics Australian public lecture at Walter

Burley[74] and Marion Mahoney Griffin's home in Castlecrag and the first wave culminating in the international launch of Ehrenfried Pfeiffer's *Bio-Dynamic Farming and Gardening* in 1938. Sydney was also the birthplace of *The Australian Organic Farming and Gardening Society* (AOFGS)—the world's first—founded in October 1944.[75] Another first was an organic farming periodical launched in April 1946,[76] though as Paull notes, Australia's organic agriculture movement was a "fast follower" of ideas that originated elsewhere.[77] Despite most agriculture prior to the period effectively qualifying as 'organic', it would become an increasingly minor component in the overall makeup of both metropolitan agriculture and the national food system.

The 11 Typologies: 1901–1945

Period Three (P3) solidified the *Period Two*'s transition from written accounts and illustrations to photographic sources. The eventual availability of aerial photography—captured from aeroplanes—helped to illuminate Sydney metropolitan agriculture's spatial reach, patterning and makeup. A high-quality 1943[78] aerial photography dataset of Sydney resembled today's satellite imagery and resulted in a flush of Map Database entries situated late in *Period Three*. It also supported the verification of written accounts with greater accuracy, particularly by year or month and spatial extent, albeit lacking finer details such as plant species identification (Fig. 2.5). Like most satellite imagery today, aerial photography focuses on the city's built footprint, strategically important sites and its expanding urban fringes, rather than the rural hinterlands where much agriculture is regularly located. In a similar way to land use mapping, aerial imagery does not always allow for clear distinctions between commercial, hobby farming, or non-productive agricultural properties maintaining rural character.[79] Nonetheless, aerial imagery's availability is a significant step greatly aiding understanding the city's makeup.

Agricultural Facilities and Education (AF): Investing in Education

Agricultural education that commenced in *Period Two* grew throughout *Period Three*. Agriculture's increasingly scientific, technical and mechanistic nature lent itself to formalised and longer durations of study and to increasingly diverse specialisations. Hardship and scarcity following the 1890s and 1930s depression and during both world wars translated to government and independent training and education initiatives geared to bolstering food production. This included prisons and asylums.

Vegetable production within prisons therefore continued in the period. Long Bay Gaol opened for female prisoners in 1909, and males from 1914. The prison developed a large farm as part of its "surprisingly ambitious" landscape modifications, which included a windmill powered irrigation system pumping from ponds.[80] Figure 5.5 (top) shows part of Long Bay Gaol's extensive farm fields that were hewn out of improbable ground, as evident in the stony foreground. The sandy soil was enriched through addition of significant quantities of manure supplied by the City Council. A detailed 2004 conservation management plan did not uncover what crops were grown in these fields, however, closely spaced rows of seedlings are separated by sparse hedges running along drainage channels feeding into a large circular pond. Although changing in size, the farm was approximately 10 acres (four hectares) until at least 1970. This scale made farming profitable for the prison, as well as facilitated outdoor exercise.[81] The farm included livestock, and in 1914–1915 a livestock shed was built in the outer grounds. Parramatta Gaol's intensive productive vegetable farm also continued to operate within its enclosed northern portion, producing considerable amounts of food (Fig. 5.5, middle). The site was earmarked for a building that did not eventuate until the 1970s and farming continued prior to its construction (see *AF* in *Period Four*). From 1915, the prison farm also housed pigs.[82] Both Parramatta and Long Bay gaol's highly cultivated farm plots are visible in 1943 aerial imagery.[83]

Agricultural training facilities also continued to diversify. In 1907, the Department of Agriculture purchased a run-down 40-acre orchard in Galston Road, Dural[84] and established the Government Orchard and Training Farm (Fig. 5.5, bottom left). The Galston district at the time was a major hub for citrus and fruit production and the site was to be

Fig. 5.5 A low east-north-east facing aerial photograph shows part of Long Bay Gaol's extensive farm fields c. 1920s (top) (See CLS&P, 2004). A 1931 photograph

used as an experimental property for research into fruit disease and good orcharding practices.[85] In 1907, Hurlstone Agricultural High School (briefly known as 'Macarthur Agricultural High School' in the 1940s) was established in Hurlstone Park (Strathfield). In 1926, it relocated to its current 112-hectare (277-acre) site in Glenfield that included an operational farm. Wartime sites also sometimes included food production. An example is Liverpool's World War II Internment Camp and its vegetable garden (Fig. 5.5, bottom right). The five-acre (approximately 2 hectare) garden produced tonnes of fresh vegetables serving not only the camp's needs but also contributed to the nearby Holsworthy Barracks.[86]

From 1910, the former Casual Labour Farm at Scheyville was turned into a Government Agricultural Training Farm, operating until the 1930s (Fig. 5.6). Also in 1910, the Hawkesbury Agricultural College's previously two-year full-time residential course in general agriculture expanded to 3 years. It subsequently became known as the Hawkesbury Diploma in Agriculture, with a dairying diploma also introduced in the same year. More widely, Hawkesbury Agricultural College continued its diverse range of educational training, with common activities during the period including animal husbandry, orcharding, spraying, fumigating, operating machinery, demonstrations and vegetable production/market gardening. Figure 5.6 (bottom) shows grazing animals in the foreground, orchards in the centre and right, and field plots.

Around the same time, the University of Sydney established Chairs in Agriculture and Veterinary Science in 1909.[87] Robert Dickie Watt became the first Australian Professor of Agriculture in 1910,[88] initially placed in the

Fig. 5.5 (continued) depicts Parramatta Gaol's verdant crops (centre). A 1909 image of the Government Orchard & Training Farm in Dural shows deciduous (stonefruit) and evergreen (citrus) fruit trees (bottom left) (A 1927 image shows spraying peach trees, sans mask: https://hornsbyshire.recollect.net.au/nodes/view/594?keywords=dural%20orchard&all=1&whole=2). The 1944 photograph (bottom right) reveals internees of the Liverpool prisoner of war and internment camp working in a vegetable field. Sources: Museums of History New South Wales (Museums of History New South Wales—State Archives Collection: Department of Corrective Services, RNCG 537, Glass negatives. RNCG-537-5-[GLASS NEG. BOXES 165 BOX 41] 165.15 CS1294 | State Penitentiary—ornamental ponds and prisoners at work in fields [no date]); National Library Australia, Bib ID: 6300099; Hornsby Shire Recollects 00534; and Geoffrey McInnes, Australian War Memorial

Fig. 5.6 A horse-drawn plough at the Scheyville Training Farm in 1926 (top). Hawkesbury Agricultural College c. 1900 (middle left), a demonstration in the orchard at the winter school for farmers in 1916 (middle right) and aerial view in 1939 (bottom). Sources: Museums of History New South Wales (Museums of History New South Wales—State Archives Collection: Government Agricultural Training Farm,

Faculty of Science, then from 1920 in a dedicated agricultural faculty. The Bachelor of Science in Agriculture curriculum was geared towards graduating agricultural scientists rather than farm managers, reflecting a scholarly rather than purely vocational education context. The first male students graduated in 1914, with the first two female students awarded agricultural science degrees in 1920.[89] Despite having its own faculty, student numbers in the agriculture programme were modest by today's standards, only having grown to 14 by 1933.[90] In 1936, the University acquired its first of many subsequent farms, McGarvie Smith Farm near Badgerys Creek.

From 1932 to 1940, St Matthew's training farm operated on South Creek Road in Cromer on the Northern Beaches. It was established by the Reverend A.R. Ebbs during the 1930s depression. The three-hectare property contained vegetable gardens, a poultry run, a small piggery and living quarters. The farm's aim was to train unemployed boys for agricultural work for several months, then secure positions on farms around New South Wales. By its final year of operation in 1940, more than 800 students had moved through the programme.[91] Finally, within the city proper, Australia's first Horticulture Certificate course was created at Sydney Technical College's gritty Ultimo site, with its first students graduating with a Gardening and Horticulture Certificate in 1938. *Period Three* thus included a range of higher, further, civic and institutional activities—regularly government initiated and/or supported—in agricultural educational and training endeavours in the Sydney region.

Mixed Farming (MF): More Specialisation, Less Diversity

The food system's industrialisation marked a significant agricultural transformation. Mechanisation and the adoption of synthetic farming methods became widespread, leading to a shift towards more specialised, focused

Fig. 5.6 (continued) Scheyville; NRS-5529, Photographs of Scheyville Training Farm. NRS-5529-1-[5/164.43]-43 | [Horsedrawn plough at Scheyville Training Farm], 1926); Western Sydney University (Archives Collection: Ref. P633, P1036) and National Library Australia (Commonwealth of Australia, National Library Australia, Bib ID: 4329539) and monocultural farming practices. This transition resulted in larger, more professionally managed farms, particularly in non-metropolitan areas where land was more affordable than in urban centres like Sydney. The emergence and adoption of technologies conducive to increasingly industrial scales of production further facilitated this shift. Improved transport played a crucial role in enabling farms to relocate to more remote areas, beyond those previously limited to producing goods suitable for transportation by train, boat and horse and cart. As a result, the traditional focus on metropolitan areas for the production of fresh and perishable foods began to gradually shift towards more distant rural regions.

The shift towards farming specialisations reduced individual properties' and local districts' agricultural diversity. This reinforced a transition from localised, diversified farming to a more commodity-driven and specialist-oriented food system, indicating the early stages of industrial 'competitive productivism' in Australian agriculture.[92] Consequently, farming communities were becoming increasingly similar in their complexion. By the end of *Period Three*, individual farms regularly focused on cultivating one or a limited number of complementary agricultural commodities, such as solely citrus (or just one type of citrus) or chickens raised for meat production (broilers). This trend often resulted in consolidation, whereby farms were sold to adjacent or competing operations, resulting in larger-scale production. Smaller-scale farms, including hobby and backyard, nonetheless persisted alongside dominant commercial and professional enterprises. These played a crucial role in sustaining Sydney throughout the challenges of two world wars.[93]

Market Gardening (MG): A Cultural Injection

Market gardening is preeminent in *Period Three*'s Mapping Database. Some entries indicate localised clusters, while countless more remained uncaptured.[94] The database therefore offers a sample of a significantly larger number of gardens within the book's boundaries. The period also witnessed new dimensions of food production emerge and/or expand. These included poultry and pig farming, viticulture, flower cultivation and mushroom production. Oakville became a significant hub for the latter,[95] initially conducted outdoors but later transitioning to indoor cultivation within sheds. Over time, mushroom cultivation would become one of Sydney's only expanding agricultural industries.

Between 1901 and 1921, the market garden trade amounted to approximately 0.5 million pounds annually, with Chinese market gardeners, hawkers, greengrocers and wholesalers accounting for 50 per cent.[96] New South Wales market gardening statistics from 1913 indicate that one-third of the state's production occurred within the metropolitan Sydney area of the County of Cumberland, encompassing 9847 acres (3985 hectares), an increase from 7764 acres at Federation in 1901.[97] From 1891 to 1911, around 30 per cent of the state's Chinese population were engaged in market gardening, compared to only 0.2 per cent of European-Australians.[98] Consequently, a significant proportion of Sydney's market gardens continued to be tended by Chinese-Australians. The primary cluster of Chinese market gardens was concentrated in Botany's rich marshland (Fig. 5.4),[99] with sites extending to unexpected locations such as Maroubra Beach (Fig. 5.7, top) and Rose Bay.[100] As in *Period Two*, Chinese market gardeners continued to demonstrate exceptional proficiency in cultivating vegetables in sandy soils with access to freshwater—a skill not easily matched by other cultural groups across Sydney.

As mentioned earlier, *Period Three* proved devastating for the Chinese community, market gardens and Sydney's fresh vegetable supply due to the White Australia Policy's exclusion of Chinese immigrants from Naturalisation and cessation of migration.[101] Most Chinese market gardeners were male and unmarried, thus did not have children to continue their work. Accordingly, an ongoing supply of workers—which effectively ceased—were needed for the Chinese model of market gardening to intergenerationally thrive at scale.

Existing Chinese immigrants were subject to various propaganda and exclusionary campaigns.[102] These efforts were often directed at their living conditions, commonly described as modest or labelled as 'shanties', 'hovels' or 'shacks'.[103] Critiques levelled by influential individuals and groups were further influenced by the period's aforementioned city beautification trend that did not view the gardens and their dwellings as elements of 'beautility'. Consequently, white Australian landowners terminated Chinese market garden leases, leading to the gradual displacement and subdivision of former market garden lands for urban development. This notably occurred in areas such as the eastern suburbs (between

Fig. 5.7 This 1910 watercolour of Maroubra Beach shows a Chinese Market Garden (top). Similarly verdant Chinese-Australian-operated market gardens

Old and New South Head Roads) and Botany. Many market gardens transitioned to golf courses and other recreational uses.

From its inception, many of Sydney's key proponents envisioned a significant portion of its population embracing agrarian pursuits. As discussed earlier, the 1919 Carruthers 'Million Farms' scheme sought to advance this goal, but solely for the white community primarily of British and Irish descent. This disparity underscores a poignant irony wherein Sydney's Chinese-Australian community exemplified agrarian values through their resourcefulness and diligent work ethic yet were excluded from the very activities that the country sought to promote. Consequently, extensive areas of market gardens were lost along with their well-established agrarian practices and traditions.

Cultural diversification before and after World War I saw immigrants arrive from Albania, Dalmatian Yugoslavia, northern Italy, Calabria, Sicily, Bulgaria, Malta, Macedonia and Greece.[104] These groups established themselves in both existing and new market garden locations, spanning the western suburbs such as Fairfield, Cabramatta, Holyrood, Canley Vale and Blacktown, as well as areas such as Brookvale and Warriewood on the Northern Beaches.[105] Lebanese immigrants also entered the market gardening sector, founding operations in Thornleigh during the 1930s.[106] The Italian community further played a crucial role in the establishment of new agricultural districts.[107] In Sydney, Italians were primarily engaged in the fruit trade and retail, although there were also diligent growers in the Sydney region who, like the Chinese, faced discrimination (Fig. 5.8).[108] A 1928 *Lismore Star* article, for example, criticises Chinese and Italian market gardeners for violating working hour limits, revealing both the period's racism and immigrant work ethics benefitting market gardening pursuits (Fig. 5.8).

By the 1920s, all market gardeners and smaller agricultural producers encountered mounting challenges from industrial agriculture's expanding scope.[109] Technological advancements, mass production and

Fig. 5.7 (continued) glisten in the morning sun on adjacent banks of Wolli Creek at Turrella in the 1930s (middle). Undercliffe also had market gardens, here captured in the 1940s (bottom). Sources: Mitchell Library, State Library of New South Wales and courtesy Myles Dunphy; and courtesy Justin Cahill (Courtesy Roy Dunkley and Dorothy Powell, courtesy Jean Young)

COMPETITION

Foreign Market Gardeners

ITALIAN MENACE

PROTEST BY AUSTRALIANS

SYDNEY, Thursday.

A deputation from the Sydney Market Gardeners' Association waited on the Minister for Labour and Industry (Mr. Farrar) to-day and requested that legislation be introduced which would limit the working hours of foreign gardeners.

It was stated that Chinese and Italian gardeners worked long hours in their gardens and Australians were unable to compete against them.

Italians, it was further stated, were a growing menace and would soon be worse than Chinese at the present rate of influx. Chinese, also, paid no income tax because it was not possible to check their profits.

One alleged case was cited of a Chinese who made £1100 from a crop of tomatoes, yet he was stated to have paid no tax.

The Minister promised to go into the question, but if the case warranted special legislation it could not be presented to Parliament this year.

Regarding a complaint that civil servants were growing vegetables and fruit in their spare time in competition with bona-fide gardeners, Mr. Farrar asked for the names of these men and the practice would be stopped.

Fig. 5.8 Urban encroachment onto market gardens in Abbotsford in 1933 (top). A photograph from a suite in a 1941 article titled 'Yugoslavs' farm' showing a economies of scale contributed to market gardens' decline. This was particularly evident in areas near Sydney's burgeoning suburban peripheries where urban expansion encroached upon agricultural land. Regions once renowned for their fertile and well-tended soils supporting intricate patchworks of market gardens were transformed by light industry and suburban housing. Evidence is visible in locations such Chullora and Granville around 1935–45[110] and in Abbotsford and Greystanes/Pemulwuy (Fig. 5.8). Losses are particularly devastating in the latter's case, given its proximity to the Prospect Hill region with its rare 'black earth' soils from basalt formations.

In 1927, Warringah on Sydney's North Shore consisted of several small settlements characterised by weekend cottages and small-scale agricultural activities. Some areas of fertile land were utilised for dairy farming, orchards, market gardening, poultry farming and flower cultivation.[111] Market gardening activities flourished in the Warriewood valley, initially dominated by Anglo-Celtic growers but later attracting a multi-ethnic population, particularly immigrants from Italy, Dalmatia and Yugoslavia (the first Yugoslav club in Australia was established in Warriewood in 1939),[112] alongside Chinese growers.

The adoption of commercial glasshouses was one technological and mass production element that benefitted market gardening. Glasshouses extended the growing seasons, allowing greater availability of produce over a longer duration that could not yet viably be shipped from afar. They became increasingly prevalent in Sydney's market gardens. Warriewood emerged as a considerable focal point, boasting a concentration of commercial production facilities.

Despite facing racial discrimination and urban growth pressures, metropolitan market gardeners remained significant at the end of *Period Three*. Over 40 per cent of Sydneysiders' vegetables originated from farms within a 40-kilometre radius of the Central Business District (CBD).[113]

Fig. 5.8 (continued) share farmer at Slavich farm in Harris Street Warriewood (centre) (Mitchell Library, State Library of New South Wales and Courtesy ACP Magazines Ltd). The fecund regions of Greystanes and Pemulwuy in 1943 (bottom left) (In Kirkpatrick, 2022: 24. Also see Elias & Coppins, 2013). A 1928 article in the *Lismore Star*. Sources: State Library of New South Wales, Mitchell Library and Courtesy ACP Magazines Ltd., Spatial Services NSW

Fig. 5.9 On the outer edges of this book's boundary are fruit growing regions including at Kurrajong (citrus) in the early 1900s (top); Cambridge Reach on

This achievement is particularly noteworthy considering that the prominent agricultural region of the Hawkesbury region is not included in this calculation.[114] Regardless, the disappearance of market gardens from Sydney's middle suburbs was apparent.[115] Analysing the characteristics of Sydney's then rural-urban fringe in 1945, Neville Wills characterised Botany-Mascot as "a striking example of a declining rural-urban area", that was "once a thriving hub of small vegetable gardens but is now rapidly assimilating into the industrial belt of South Sydney" (Fig. 5.4).[116] Regrettably, such loss of Sydney's market gardens would escalate significantly in the post-World War II period.

Fruit (FR): Sydney's Last Fruitful Period?

Fruit cultivation is the second largest typology in the Map Database.[117] *Period Three*, however, witnessed Sydney's peak fruit production. While some new orchards appeared, overall, fruit cultivation was diminishing by the period's close. As discussed in *Period Two*, Mediterranean fruit fly's (*Ceratitis capitata*, or 'medfly') emergence in Perth in 1895[118] and spread led to Sydney's thinner-skinned stone fruit production demise. The challenging pest had declined in numbers towards the end of the period and the last sighting in New South Wales was reported in 1941,[119] though possibly due the Queensland fruit fly's increasing presence. Less affected regional irrigation districts also facilitated transportation of fruit to Sydney's markets, further contributing to the loss of orchards within the metropolitan area.

Kurrajong was a prominent fruit-growing region in the north-western reaches of the book's boundary. The cooler, elevated area specialised in stone fruits and apples (Fig. 5.9, top). Peaches and various other fruits

Fig. 5.9 (continued) Hawkesbury River flats c. 1920–1925 (middle); and Maroota 1942. Sources: Museums of History New South Wales (Museums of History New South Wales—State Archives Collection: Department of Railways; NRS-12932 | Original Prints of Photographs used in NSW trains. NRS-12932-1-[X2450]-10-107 | [Citrus Orchards, Kurrajong (NSW)]); National Library of Australia (Cambridge Reach showing orchards, Hawkesbury River, NSW Bib Id: 791584); State Library of New South Wales

Fig. 5.10 Fruit growing in *Period Three*: the panorama (top) shows Waratah Farm Orchard Ingleside c. 1917–1946 where the Narrabeen plum variety was registered;

were cultivated around Wiseman's Ferry and Maroota, where an iconic photograph depicts gridded orchard geometries carved into undulating bushland (Fig. 5.9, bottom). A diverse range of fruits including pears, plums, currants, gooseberries, strawberries, cherries and melons were grown across the broader Hawkesbury district, often transported via riverboat.[120] The 1930s fruit fly outbreak significantly affected the area, leading to numerous orchards' closure. Despite these challenges, the Hawkesbury district maintained a substantial citrus cultivation, with 286,000 citrus-bearing trees recorded in 1944, constituting approximately 20 per cent of the total citrus trees in New South Wales.[121]

Maroota's fruit production suffered from sand and gravel mining later in *Period Three*. Quarrying earlier impacted orcharding in Hornsby's Old Man's Valley from 1909.[122] In Ingleside, meanwhile, Isaac Larkin of Waratah Farm Orchard registered the now popular Narrabeen Plum variety with the Department of Agriculture as a hybrid of the Japanese variety *Prunus salicina* (Fig. 5.10, top). Situated near the current Baha'i temple (opened in 1961), goods from the farm were transported by horse and dray to Manly Wharf, where they were consigned by a produce agent to Sydney markets and subsequently ferried there for distribution.[123]

The greater Epping and Carlingford districts served as hubs for various agricultural activities, especially fruit growing, and a private rail line serviced growers (Fig. 5.10). In some orcharding areas, income was initially generated through clearing timber to offset delays until trees reached maturity and bore fruit (Fig. 5.10, bottom right). Glenorie, located 32

Fig. 5.10 (continued) three c. 1911–1916 photographs show loading fruit Pennant Hills train station (Rex Hazelwood, Hornsby Shire Recollects), packing fruit at Sonter Orchards (Mitchell Library, State Library of NSW, ON 151/1-19, FL980560 and FL980566), and apricots at Mobbs' farm (Rex Hazelwood, Hornsby Shire Recollects); a Baulkham Hills orchard in 1914 (McClymont, 2013: 31); hoeing the soil by hand in Les Smith's orchard in West Pennant Hills (bottom left); and removing forest to grow fruit in Carlingford c. 1920 (bottom right). Sources: National Library of Australia (Panorama of Waratah Farm Orchard, Ingleside, New South Wales/EB Studios. Bib ID: 6248299); State Library of New South Wales (Mitchell Library, State Library of NSW, ON151/1–19, FL980560 and FL980566); Hornsby Shire Recollects; and Museums of History New South Wales (Museums of History New South Wales—State Archives Collection: Government Printing Office; NRS 4481, Glass negatives. NRS-4481-3-[7/15963]-St4397 | Government Printing Office 1 - 16823 - Baulkham Hills: distant view of orchard, 1914) kilometres north-west of Sydney city, established itself as a significant fruit-growing district renowned for citrus cultivation (Fig. 6.1). In 1922, the district boasted a sprawling expanse of approximately 1000 acres (800 bearing) dedicated to citrus fruits, with an estimated 100,000 trees yielding an average annual crop of 90,000 cases.[124] The district produced a symphony of citrus fruits including Navel, Valencia, Parramatta and Paterson River oranges, as well as Lisbon and Sweet Rind lemons and various types of mandarins. The agricultural bounty did not end there, with 10,000 boxes of passionfruit, 10,000 boxes of fruit, apples and stone fruit, 1500 boxes of loquats, amongst tomatoes, peas and beans.[125]

In the Camden and Elderslie regions, Spring Farm was one of several properties dedicated to cultivating table grapes for Sydney markets in the 1930s and 1940s.[126] At its zenith, the formidable Camden Park Estate boasted an impressive array of 7200 fruit trees, comprising 56 varieties of apples, 31 varieties of pears, 23 varieties of citrus fruits, as well as apricots, plums, cherries, figs and quinces. Even fallen fruit found a purpose, as pigs housed nearby were permitted to feast upon the windfall.[127]

As *Period Three* drew to a close, many of Sydney's orchards faced the threat of (sub)urban expansion. Some succumbed to encroaching development while others, anticipating inevitable urbanisation, were allowed to fall into neglect. Other landowners awaited opportunities to profit from the subdivision of orchard blocks into smaller parcels for residential housing.

Cropping (CR): Feeding Livestock

No dedicated statistics for cropping in *Period Three* were identified specifically for Sydney. However, a comprehensive 1948 survey (discussed in *Period Four*) revealed cropping as the primary horticultural land use in the County of Cumberland, albeit to a smaller degree than grazing in terms of overall land usage. Much of the cropping in the region was geared towards producing local feed for livestock (for example, hay, pasture and maize) rather than commercial cultivation of broadacre crops for direct human consumption. Crops, either harvested or foraged by livestock, served as animal feed supporting various intensive operations across Sydney, including abattoirs, fattening operations and dairies.

William Farrer's groundbreaking development of the rust-resistant wheat variety 'Federation' revolutionised Australian agriculture on its release in 1903. Despite flour millers' initial resistance towards this 'hard' wheat variety, Federation swiftly rose to prominence and became the leading variety planted nationwide from 1910 to 1925. Its widespread adoption played a pivotal role in tripling the national wheat harvest during the period.[128] While Federation held promise for improving wheat cultivation in Sydney's climate, by *Period Three* the agricultural landscape had shifted. Cropping activities had largely transitioned to expansive rural districts serviced by robust rail infrastructure, rendering metropolitan cropping's financial feasibility beyond supplementary pursuits increasingly challenging.

Intensive Livestock (IL): The Last True Free-Range Era

Sydney's intensive livestock production underwent upheaval, gradually being displaced from built-up areas and transitioning towards more regulated, commercial and larger-scale operations on separated lands. This shift mirrored patterns observed in livestock processing (see *Agricultural Infrastructure*). Livestock's removal from urban life catalysed an incremental divide between production and consumption processes, distancing urban populations from the smells, dusts, squeals and wastes of livestock production, slaughter and processing.

As discussed earlier, health and hygiene concerns were a central reason for urban livestock's decline, exacerbated by Sydney's typhoid fever, plague and Spanish influenza outbreaks. Consequently, increasingly formalised bylaws and planning processes were established to regulate and enforce parameters for livestock practices, leading to their incremental expulsion from built-up areas.[129] Sydney's ballooning population also clashed with intensive livestock keeping in regions with growing densities. Livestock intensity was not solely determined by on-site stocking rates but also by adjacent land uses, spatial proximities and public perceptions. Women's entry into the workplace also shifted domestic practices, reducing levels of animal husbandry at home.

As urban densities increased and backyard sizes shrunk, complaints about neighbouring pigs, dairy cows, roosters or beehives prompted local government regulations banning livestock in suburban areas. Animals wandering across roads increasingly became a nuisance and danger as motor vehicles increased (Fig. 5.2, top). Historical evidence also reveals a recurring conflict between intensive animal operations and suburban development, exemplified by land developers like Sir Arthur Rickard who promoted the same area of swampy land in 1924 for both rural farmlets for pig raising as well as "peerless" residential land for young home buyers to live in healthy surroundings.[130] A similar pitch had been made in 1906 in an advertisement for a Warriewood/Narrabeen subdivision[131] and undoubtedly in many other instances across the city. Such conflicts persist in today's peri-urban Sydney regions, underscoring regular ineffectiveness of planning measures in avoiding and mitigating conflicting land uses, both before and after the official legislature in 1945 (see *Periods Four*, *Five*, and *2021+*, and Fig. 7.3).

As a commercial hub, Sydney contained animal studs to supply breeding stock. Livestock breeding and husbandry's significance to the period is highlighted by publications such as *The Farmer and Settler*, a broadsheet newspaper that circulated primarily on a weekly basis between 1906 and 1957. A feature article and advertisement published in 1936 noted Sir Frederick Stewart's newly knighted status and his St Cloud Pig and Cattle Stud in Dundas.[132] The article offered insights into the operations of the farm and Stewart's advocacy for the advancement of livestock breeding for the wealthy Australian gentleman. Another proximate piggery was located on the site of the now Northern Beaches Secondary College in Brookvale.

Poultry

Metropolitan intensive livestock's marginalisation during *Period Three* was significantly contrasted by the growth of Sydney's poultry sector. Poultry dated back to the First Fleet but the influx of new migrants, particularly from southern Europe, significantly catalysed the sector, especially for meat production. Much of the period's poultry-keeping was characterised by birds enclosed at considerably more humane densities than the industry from *Period Four* onwards (Fig. 5.11). This reflects evolution from domestic, hobbyist and small-scale operations.[133] 'Layer hens' eggs were the poultry industry's primary output, with meat production still a distant aspect largely prompted by hen turnover to maintain egg-laying productivity. This would reverse in *Period Four*, when Sydney's commercial broiler (meat) industry experienced phenomenal growth.

The Grantham Poultry Stud in Seven Hills exemplified best practices in commercial poultry-keeping. Established in 1917 on part of the former Prospect Common as a governmental research station and model poultry farm under the WWI Soldier Settlement Scheme, Grantham later evolved into the Grantham Poultry Research Station and continued operating until 1988. It was the focus of the New South Wales' poultry industry for over 70 years.[134] Bird stocking densities in 1917 (Fig. 5.11, second row) were notably modest compared to contemporary practices, with the surrounding landscape still bearing remnants of Cumberland Plain Woodland and lacking significant urban development.

Prior to the dominance of broiler and layer chickens in the latter part of the twentieth–century, poultry-keeping was more varied in terms of types and breeds, which ranged from chickens, ducks,[135] geese and even ostriches. An intriguing example is the Diamond Bay ostrich farm, 'Barracluff's', which operated from 1889 to 1918 near South Head on the threshold of Rose Bay and Vaucluse (Fig. 5.11). Barracluff's housed a stock of 100 birds, originally sourced from Morocco and Egypt, and thrived on the demand for ostrich feathers used in fashion accessories. However, the outbreak of World War I precipitated a decline in demand for frivolous feathers. Despite its closure, the farm enhanced the profile of its owner Joseph Barracluff, who served as an alderman of Waverley for a decade and mayor from 1914 to 1915.[136]

While domestic poultry-keeping was prevalent across the period, it gradually diminished across the expanding city. In some instances, poultry were integrated in runs coupled with citrus trees. Some citrus growers opted to abandon their orchards entirely in favour of poultry farming, driven by the promise of quicker capital returns and a smaller land footprint. Increasing commercialisation and intensification necessitated greater separation between facilities and residential areas and increased

Fig. 5.11 McCombs Poultry farm, The Spit 1906 (top); Grantham Poultry Institute (then stud) 1917 in Seven Hills; Barracluff's ostrich farm c. 1905 near South Head;

stocking densities to remain viable. Instances of oversupply (chiefly eggs) led to closures in the 1920s.[137] Regardless, commercial operations did not bode well for domestic poultry-keepers in suburban and urban regions. Towards the end of the period, polyculture (poultry and fruit) practices became increasingly rare commercially, with poultry transitioning into sheds in *Period Four* for both egg and meat production. Moreover, a significant cultural shift occurred towards chicken meat consumption, hence broilers' production next proliferated (see *Period Four*, *IL*).

Dairying

Dairies were prevalent throughout Sydney in *Period Three* in all regions and most suburbs. Although diminishing over the course of the period, their number peaked in *Period Three*. In 1911 there were 350 registered dairies in Sydney and by 1921 this had climbed to 427.[138] The Metropolitan Milk Board was established in 1929 and its 1932 annual report identified 517 dairies in the metropolitan area. By 1940, this number had dropped to 289.[139]

As refrigeration was not yet widespread, especially domestically, local dairies were still an integral means to supplying Sydney with milk. These extended from inner urban areas to concentrations in suburbanising regions like Marrickville, further out in Ryde,[140] coastward to Curl Curl (Fig. 5.11, bottom) north to Narrabeen, Mona Vale[141] and even Palm Beach and Currawong Beach.[142] Sydney's dairy industry also experienced notable shifts during the period.[143] In 1903, 55 per cent of Sydney's milk was supplied from country areas, equating to an impressive 45 per cent from the city itself. By 1912, country areas supplied 69–70 per cent.[144] This remained steady at 70 per cent up to 1934.[145] Other figures for 1920

Fig. 5.11 (continued) a Mona Vale goat farm in 1942 (third row, right); and a dairy in Curl Curl 1910 (bottom). Sources: Museums of History New South Wales (Museums of History New South Wales—State Archives Collection: Government Printing Office; NRS 4481, Glass negatives. NRS-4481-3-[7/15903]-M5371 | Government Printing Office 1—30566 —Seven Hills Soldiers Settlement, Grantham Stud Poultry Farm, 1917); National Museum Australia; Mitchell Library, State Library of New South Wales and Courtesy ACP Magazines Ltd correlate, with about one-third of Sydney's milk requirements being produced within the metropolitan area.[146] Correspondingly, the per capita milk consumption was estimated at half a pint per day, translating to 415,000 pints per day for Sydney's population of 830,000 in 1920.[147]

Dairying for milk production operated on a smaller scale than butter production, which could be stored more easily.[148] In 1912, the New South Wales Board of Trade described,

> A great number of dairymen keep herds of less than ten cows. Some [...] dairymen have large herds. There is one herd of 250 cows at Waterloo, another of 140 cows at Zetland, another of 130 cows at Greenwich, another of 126 cows at North Sydney, another of 125 cows at Willoughby, another of 125 cows at Concord, and other of 100 cows at Enfield, Canterbury and Woollahra.[149]

The proportion of urban dairies only for milk distribution depots is unclear. For example, in 1928 one Mosman dairy claimed to be the sole one in its municipality to keep cows.[150] Stephen Codrington's data reveals a rise in numbers, with 350 registered dairymen and 7345 cows in Sydney in 1911 to 427 and 7856 respectively in 1921.[151] The average farm size was approximately 1.2 hectares and herds ranged from 40 to 50 cows. As such, dairies relied on supplies of cropfeed grown offsite.

Urban dairies supplied raw, untreated milk.[152] In 1934, less than 5 per cent of the metropolitan milk supply was bottled.[153] Sydney's urban dairies therefore faced continued concerns over milk quality and hygiene. Following the 1886 tracing of milk-borne typhoid fever to a Leichhardt dairy, a 1901 report by the Metropolitan Medical Officer of Health classified 12 per cent of Sydney's metropolitan dairies as being in a bad condition, with only 20 per cent having a hot water supply for cleaning. By 1910, the number of dairies classed as 'bad' had dropped to 4 per cent, likely due to the efforts of the central health authority.[154] A 1903–1904 Royal Commission highlighted poor hygiene standards, adulteration and chemical preservatives, contributing to a perception of Sydney's "injurious quality of milk".[155] Country producers leveraged this to their gain, marketing their milk as wholesome and clean.[156] The Milk Board introduced a registration for dairies and vendors, as well as a regular system of testing, which from 1937 became annual.[157] Efforts to address concerns led to improvements, with a decrease in poorly maintained dairies from 12 per cent in 1901 to 4 per cent by 1910. Raw milk declined in the late 1930s and became rarer following World War II.[158]

Sydney's dairying industry also grappled with the ever-expanding city. As the residential population burgeoned, so did grievances from inhabitants regarding dairies' presence. Marrickville offers a case in point during the 1920s and 1930s. Despite a decline from 40 registered dairies in 1893 to merely nine by 1923, complaints persisted.[159] The Council impounded 171 straying cows from a herd of 370, refusing to release them until fines were paid. By 1926 the Council disallowed new dairy registrations and by 1928 stopped re-registration of existing dairies.[160] This exemplifies how local governmental regulations were employed to desist urban dairying, ostensibly to placate new residents who encroached upon pre-existing agriculture—a pattern still occurring in Sydney today.

Grazing (GZ): Town-Scape Pastoralists

Although grazing constituted Sydney's largest agricultural land use during *Period Three*, metropolitan grazing operations were regularly at hobbyist scales compared with the vast pastoral ventures beyond the city's limits. Smaller-scale grazing was nonetheless widespread, albeit extensive rural production scales limited commercial scope within Sydney's constrained landholdings.[161] Cattle grazing remained a common sight across various Sydney areas long-since suburbanised, with cows' hard-hoofs eroding the banks of Sydney's waterways just as they continue to do today in regional areas. Historic photographs capture cattle grazing the banks of Cooks River in Earlwood in 1904[162] as well as in estuarine lagoons around 1906.[163] The immediate landscape was visibly impacted, creating a temporal landscape succession from forest to woodland and then open pastureland. Once denuded and encroached, pastures transitioned to urban and residential development.

A 1914 photograph showed 'Macarthur Merinos' as "pure-bred descendants of [the] original flock" at Camden Park Estate, offering a glimpse into the fine wool industry and its linkages to Sydney.[164] The

Macarthur property, situated in the heart of the Cowpastures near Camden, maintained its significance throughout *Period Three* as evidenced by Hardy Wilson's beautifully executed 1919 map of the 'Cow Pasture Road' (Fig. 5.12). His depiction focuses on Parramatta to Camden, providing rich visual interest and cartographic creativity. Amongst indigenous fauna, the map includes butter, mutton, spatchcock, honey and wine production within its landscape mosaic.

The widespread metropolitan presence of cattle during this period directly ties into the 'processing' at Glebe Island (1830s–1915/16) and Homebush (1907–1988). As such, some of Sydney's grazing lands played a role in the 'finishing' process of cattle transported to the city from other regions. This equated to sites where cattle were rested and fattened prior to slaughter, processing and ultimately, distribution within and beyond city limits.

Vineyard Wineries (VW): Never to Fully Recover

Some resilient grape growers and wineries managed to ride out the devastation wrought by *Phylloxera* in *Period Two* and re-establish their vines. In *Period Three*, Sydney boasted several grape-growing regions, extending beyond greater Liverpool to areas along the Georges River, Cabramatta West, Castle Hill and greater Kingswood. Throughout metropolitan Sydney, numerous wineries dotted the landscape, with standout examples including 'Leonay' in Emu Plains and 'Vicary's' in Luddenham, amongst many others.

As mentioned in *Period Two*, Minchinbury began producing its sparkling wines in 1908, utilising grapes from their 1903 vintage in a venture that would secure their renown.[165] Under the expertise of esteemed winemaker Leo Buring and with state-of-the-art equipment imported from France, Minchinbury employed the traditional bottle-fermentation method to craft its sparkling wine. In 1912, Penfolds Wines acquired Minchinbury and initiated a phase of expansion. The cellars were enlarged to accommodate over 1.25 million bottles, while the vineyards were expanded to cover more than 400 acres (162 hectares). Additionally, the

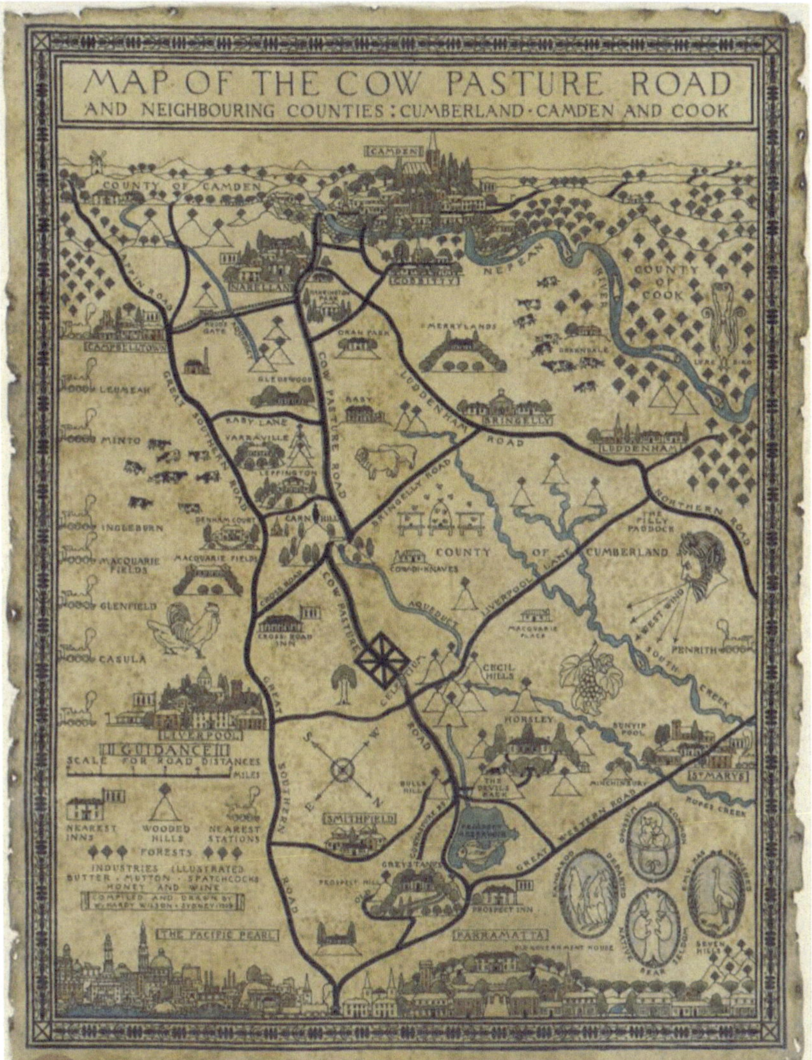

Grazing

Fig. 5.12 An iconic 1919 depiction of the Cow Pasture Road. Source: W. Hardy Wilson, National Library of Australia estate saw the planting of over 90 Kalamata olive trees.[166] The grape varieties cultivated in their Eastern Creek vineyard included Verdelho, Riesling, Cabernet Riesling, Pinot Noir, Hermitage, Traminer and Pinot Blanc. The absence of cover crops on the soil eventually led to soil exhaustion.

The ever-expanding metropolis nevertheless encroached upon vineyards' borders. Early remnants managed to persist on still productive farming lands but many ultimately succumbed to the relentless pressure of urban growth. Some wineries already faced this threat during *Period Three* but for others, it became an inevitable reality in *Period Four*.[167]

Agricultural Infrastructure (IN): Consolidation and Industrialisation

The Map Database's total *Agricultural Infrastructure* entries declined from 35 in *Period Two* to 27 in *Period Three*. This primarily reflects escalating industrialisation, as well as the concentration and clustering of operations across agri-food supply chains. This phenomenon is evident in Sydney's abattoir sector. In 1912, a Royal Commission investigating the supply and pricing dynamics of meat, fish, fruit and vegetables identified a total of 48 abattoirs distributed throughout the city and its suburbs. Among these, the largest facility was situated at Glebe Island, while Enfield, Mascot and Hornsby each accommodated five abattoirs, and Granville, Liverpool and Prospect contained four.[168] During the same period, the Labor government established the Metropolitan Meat Industry Board, which led to the consolidation of livestock processing operations at a specialised facility located in Homebush. This state venture created a near monopoly of slaughtering activities within the Sydney region.[169]

Abattoir consolidation largely stemmed from health and hygiene concerns. Following the outbreak of plague 1900 traced to waterfront rats near The Rocks and Millers Point, impetus increased to close the adjacent public abattoir at Glebe Island. The aged facility had been operating since the 1830s (see *Period Two, IN*). A 1902, the Parliamentary Standing Committee recommended the establishment of an abattoir at Homebush. The state resumed 367 hectares of Homebush Estate in 1907. In 1910, construction of a significant complex commenced, with 44 slaughter-houses, administration buildings, stabling, by-products treatment buildings, latrines, drafting yards and roadways.[170] 1914 marked the initiation of large-scale livestock slaughtering and the production of meat and associated by-products.

Homebush Abattoir had become the largest facility of its kind within the Commonwealth by 1923, employing a workforce of up to 1600. Its grim infrastructure facilitated an immense killing capacity, with the ability to process approximately 18,000 to 20,000 sheep, 1500 cattle, 2000 pigs and 1300 calves daily.[171] The operations yielded a diverse array of by-products including tallow, dripping, fertiliser, oil, sinews, hoofs, hair, glue pieces, bones and horns, all of which were monetised. In the 1930s, algal blooms developed in Homebush Bay from the dumping of carcass' waste material.[172] The behemoth complex nevertheless continued its gruesome operations well into *Period Four*.

The Davis Gelatine Factory in Botany was another notable livestock processing facility and part of a growing industrial complex supplanting the area's market gardens (Fig. 5.4). The factory opened in 1917 and closed in 1990. It used animal offcuts to make gelatine, glue and other products (Fig. 5.13, second row). Today, such processes are usually concealed in their production but they were far more visible in early twentieth-century Sydney, and indeed, odorous, even from a distance.

Flour mills also persisted throughout the period, exemplified by the enduring presence of John Darling and Sons in Rhodes.[173] Opening in 1919,[174] Darling's was an early adopter of industrial-scale production that would come to dominate the Rhodes Peninsula, later becoming Allied Feed Mills (Fig. 5.13, third row left). Sydney's formidable industrial food manufacturing establishments were also noteworthy. Arnott's boasted a substantial biscuit factory situated in North Strathfield, while Nestlé operated a chocolate factory established in Abbotsford in 1918 that employed over 1000 staff (Fig. 5.13, third row right).[175] Both enterprises proudly laid claim to being the largest of their kind in the southern hemisphere. Arnott's leveraged its advanced industrial and technological capabilities as a means to captivate and attract clientele[176]—a marketing approach no longer in vogue given their website today omits any details or imagery of historical or current food manufacturing facilities.[177]

Fig. 5.13 Farm machine from Dangar, Gedye & Co Ltd., Sydney c. 1900 (top left); a horse grinding wheel at Rouse Hill c. 1908 (top right); Davis Gelatine advertising

As mentioned earlier in the chapter, *Period Three* witnessed the transition from horse-drawn urban transport and agricultural usage to machines stemming from the enthusiastic uptake of motorised vehicles and tractors. From the early 1900s, an array of experimental farm machinery became available (Fig. 5.13, top left), evident in, for example, in a 1916 machinery catalogue[178] and demonstrations by importers, manufacturers and sellers.[179] Despite availability, the horse remained the main source of farm power nationally until the 1930s (Fig. 5.13, top right), with tractor-drawn ploughs only becoming normalised from 1945.[180]

Sydney utilised a variety of transportation infrastructures to move its food to market, often tackling challenging geography. After harvest, fresh produce regularly travelled by various modes including horse and cart, rail, boat, barge or ferry, usually to agents who sold to retailers, to then be taken home by consumers by tram, horse, cart, car, bicycle, or on foot. Food was therefore subject to multiple points of handling and prone to damage, especially delicate fruits. Figure 5.10 shows various stages of packing, loading and transporting fruit in the Pennant Hills/Carlingford district. This railway line was initially privately owned and provided improved transport options for fruit growers from the districts around Baulkham Hills, Castle Hill and Dural.[181] It also serviced residents in new housing subdivisions which were paradoxically diminishing agricultural production lands in the region.

The advancement of machinery, highlighted by the introduction of harvest grading machines used for wheat between 1900 and 1930[182] and subsequently for fruit,[183] played a pivotal role in streamlining various agricultural practices. Such technological innovations markedly enhanced the efficiency of growing, harvesting and processing operations,

Fig. 5.13 (continued) (Courtesy of Bayside Library Service); Darling flour mill in Rhodes 1937; Nestlé Factory Abbotsford, 1940; Arnott's Biscuit Factory advertisement showing their North Strathfield factory; and Sydney (now Paddy's) Markets on Quay and Hay Streets c. 1911–1916 (bottom right). Sources: Trove; Caroline Simpson Library, Museums of History NSW; Courtesy Bayside Library Service; Royal Australian Historical Society (Set 72157640448188933, ID 29885439752, Original title John Darling & Sons Flour Mills Rhodes—27 July 1937); City of Canada Bay Local Studies Collection; and Rex Hazlewood, State Library of NSW consequently diminishing the labour intensity associated with food production while enabling greater yields.

Sydney's fresh food markets continued to fluctuate from *Period Two* into much of *Period Three*. Despite the significant expenditure of over £400,000 on city markets during the final decade of the nineteenth century, there was a decline in genuine market space.[184] As outlined in the preceding chapter, the expensive Queen Victoria 'Market' Building (QVB) failed to fulfil its original purpose as a fresh food market, offering ornamental grandeur over the functionality required for the fresh food trade. The QVB accommodated primarily value-added products, functioning as a massive wine cellar where merchants like Lindemans and Penfolds stored over 0.5 million bottles.[185] The QVB's trajectory epitomises a prevalent theme in Sydney's historical evolution, namely, the pre-eminence of high commerce usurping local physical activities and the processes underpinning it.[186]

Between 1909 and 1914, a new market complex emerged close to Darling Harbour, supplanting the congested Belmore market. This contemporary facility encompassed provisions for the sale of fruit and vegetables, fish, poultry and rabbits.[187] In response to a scarcity of stall space, a new producer's market, 'No.6 Market', was erected in 1936.[188] By the early 1900s, the prevailing conditions at the Belmore markets had deteriorated to the extent of being deemed intolerable. Consequently, during a period of urban reform that culminated in the 1908–1909 Royal Commission, the markets underwent relocation to a more spacious locale near the Darling Harbour railway goods yards situated west of George Street. The ensuing transformation gave rise to brick buildings complemented by banana-ripening towers and loading bays, defining the landscape of the Haymarket district and remaining until they were located to Flemington in 1975.[189]

With respect to the period's refrigeration, the 'Coolgardie safe' of *Period Two* was a relatively short-lived domestic refrigeration solution, replaced by the icebox in early *Period Three*. Similarly, the 1920s kerosene refrigerator was similarly briefly used, largely superseded by electricity's availability. Exceptions existed in areas not serviced by electricity like rural Australia or for those unable to afford the running an electric or kerosene refrigerator.[190]

Aquaculture (AQ): Teach a (Hu)man to ~~Fish~~ Trawl…

Sydney's produce market flux was largely mirrored by its seafood markets. In 1914, the Woolloomooloo fish market closed and sales were moved to Haymarket's Municipal Markets. Redfern's private fish market also closed in 1923 when the Sydney City Council took control of marketing.[191] In 1945, under the Fisheries and Oyster Farms Act, the state government assumed control over fish marketing and introduced a regulated sales system.[192] Regulation at the time was not particularly effective at managing the pollution of aquatic environments, nor that of fish stocks. Figure 5.14 shows a substantial catch of sea garfish (*Hemiramphus*) at Thompson's Beach[193] (near Sydney) in 1911, demonstrating that even before ocean trawling vessels, the use of fine nets could deplete fish stocks.

As seen in *Period One*'s pollution of the Tank Stream, aquatic systems were regularly utilised as repositories for wastes. In *Period Three*, chemical and heavy industries and abattoirs significantly ramped-up their activities, widely impacting Sydney Harbour. Technology also facilitated greater consumption. In the late 1920s, for example, Tiger Flathead had been exhausted in Sydney's south after only a decade of ocean trawl fishing, and Flathead reduced from a catch of 2.3 million tonnes in 1928 to 0.2 million in 1937.[194] Finally, although not for food, seaweed was harvested in Botany Bay during World War II, dried in Kurnell, and sent to Newcastle to make gelatin.[195]

Miscellaneous (MC): Roll Up Your Sleeves

Several Agricultural Shows took place during *Period Three*, including the St Ives Show, which commenced in 1921 at Hassall Park but shortly thereafter moved to St Ives Showground. The most prestigious of these was The Royal Easter Show (Fig. 5.14, centre right).

In 1927, *Period Two*'s Iliffe Nursery was sold and subdivided, with local streets christened after members of the Iliffe family.[196] Meanwhile, the year 1932 marked the inauguration of Hammondville, a new suburb designed as a resettlement area for former inhabitants of inner-city slums. This development was underscored by a commitment to fostering self-sufficiency

Fig. 5.14 Sea Garfish catch near Sydney, 1911; one of Maze's agricultural land use studies of the Kurrajong-Windsor District, 1943; The Governor General (centre) ostensibly inspecting agricultural products at the 1935 Royal Easter Show; and a *'Grow your own'* illustration from Arthur Yates & Co's Sydney 1944 seed book.

in food production.[197] Notably, the distribution of food retail remained markedly decentralised in comparison to *Period Four*, which witnessed the emergence of supermarkets. A testament to the period's productivity can be found in Mosman, which boasted an impressive tally of 33 fruit and vegetable shops in 1934.[198] Even amidst the challenges of wartime rationing in 1941, retailers such as Grace Brothers (Myer) continued to maintain a robust retail fruit display at their Broadway store.[199]

Domestic food production during World War II was an important component amidst wartime hardships. A 'Dig for Victory' campaign was launched in Australia in January 1942 by Prime Minister John Curtin, drawing on successful similar initiatives in the US and UK. The campaign encouraged people to produce as much food of their own food as possible during World War II's fresh food shortages,[200] as visible in William Dawson's *My victory garden: Simple rules for growing your own vegetables* published in 1944.[201] Arthur Yates and Company's seed book describes food as an important 'weapon of war' (Fig. 5.14, bottom right).[202] Several examples of seed packets for domestic vegetable gardeners from the 1920s and 1930s were shown at a Sydney Living Museums 'Eat Your History' Exhibition.[203] These show interesting varieties of heirloom vegetables, along with culinary implements that reveal practices, styles and approaches to growing and eating at the time.

On a bigger scale, arable farming on the Hawkesbury River's alluvial flats also supported thriving citrus and orange orchards, ample pastureland and upland poultry farms throughout the period. W.H. Maze's 1943 study of the Kurrajong-Windsor district revealed that farm production was not only oriented to the central Sydney market but also Parramatta and locally. He concluded that "it would seem likely that the area is approaching its maximum agricultural development".[204] Was this just a wartime boom or would such agricultural productivity continue in Sydney's future in *Period Four*?

Fig. 5.14 (continued) Sources: David Stead, National Library of Australia; Taylor & Francis (Fig. 6 from Maze, W. H. (1943). Land utilization surveys in the Kurrajong/Windsor District, New South Wales. *Australian Geographer*, 4(6), 155–174. https://doi.org/10.1080/00049184308702232. Copyright © Geographical Society of New South Wales Inc., reprinted by permission of Taylor & Francis Ltd); Mitchell Library, State Library of New South Wales; and Caroline Simpson Library, Museums of History NSW

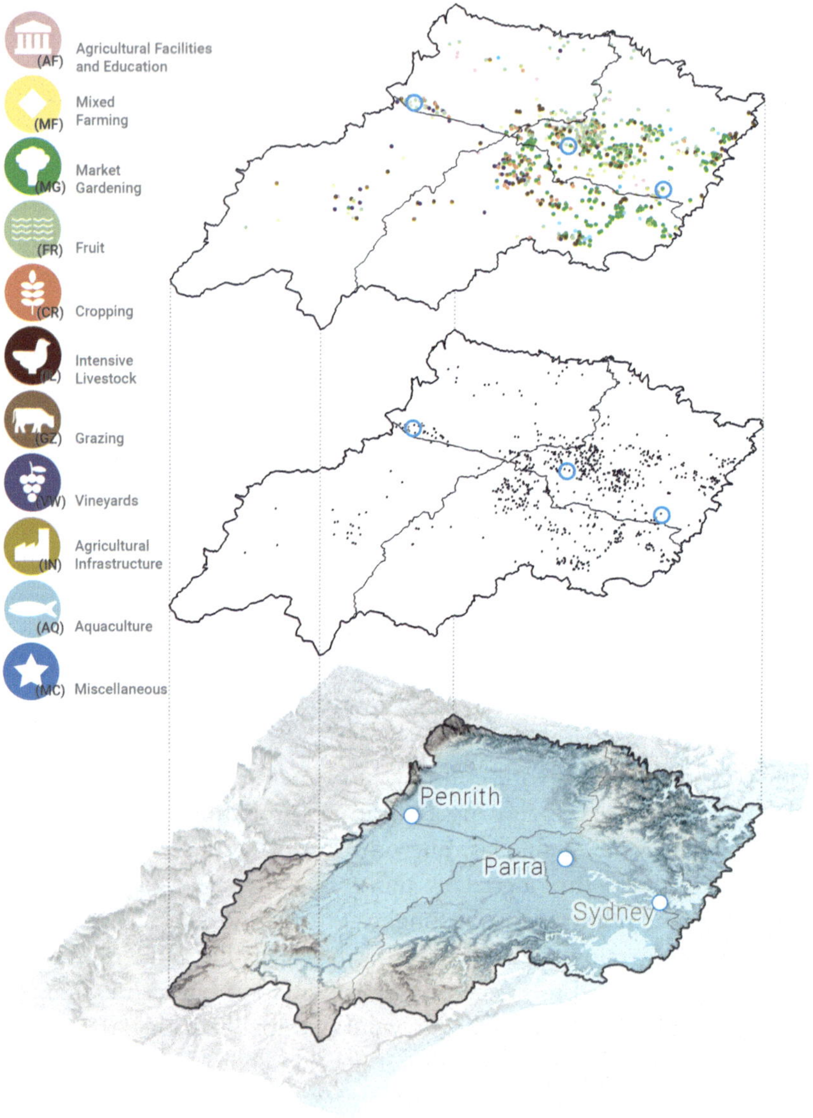

P3
1901-1945

Fig. 5.15 *Period Three*'s Map Database: 3D. While *Period Three*'s database maps contains fewer total entries than *Period Four*, this results from *Period Four*'s greater availability of satellite aerial imagery (*Period Three* only had 1 year (1943) of viable coverage to populate the Map Database via aerial imagery). It is unclear when Sydney's per capita agricultural activity peaked, but the city was undoubtedly a powerhouse of local food production between 1901 and 1945. Source: Joshua Zeunert, Josh Gowers and Stephanie Stankiewicz

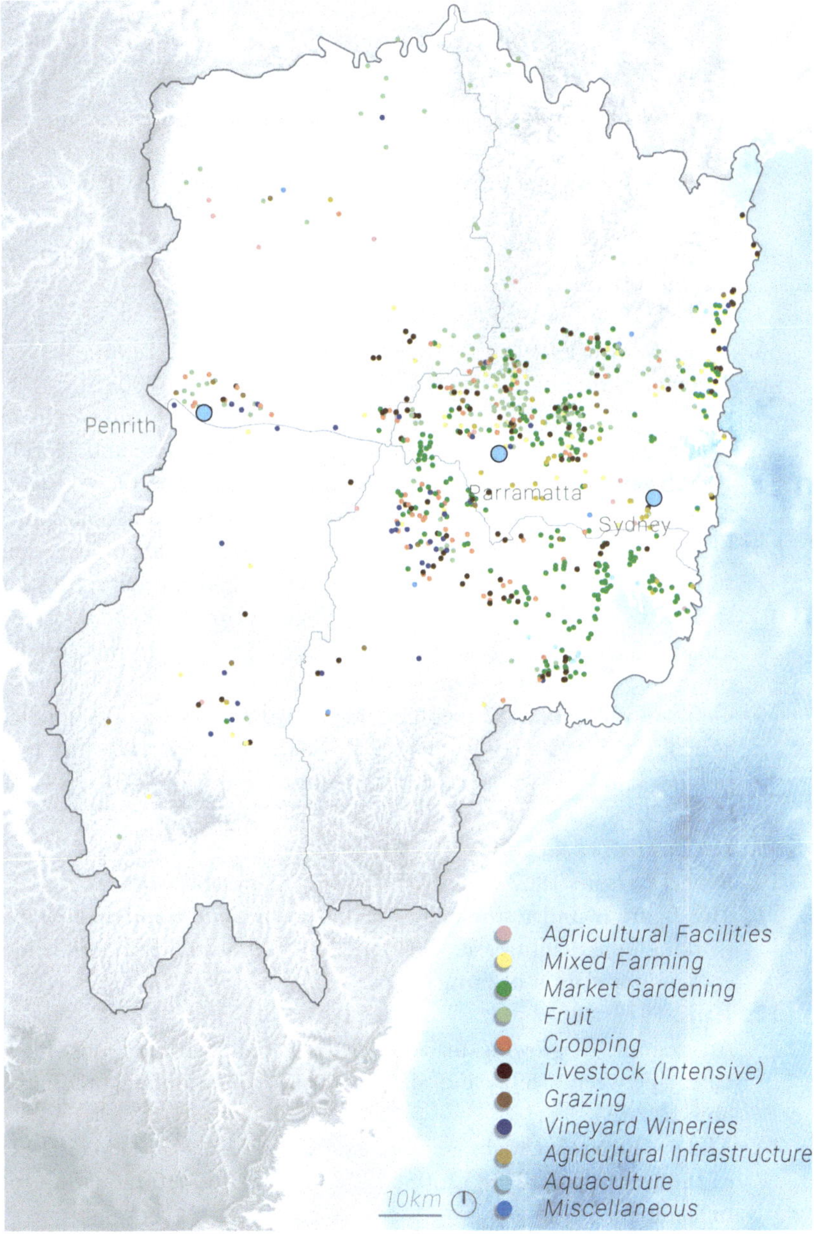

Fig. 5.16 *Period Three*'s Map Database: Plan. Source: Joshua Zeunert, Josh Gowers and Stephanie Stankiewicz

Notes

1. Sydney's April 1901 population: https://trove.nla.gov.au/newspaper/article/24730043.
2. https://www.abs.gov.au/AUSSTATS/abs@.nsf/Lookup/3105.0.65.001Main+Features12014?OpenDocument.
3. See e.g. https://www.nma.gov.au/defining-moments/resources/bubonic-plague.
4. McGill, 2017.
5. Western Sydney University Archives Collection: Ref. P1496.
6. To manufacture synthetic nitrogen, see Erisman et al., 2008.
7. Clark (1988: 635–636) states: "In 1897 Cuming Smith and Company commenced importing rock phosphate from Tennessee and Florida. From 1900 rock phosphate supplies were imported from Ocean Island, Nauru and Christmas Island. Erection of conventional phosphate plants continued, as new companies emerged to make use of increasing supplies of byproduct sulphuric acid and as agricultural cultivation spread. By 1930 there were some eighteen plants in operation with capacity in excess of one million tonnes". See: http://www.austehc.unimelb.edu.au/tia/604.html.
8. German philosopher Rudolph Steiner issued a call for a differentiated agriculture in 1924 in Poland. In 1925 he wrote his treatise, which later became known as 'biodynamic' agriculture. See Paull, 2013: 38 and Code, 2018.
9. See Erisman et al., 2008.
10. Rachel Carson's 1962 book Silent Spring is a notable work.
11. This figure is underestimated as it did not include Aboriginal people working in agriculture, see O'Connell: https://australianfoodtimeline.com.au/population-of-farmers/.
12. Jeans, 1972: 295.
13. An example of growing maize in Sydney is shown here, most likely taken between 1892 and 1917: https://collection.maas.museum/object/28824.
14. ABS, https://www.abs.gov.au/ausstats/abs@.nsf/94713ad445ff1425ca25682000192af2/3852d05cd2263db5ca2569de0026c588!OpenDocument.
15. Australia exported 72 per cent of its total agricultural output by value from 2017 to 2020 (ABARES, 2022).

16. O'Connell, '1893 Mediterranean diet recommended': https://australianfoodtimeline.com.au/1892-mediterranean-diet-recommended/.
17. O'Connell, 2017; Hunwick, 2018.
18. This is the title of Hoorn's (2009) book focusing on Australian pastoralism, its depiction in art and its shaping of the landscape and perceptions of it.
19. Burnley, 2001, James, 2016: 12.
20. Not insignificant considering that in 1883 Australians were apparently the highest per-capita consumers of tea in the world, almost one-third above that of British consumers. See O'Connell: https://australianfoodtimeline.com.au/griffiths-australian-grown-tea/.
21. Birmingham, 1999: 106.
22. As discussed in the remaining chapters. See also Zeunert & Freestone, 2022.
23. ABARES, 2021: 2.
24. Ibid. In 2021, 55 per cent of Australia's land is directly devoted to agriculture and more again to the resources it requires. About 2 per cent of Australia's population works in agriculture, but many of these do not own or lease the land.
25. All 200 pages can be viewed here: https://recordsearch.naa.gov.au/SearchNRetrieve/Interface/ViewImage.aspx?B=83899.
26. See Lewis (1984) for a critique of Carruthers.
27. Lewis, 1984: 71.
28. Lewis, 1984: 72.
29. City Plan, 2014: 27–28.
30. O'Connell, 2017; n.d.
31. City Plan, 2014: 4.
32. City Plan, 2014: 4.
33. Birmingham, 1999: 311.
34. Cashin & Mcdermott, 2002.
35. City Plan, 2014: 26.
36. City Plan, 2014: 27.
37. In the 1947 census. Nationally this was a quarter of the population. See City Plan, 2014: 9.
38. City Plan, 2014: 9.
39. A prime example is groundwater pollution by the company 'Orica' (formerly ICI Australia). The NSW EPA characterise this as, "Poor environmental management and waste disposal practices [from 1942]

up to the 1980s resulted in significant amounts of chemicals contaminating the underlying soil and groundwater". See EPA NSW (n.d.).

40. Curby, 1997: 55.
41. City Plan, 2014: 9
42. Spearritt, 2000: 144.
43. The photograph is also prescient of how expansion of aviation and industrial development would displace metropolitan agriculture in Sydney's backyard vegetable garden in Mascot, Botany and their surrounding areas.
44. Hovenden, 1981: ii.
45. City Plan, 2014: 9.
46. Ashton & Waterston, 2000: 50.
47. Spearritt, 2000: 144.
48. Hovenden, 1981: ii.
49. Spearritt, 2000: 144.
50. City Plan, 2014: 9.
51. Spearritt, 2000: 144.
52. New South Wales Parliament, 1920.
53. See: https://mhnsw.au/stories/general/spanish-flu-1919/.
54. Spearritt, 2000: 24.
55. Spearritt cites city historian Shirley Fitzgerald, see Spearritt, 2000: 14.
56. Spearritt, 2000: 69–72.
57. Gaynor, 2006: 26.
58. Freestone, 2000.
59. Freestone, 2000.
60. In this period Ebenezer Howard's Garden Cities recognised the essential need for food production in the hinterland and adjacent to satellite villages and towns. Frank Lloyd Wright's vision for the Broadacre City in the US clearly acknowledges the central role of metropolitan agriculture (see Zeunert, 2017: 177). Also see discussion of Professor of Planning Denis Winston's book, *Sydney's Great Experiment* (1957) in *Period Four*.
61. Town planning as a discipline was formalised in America with the founding of the American City Planning Institute in 1917, followed later by the American Society of Planning Officials in 1934. The Planning Institute of Australia (PIA) was founded in 1951.
62. See Zeunert & Freestone, 2022.
63. Brown, 1937.

64. Ashton & Waterson, 2000: 48.
65. Birmingham, 1999: 223–224.
66. See for example, images of Chullora and Granville here, and 1940s images here.
67. Birmingham, 1999: 223–224.
68. Spearritt, 2000: 3.
69. Wills, 1945: 35.
70. See a 1927 painting of Castle Hill by Robert Johnson: https://www.sl.nsw.gov.au/collection-items/vineyard-castle-hill. Although titled 'vineyard', an orchard appears adjacent to a farmhouse, with a vineyard likely in the foreground on the far right. The nature of the expanding city is evident through farming properties extending into the then hinterland of blue gum wet sclerophyll forest. The painting faces west, with the Blue Mountains veiled in the background.
71. Christie, 1988.
72. Christie, 1988.
73. Although Paull (2013) makes no mention of the contribution of the Australian founded movement of Permaculture by Bill Millison and David Holmgren. See Code, 2018.
74. Walter had died in India the year before, see Paull, 2012.
75. O'Connell: https://australianfoodtimeline.com.au/1944-worlds-first-organic-farming-society-in-sydney/.
76. Paull, 2013: 37.
77. As well as Australia providing the first statement of organic farming principles, see Paull, 2013: 37.
78. For the purposes of this book this imagery was initially (and still is at the time of writing) available on the 'SIX Maps' website and in the latter research stages, on the Historical Imagery Viewer website.
79. Aslin et al., 2004; Willis, 2007.
80. CLS&P, 2004: 22.
81. Pers.Comm. Colleen Morris, 17 March 2020.
82. TKDA, 2017: 38.
83. And later imagery, search -33.97194, 151.24514 at: https://portal.spatial.nsw.gov.au/portal/apps/webappviewer/index.html?id=f7c215b873864d44bccddda8075238cb.
84. Situated where Uralla Road is now located.
85. https://hornsbyshire.recollect.net.au/nodes/view/584?keywords=dural%20orchard&all=1&whole=2.

86. https://dictionaryofsydney.org/entry/liverpool_internment_camp_during_world_war_ii.
87. The first appointment commenced in 1910. See Copeland, 2020: 59.
88. At the age of 28 years old. See Copeland, 2020: 59.
89. Copeland, 2020: 59.
90. Copeland, 2020: 60.
91. https://dictionaryofsydney.org/entry/cromer.
92. Dibden & Cocklin, 2005; Wolf, 2014.
93. See Gaynor, 2006 and the Yates illustration, 'From the Idle Plot…Grow your Own': https://mhnsw.au/whats-on/exhibitions/eat-your-history-library/.
94. Only 1943 aerial photography was available for survey—earlier aerial imagery was not available.
95. Nichols, 2016: 77.
96. Christie, 1988: 97.
97. Christie, 1988: 92–93.
98. Christie, citing Yong. See Christie, 1988: 74–75.
99. Ibid.
100. A 1920 R.M. Ronald watercolour of Rose Bay in the State Library NSW shows large areas of Chinese market gardens still present on O'Sullivan Road. During the second half of the nineteenth century, leases were taken up by Chinese market gardeners on low lying areas referred to as the 'Rose Bay Flats', centred around the Rose Bay Creek which ran inland from Lyne Park. They were gradually consumed by residential development, the Woollahra and Royal Sydney Golf Courses and Woollahra Park. By 1926 the last of the gardens disappeared. See: https://archival.sl.nsw.gov.au/Details/archive/110369436.
101. See: https://www.aph.gov.au/About_Parliament/Parliamentary_Departments/Parliamentary_Library/pubs/APF/monographs/Within_Chinas_Orbit/Chapterone#_ftn1.
102. See for example Christie, 1988: 95–98; Fitzgerald, 2008.
103. Migration Heritage Centre, 2011.
104. Christie, 1988: 93–94; James, 2016: 12.
105. Ibid.
106. James, 2016: 12, citing Burnley, 2001.
107. Christie, 1988: 100–101.
108. Also see: https://trove.nla.gov.au/newspaper/article/175814769?searchTerm=Italian%20Market%20Gardeners%20sydney&searchLimits=.

109. NSW Migration Heritage Centre, 2011.
110. See Milton Kent aerial photography at the SLNSW here.
111. Spearritt, 2000: 43.
112. See Gojak, 2009; Macleod, 2010: 67.
113. Wills, 1945.
114. Wills, 1945.
115. Zeunert & Freestone, 2022.
116. Wills, 1945: 35.
117. Albeit orchards are an easy typology to identify in aerial photography due to their gridded patterning.
118. Bonizzoni et al., 2004: 3846.
119. Meats, 1989.
120. Nichols, 2015.
121. Nichols, 2015.
122. Higgins, 2016: 42. The Higgins orchard later featured in the Cartoon *Ginger Meggs*, see https://hornsbyshire.recollect.net.au/nodes/view/2213.
123. See http://nla.gov.au/nla.news-article190540321; http://www.pittwateronlinenews.com/Waratah-Farm-Ingleside-The-Narrabeen-Plum.php; and https://files.northernbeaches.nsw.gov.au/sites/default/files/documents/general-information/pittwaters-past/lh-ingleside.pdf.
124. McClymont, 2013: 76.
125. McClymont, 2013: 76.
126. https://dictionaryofsydney.org/entry/spring_farm.
127. https://www.belgennyfarm.com.au/history/agriculture-at-camden-estate/horticulture-and-viticulture.
128. O'Connell, n.d., 1903.
129. See pig carcasses piled in carts in Liston, 2009: 59.
130. Birmingham, 1999: 226–227. Also see Fig. 7.3.
131. See: https://www.facebook.com/NarrabeenReunions/photos/a.1848129648548071/1442619835765723/?type=1&theater and https://www.facebook.com/NarrabeenReunions/photos/a.1848129648548071/1442619835765723/?type=1&theater. The 1906 advertisement and subdivision plan at Warriewood Estate claims that its 66 farm blocks are 'all first class', with 'deep fertile soil', and the 'absolute pick of the district'. The Sydney Morning Herald reported on 3 October that 13 farm blocks had been sold on the day at £20,10s to £35 per acre. Any potential exaggerations aside, the greater Warriewood valley indeed became a very notable production district, particularly for market gardening and greenhouse-based food production (see P4, *MG*).

132. https://trove.nla.gov.au/newspaper/article/117213779.
133. As opposed to contemporary standards, routinely criticised as not representative of such conditions. See Zeunert, 2024.
134. Davies & Raymond, 2005.
135. Hothams in Gymea, for example, specialised in the supply of ducks. See Curby, 1997: 55.
136. See: https://dictionaryofsydney.org/entry/barracluffs_ostrich_farm.
137. McClymont, 2013, 128.
138. Marrickville Unearthed, 2019.
139. Spearritt, 2000: 23–24.
140. Search 'Hoppings' 'Goodchild' and 'Muraine' (dairy) at: https://ryde.spydus.com/cgi-bin/spydus.exe/MSGTRN/WPAC/HOME.
141. https://northernbeaches.recollect.net.au/nodes/view/27835.
142. http://www.pittwateronlinenews.com/st-clouds-elanora-heights-the-stewart-legacy.php.
143. See Lewis, 1980 and Wilkinson, 1999.
144. Marrickville Unearthed, 2019.
145. Lewis, 1980: 196.
146. Codrington, cited in Wilkinson, 1999: 3.
147. https://trove.nla.gov.au/newspaper/article/28094787 and https://www.ausstats.abs.gov.au/ausstats/free.nsf/0/2DB373BF5D0CD9D0CA257AEF00193903/$File/13010_1901_1919%20section%204.pdf.
148. Wilkinson, 1999: 3.
149. Codrington, cited in Wilkinson, 1999: 3.
150. Spearritt, 2000: 24.
151. In Wilkinson, 1999: 3.
152. Lewis, 1980: 199.
153. Lewis, 1980: 203.
154. Lewis, 1980: 197.
155. Lewis, 1980: 203–204.
156. See Lewis, 1980: 202.
157. Lewis, 1980: 201.
158. Lewis, 1980: 203.
159. Smart, 2019 and Marrickville Unearthed, 2019.
160. Smart, 2019.
161. To an extent greater than the total 'Grazing' database entries for *Period Three* suggest. Available aerial photography (1943) in P3 covered little of this book's outer regions (Fig. 2.6) where grazing was widespread.

162. Madden & Muir, 1989: 29.
163. Albeit without a specified location, see: https://collection.maas.museum/object/29680.
164. https://www.belgennyfarm.com.au/history/agriculture-at-camden-estate. See Garran & White (1985) for a critique of the significance of the Macarthurs' claims.
165. Sydney Vignerons Association, 2011.
166. The grove of over 90 olive trees was planted in 1912.
167. Sydney Vignerons Association, 2011.
168. Spearritt, 2000: 23.
169. Spearritt, 2000: 23.
170. Sydney Olympic Park Authority (n.d.).
171. The State Library has a range of images such as here, here, here and here.
172. Sydney Olympic Park Authority (n.d.).
173. See here and here.
174. https://www.canadabay.nsw.gov.au/community/community-services/the-rhodes-peninsula/rhodes-history.
175. See Elias & Coppins, 2013: 112 and https://www.flickr.com/photos/canadabayconnections/5910971542/in/pool-eat_history/, https://www.flickr.com/photos/canadabayconnections/4712764173/in/pool-eat_history/, https://www.flickr.com/photos/canadabayconnections/4705635702/in/pool-eat_history/.
176. https://www.flickr.com/photos/canadabayconnections/5910728191/in/pool-eat_history/.
177. https://www.arnotts.com/our-story.
178. See this Dangar, Gedye & Co Ltd. 1916 catalogue: https://archive.org/details/dangargedyecogeneralcatalogue1916/page/n13/mode/2up.
179. E.g. the Clyde Engineering Co were based in Granville and Australian agents for Fowler products, demonstrating a crane road locomotive and balance plough (made in Leeds, England from 1900 to 1940), see here: https://collection.maas.museum/object/367697.
180. Citing Jeans, 1979, see Pratley & Rowell, 2003: p. 1 of Chap. 1.
181. Sahni, 2019.
182. https://collection.maas.museum/object/375492.
183. https://collection.maas.museum/object/373632.
184. Christie, 1988: 83.
185. Christie, 1988: 83.
186. McIntyre, 2009: 4–5.

187. See photography: https://www.flickr.com/photos/statelibraryofnsw/5935600839/in/album-72157608284326799/.
    https://www.flickr.com/photos/statelibraryofnsw/5936098692/in/album-72157608284326799/.
    https://www.flickr.com/photos/statelibraryofnsw/9909574183/in/album-72157608284326799/.
188. Christie, 1988: 152.
189. Spearritt, 2000: 24.
190. O'Connell, n.d., late 1890s.
191. https://trove.nla.gov.au/newspaper/article/28080709.
192. O'Connell, n.d.: https://australianfoodtimeline.com.au/first-sydney-fish-market/.
193. Unknown location, but possibly Gordons Bay or Thompson's Bay.
194. Clark, 2017.
195. Curby, 1997: 97.
196. Rathbone, 2006: 39.
197. Liston, 2009: 96–97.
198. https://www.flickr.com/photos/mosmanlibrary/6098284301/in/pool-eat_history/.
199. See: https://www.flickr.com/photos/statelibraryofnsw/4066602857/.
200. https://sydneylivingmuseums.com.au/exhibitions/eat-your-history-library.
201. https://sydneylivingmuseums.com.au/exhibitions/eat-your-history-library#item-details-1.
202. https://sydneylivingmuseums.com.au/exhibitions/eat-your-history-library.
203. Some is now displayed online, see https://sydneylivingmuseums.com.au/exhibitions/eat-your-history-library.
204. Maze, 1943: 168.

References

ABARES. (2021). *Snapshot of Australian agriculture 2021*, ABARES insights issue 2 2021, Australian Bureau of Agricultural and Resource Economics and Sciences, Australian Government Department of Agriculture, Fisheries and Forestry. https://www.awe.gov.au/abares/products/insights/snapshot-of-australian-agriculture-2021

ABARES. (2022). Snapshot of Australian agriculture 2022. Australian Bureau of Agricultural and Resource Economics and Sciences, Australian Government Department of Agriculture, Fisheries and Forestry.

Ashton, P., & Waterson, D. (2000). *Sydney takes shape: A history in maps*. Hema.

Aslin, H., Kelson, S., Smith, J., & Lesslie, R. (2004). *Peri-urban landholders and bio-security issues—a scoping study Australian Government*. Bureau of Rural Sciences.

Birmingham, J. (1999). *Leviathan: The unauthorised biography of Sydney*. Knopf.

Bonizzoni, M., Guglielmino, C. R., Smallridge, C. J., Gomulski, M., Malacrida, A. R., & Gasperi, G. (2004). On the origins of medfly invasion and expansion in Australia. *Molecular Ecology, 13*, 3845–3855.

Brown, A. J. (1937). City of the future. *Sydney Morning Herald*, 13.

Burnley, I. (2001). *The impact of immigration on Australia: A demographic approach*. Oxford University Press.

Carson, R. (1962). *Silent spring*. Houghton Mifflin.

Cashin, P., & Mcdermott, J. (2002). Riding on the sheep's back: Examining Australia's dependence on wool exports. *Economic Record, 78*(242), 249–263.

Christie, M. (1988). *The Sydney markets 1788–1988*. Sydney Market Authority.

City Plan. (2014) Report on City of Sydney Industrial & Warehouse Buildings Heritage Study. City Plan: October 2014: H-13070, Accessed 10 April 2020. https://www.google.com/url?sa=t&rct=j&q=&esrc=s&source=web&cd=&ved=2ahUKEwianbmfqL7wAhVg73MBHft6Be4QFjAAegQICBAD&url=https%3A%2F%2Fwww.cityofsydney.nsw.gov.au%2F-%2Fmedia%2Fcorporate%2Ffiles%2F2020-07-migrated%2Ffiles_c-1%2Fcos-warehouses-industrial-buildings-heritage-study-report-final-2-13oct2014-accessible-version.pdf%3Fdownload%3Dtrue&usg=AOvVaw1h7o7uiw0KxM2GAvdoDJfy

Clark, L. (1988). *Technology in Australia, 1788–1988: A condensed history of Australian technological innovation and adaptation during the first two hundred years*. Australian Academy of Technological Sciences and Engineering.

Clark, A. (2017). Plenty of fish in the sea? Not necessarily, as history shows. *The Conversation*. Accessed October 20, 2020. https://theconversation.com/plenty-of-fish-in-the-sea-not-necessarily-as-history-shows-84440

CLS&P. (2004). *Long bay correctional complex conservation management plan*. Clive Lucas Stapleton & Partners. https://csa.intersearch.com.au/csajspui/bitstream/10627/895/1/LongBayCMP04105%20-%20289p.pdf

Code, J. (2018). Alternative agriculture: Innovations for growing and cultivating diverse ways of knowing. In J. Zeunert & T. Waterman (Eds.), *Routledge handbook of landscape and food* (pp. 127–139). Routledge.

Copeland, L. (2020). A century of agriculture in the University of Sydney. *Agricultural Science, 31*, 58–65.

Curby, P. (1997). *Sutherland Shire: Pictorial history*. Kingsclear Books.

Davies, P., & Raymond, L. (2005). *Grantham poultry Research Station conservation management Plan*, prepared by the office of Paul Davies Pty Ltd, September 2005.

Dibden, J., & Cocklin, C. (2005). *Sustainability and change in rural Australia*. University of New South Wales Press.

Elias, J., & Coppins, S. (2013). *Pictorial history Holroyd*. Kingsclear Books.

EPA NSW. (n.d.) Orica botany. 9 September 2022. https://www.epa.nsw.gov.au/working-together/community-engagement/updates-on-issues/orica-botany-bay-incident/orica-botany

Erisman, J., Sutton, M., Galloway, J., Klimont, Z., & Winiwarter, W. (2008). How a century of ammonia synthesis changed the world. *Nature Geoscience, 1*, 637.

Fitzgerald, S. (2008). *Red tape gold scissors: The story of Sydney's Chinese*. Halstead Press.

Freestone, R. (2000). From city improvement to the city beautiful. In S. Hamnett & R. Freestone (Eds.), *The Australian Metropolis: A planning history*. Sydney. Chapter 2.

Garran, J. C., & White, L. (1985). *Merinos, myths and Macarthurs: Australian graziers and their sheep, 1788–1900*. ANU Press.

Gaynor, A. (2006). *Harvest of the suburbs: An environmental history of growing food in Australian cities*. UWA Press.

Gojak, D. (2009). *People in glass houses: the disappearance of 'Crystal Valley', Warriewood, NSW and implications for heritage conservation*. Accessed May 4, 2020. https://www.aicomos.com/wp-content/uploads/2009_Unloved Modern_GojaK_Denis_Crystal_Valley_Slides.pdf

Higgins, L. (2016). The family of old Mans Valley. https://hornsbyshire.recollect.net.au/nodes/view/3053

Hovenden, L. (1981). The motor car in New South Wales 1900–1937, MA Thesis, University of Sydney.

Hunwick, H. (2018). *The food and drink of Sydney: A history*. Rowman & Littlefield.

James, S. (2016). *Farming on the fringe: Peri-urban agriculture, cultural diversity and sustainability in Sydney*. Springer.

Jeans, D. (1972). *An historical geography of New South Wales to 1901*. Reed Education.

Kirkpatrick, A. (2022). Laying the groundwork for soil biodiversity. *Landscape Architecture Australia, 173*, 22–26.

Lewis, M. (1980). Milk, mothers, and infant welfare. In J. Roe (Ed.), *Twentieth century Sydney: Studies in urban & social history* (pp. 193–207). Southwood Press.

Lewis, G. (1984). Million Farms' campaign, NSW 1919-25. *Labour History, 47*, 55–72.

Liston, C. (2009). *Pictorial history: Liverpool & district*. Kingsclear Books.

Madden, B., & Muir, L. (1989). *A history of Earlwood, Undercliffe and Clemton Park, NSW*. Canterbury Municipal Council. https://issuu.com/canterburylib/docs/earlwoods_past

Macleod, V. (2010). Warriewood, 66–67, Viewed 4 May 2020. https://epress.lib.uts.edu.au/index.php/sydney_journal/issue/view/118

Marrickville, U. (2019). Marrickville's last dairyman. Viewed 4 May 2020. http://www.marrickvilleunearthed.com/209-livingstone-rd.html

McGill, J. (2017). *Pictorial history Campbelltown & District*. Kingsclear Books.

McClymont, J. (2013). *Pictorial history: The Hills District*. Kingsclear Books.

McIntyre, J. (2009). Not rich and not British: Philip Schaeffer, 'Failed' colonial farmer [online]. *Journal of Australian Colonial History, 11*, 1–20. https://search.informit.com.au/documentSummary;dn=997044012583652;res=IELHSS

Meats, A. (1989). Bioclimatic potential in fruit flies: Their biology, natural enemies and control. In A. Robinson & G. Hooper (Eds.), *World crop pests* (Vol. 3B, pp. 241–252). Elsevier.

Migration Heritage Centre. (2011). *Happy Valley, Chinese market gardens and migrant camps*. Migration Heritage Centre NSW. Accessed April 20, 2020. https://www.migrationheritage.nsw.gov.au/exhibition/atthebeach/happy-valley/index.html

New South Wales Parliament: Joint Volumes of Papers Presented to the Legislative Council and Legislative Assembly (Parliamentary Papers) (1920) Volume 1, 'Outbreak of Pneumonic Influenza in New South Wales in 1919', Section V, Part I: 153.

Nichols, M. (2015). Fruit growing in the Hawkesbury. http://hawkesburyheritage.blogspot.com/2015/

Nichols, M. (2016). *Pictorial history: Hawkesbury*. Kingsclear Books.

O'Connell, J. (2017). *A timeline of Australian food: From mutton to MasterChef*. New South Publishing.

O'Connell, J. (n.d.) *Australian food history timeline.* Accessed May 12, 2019. https://australianfoodtimeline.com.au/

O'Connell, J. (n.d.) 1893: Mediterranean diet recommended. https://australianfoodtimeline.com.au/1892-mediterranean-diet-recommended/

O'Connell, J. (n.d.) 1901: Griffiths Australian-grown tea: https://australianfoodtimeline.com.au/griffiths-australian-grown-tea/

O'Connell, J. (n.d.) 1903: William Farrer's Federation wheat: https://australianfoodtimeline.com.au/william-farrer-federation-wheat/

O'Connell, J. (n.d.) 1872: First Sydney fish market. https://australianfoodtimeline.com.au/first-sydney-fish-market/.

O'Connell, J. Late (n.d.) 1890: Coolgardie safe invented: https://australianfoodtimeline.com.au/coolgardie-safe-invented/.

O'Connell, J. (n.d.) 1901: A population of farmers. https://australianfoodtimeline.com.au/population-of-farmers/.

O'Connell, J. (n.d.) 1944: World's first organic farming society in Sydney. https://australianfoodtimeline.com.au/1944-worlds-first-organic-farming-society-in-sydney/.

Paull, J. (2012). Walter burley griffin and Marion Mahony griffin, architects of anthroposophy. *Journal of Bio-Dynamics Tasmania, 106*, 20–30.

Paull, J. (2013). A history of the organic agriculture movement in Australia. In B. Mascitelli & A. Lobo (Eds.), *Organics in the global food chain* (pp. 37–61). Connor Court Publishing.

Pratley, J., & Rowell, L. (2003). Evolution of Australian agriculture: From cultivation to no-till. In J. Pratley (Ed.), *Principles of field crop production* (pp. 1–25). Oxford University Press. Accessed February 23, 2020, from https://www.csu.edu.au/__data/assets/pdf_file/0005/2805521/Chapter1_PratleyRowell.pdf

Rathbone, R. (2006). *Cameos of Bexley*. R.W. Rathbone.

Sahni, N. (2019). *History of carlingford to clyde—T6 Rail Line*. City of Parramatta, Parramatta Heritage Centre. Accessed April 20, 2020. https://historyandheritage.cityofparramatta.nsw.gov.au/blog/2019/12/11/history-of-carlingford-to-clyde-t6-rail-line

Smart, A. (2019). *Dairy beloved building*. City news. August 14, https://cityhub.com.au/dairy-beloved-building/

Spearritt, P. (2000). *Sydney's century: A history*. UNSW Press.

Sydney Olympic Park Authority (n.d.). History & heritage: Industrial history. Accessed April 2, 2024. https://www.sopa.nsw.gov.au/learning/history-and-heritage/armory-colonial-geological-and-industrial-history#industrial-history

Sydney Vignerons Association. (2011) Historical timeline: Wine and grape history in the Sydney Region. Accessed April 13, 2020. https://www.hawkesbury.net.au/sydneywine/overview/history.html

TKDA. (2017). *Parramatta north historic sites consolidated conservation management plan*. Tanner Kibble Denton Architects report for UrbanGrowth NSW. https://www.cityofparramatta.nsw.gov.au/sites/council/files/inlinefiles/D04731377%2D%2DPNHS-CMP_Part-B_Parramatta-Gaol.pdf

Wills, N. R. (1945). The rural-urban fringe: Some agricultural characteristics. *Australian Geographer, 5*, 29–35.

Willis, A. M. (2007). From peri-urban to unknown territory. *Design Philosophy Papers, 5*, 79–90.

Wilkinson, J. (1999) *Dairy industry in NSW: Past and present*. Briefing Paper No 23/99, NSW Parliamentary Library Research Service. Accessed July 15, 2020. https://www.parliament.nsw.gov.au/researchpapers/Documents/dairy-industry-in-nsw-past-and-present/23-99.pdf

Winston, D. (1957). *Sydney's great experiment: The progress of the cumberland county plan*. Halstead Press.

Wolf, S. (2014). *The neoliberal regime in the agri-food sector: Crisis, resilience and restructuring*. Routledge.

Zeunert, J. (2017). *Landscape architecture and environmental sustainability: Creating positive change through design*. Bloomsbury.

Zeunert, J., & Freestone, R. (2022). From rural lands to agribusiness precincts: agriculture in metropolitan Sydney 1948-2018. In C. Brisotto & F. Lemes de Oliveira (Eds.), *Re-imaging resilient food landscapes—perspectives from planning history* (pp. 247–272). Springer.

Zeunert, J. (2024). Bridging rural and urban disconnections: Spatial graphic explorations of Australia's livestock landscapes. In R. Pernice & B. Chen (Eds.), *Urban regeneration and rural revitalization in Australia and China* (pp. 110–143). Routledge.

6

Period Four: 1946–1999

Box 6.1 Population 1946 and 1999

Population 1946: c. 1.7 million
Population 1999: 4 million
Increase = 2.4-fold (2.3 million)

J. Zeunert, A. Daroy, *Sydney's Food Landscapes*,
https://doi.org/10.1007/978-981-96-0710-5_6

Fig. 6.1 Glenorie, 1970. Source: Spatial Services NSW

Box 6.2 Key Events 1946–1999

- Population boom following World War II creating metropolitan growth;
- Motorised transport's domination, particularly cars, shaping city planning and urban form;
- The Green (Agricultural) Revolution, resulting in larger production scales, reduced human labour and more synthetic chemical inputs;
- Technological advancement, including space exploration and commercial aviation travel;
- The Cold War and nuclear proliferation;
- The emergence of modern environmentalism and environmental legislation;
- Relative eradication of disease;
- The end of the *Immigration Restriction Act* (White Australia Policy) in 1958 with its last remaining policy components removed in 1973;
- Shifting gender and social dynamics aided by feminism's rise and the contraceptive pill's release in Australia in 1961;
- The emergence of neoliberalism, dissolving tariffs and subscribing to a free trade ideology, particularly in the 1980s and 1990s;
- A shift from analogue to digital technologies, alongside the emergence of computers and the internet.

Box 6.3 Key Events in Sydney 1946–1999

- The development of five metropolitan-scale planning schemes;
- A property boom in the 1960s and early 1970s, supported by the *Strata Title Act* of 1961, the rise of apartments and implementation of 'rubber stamp' planning processes;
- 'Green bans' in the 1970s, namely, construction union resistance to socially and environmentally unsustainable projects;
- The growth of manufacturing industries followed by significant decline;
- The shift from manufacturing to a knowledge- and service-based economy;
- Continued population growth and spatial expansion of the urban footprint;
- City building towards the Olympic Games in 2000.

Box 6.4 Sydney's Metropolitan Agriculture Trends 1946–1999

- Metropolitan agriculture and total agricultural land decimated by urban sprawl;
- Continuing technological development replacing human labour with fossil-fuel-powered machinery;
- Widespread normalisation of synthetic applications (fertilisers, pesticides and herbicides);
- Significant cultural diversification of foodways due to immigration;
- Substantial contraction in commercial metropolitan fruit production;
- The loss of prior market gardens due to urban growth;[1]
- Further decline of metropolitan dairies;
- Livestock intensification and the shift to factory farming especially in poultry farming;
- Growth and intensification of the mushroom industry;
- Relocation of agriculture to, and beyond, the ever-creeping peri-urban fringe;
- Food production increasingly replaced by ornamental horticulture, particularly turf, cut flowers and within nurseries;
- Supermarkets' emergence and growth, diminishing local food production and retail, subsequent upscaling of supply chains, logistics networks and roads;
- Conflicts between agricultural production (farming) and processing activities (e.g. abattoirs) with encroaching land uses (residential, commercial);
- The prioritisation of recreational and sporting landscapes and restoration ecology[2] in planning and public domain design over agricultural landscapes.

After the end of World War II in 1946, Sydney was home to 1.7 million people. By 1999, the population had increased by nearly two-and-a-half times to 4 million.[3] This surge precipitated unparalleled urban expansion and growth, emblematic of a broader global trend. Sydney's environmental footprint soared. The metropolis continued to decentralise, sprawling outwards, facilitated by widespread car ownership and the availability of cheap and abundant crude oil. *Period Four* (1946–1999) was therefore a time of substantial socioeconomic change. The era was characterised by elevated living standards, the advancement of women's liberation, rapidly accelerating consumer culture and a discernible transition towards a white-collar, knowledge-based economy. Against this backdrop of economic prosperity, however, an increasingly conspicuous spectre of pollution, ecological degradation and species extinction emerged. This phenomenon was accompanied by increased global awareness of anthropogenic impacts, set against the looming threat of nuclear war throughout the first decades, gripping collective consciousness. Remarkable technological gains and prosperity were hence tempered by a shadow awareness of humans' capacity for mass destruction, bringing into question degradative modes of production and consumption.[4]

Official town and urban planning emerged, seeking to manage economic and population growth manifest in metropolitan composition and form. The city's planning arc, though, would shift from aspirational ideals and visions evident in the County Cumberland Council's (CCC) 1948 Planning Scheme (CCPS), to normalising the discipline's alignment with neoliberal government agendas of rapid economic growth.[5] In his 1957 book *Sydney's Great Experiment*, the first Chair and Professor in Town and Country Planning at the University of Sydney,[6] Denis Winston, warned that,

> The surroundings of Manchester and Pittsburgh and the suburbs of Birmingham and Chicago give us a full preview of the County of Cumberland in twenty years' time *if we don't succeed in controlling development instead of letting developments control us* (our emphasis).[7]

Winston further noted that,

> Australia has nearly three million square miles of territory, but really good growing land with adequate rainfall makes up a very small percentage of the total: where such favourable land occurs near a large population centre it is of double importance, for only from such areas can really fresh vegetables come to city markets [..] the first essential is to preserve the land where they can be grown: to raise bungalows on first class agricultural land may be profitable for the "developers" but is a serious loss for everyone else.[8]

Winston's warning would fall on deaf ears. In contrast, the 1968 Metropolitan Plan for Sydney stated that "Sydney [...] has added another million to its population in the last twenty years, and a further growth of about two and three-quarter millions is expected over the next thirty years".[9] The city's agriculture during *Period Four* is hence characterised by its erasure and usurpation by other land uses, predominantly suburban.

Sydney experienced a devastating loss of commercial agricultural production, shifting from being a productive local food powerhouse to becoming a bastion of sprawling commercial real estate development. The key production types defying decline were the intensive and spatially concentrated practices of poultry raising and mushroom farming—both chiefly in sheds and not aligned with conventional notions of agricultural landscapes. Metropolitan farmland continued to be lost to urban development but was also caused by more lucrative forms of non-edible 'amenity' production, such as turf farming and ornamental horticulture. This mirrored Sydney's shifting demographics as more dual income households used their disposable incomes to move away from supplementary food gardens, embracing consumer culture and non-productive ornamental gardens seen as higher status.[10]

Relevant Global and National Events: 1946–1999

World War II's effects continued to be felt years after its official conclusion in 1945. The extent of social upheaval cannot be overstated, especially for ex-service men and women, displaced people and families affected by trauma. Food, petrol and other rationing were widely felt and did not completely end in Australia until 1950.[11] Agriculture provided a significant transitional market for wartime chemical producers. Dichlorodiphenyl-trichloroethane (DDT), for example, extensively used by the United States to eliminate wartime pathogens, was repurposed as an agricultural insecticide. Other synthetic pesticides were also central agents for the period's 'Green Revolution' in agriculture (c. 1950–1984) that hugely increased global food production through high-yield crops.[12] The movement was facilitated by global agricultural systems increasing their use of synthetic inputs through fertilisers, pesticides and amendments. Mechanised technologies, such as tractors, trucks and crop-dusting planes, assisted their distribution as well as enabled harvesting, processing and distribution. Subsequent increased agricultural yields and food availability were unprecedented in human history. This evolution in technology and manufacturing was an ecological 'devolution' facilitating ecocide, as Rachel Carson illuminated in her landmark 1962 work, *'Silent Spring'*, which catalysed modern environmentalism.

The progressive replacement of human and animal labour with finite fossil fuel-powered machinery via diesel and petroleum (including tractors, utility vehicles ('utes'), harvesters and trucks) proliferated. Producers were able to hugely increase agricultural production scales with less labour than ever before.[13] A global abundance of crude oil made its by-products affordable. Food systems' interdependence with fossil fuel energy sources became enmeshed. Fossil fuel not only propelled movement but came to span all facets of food production, processing and distribution.[14] Expanding rural agricultural production scales meant that smaller metropolitan land parcels fell increasingly out of step with the industrialised food system. Federal agricultural philosophy embraced 'competitive productivism', which equated to 'get big or get out'.[15] Thousands of Sydney's metropolitan agricultural operations dissolved over the period, in part due to improved transport and logistical infrastructures that trucked, shipped and flew in produce from across NSW, Australia and the globe. Initially a novelty,[16] by the end of *Period Four* supermarkets had been normalised, along with the strengthening duopoly of retailers Coles and Woolworths. Supermarkets and the growing prevalence of chain stores preferred larger suppliers, driving further expansion of agricultural production scales and farm agglomerations made possible by fossil fuels and large machinery. Many smaller growers struggled and collapsed. Larger production scales proved especially problematic for Sydney's agriculture given the city's high land values and smaller parcel sizes than Australia's rural production settings.

The period was also significant for its social and cultural revolutions. Following Carson, modern environmentalism emerged globally in the 1960s. The movement was, perhaps surprisingly, assisted by space exploration, particularly its photographs such as 'Earthrise' (1968) and 'The Blue Marble' (1972) that for the first time captured the planet as a holistic entity in an abyss of deep space. Along with works including James Lovelock and Lynn Margulis' 'Gaia Hypothesis' (1974), this assisted recognition of Earth as an interconnected ecological system. Much of *Period Four*'s environmentalism focused on ecosystem and species conservation, which later shifted to global warming concerns and 'sustainable development' (see *Period Five*). Agricultural and food production impact, as well as dietary choices, were not yet well connected within the period's environmentalism, other than within the animal rights movement.[17] Protection of remnant ecosystems, however, and ecological restoration as a later impetus for their reinstatement, grew from grass-roots origins to gain momentum. Urban restoration ecology sought to re-establish native plant species and vegetation communities, as distinct from implementing First Nations' Aboriginal land management underpinned by Indigenous knowledge systems.[18] At times, this concern with reinstating indigenous plants conflicted with metropolitan agriculture and found greater widespread acceptance.

Period Four also saw a significant decline in egalitarian social aims and rise in neoliberal capitalism. The period's last 15 years particularly witnessed the emergence of neoliberal ideologies within governance, as championed by Ronald Reagan in the US and Margaret Thatcher in the UK. One of neoliberalism's key ideologies is the determination of outcomes by a supposedly 'free market' with minimal government intervention. Typically, the government takes a passive stance towards its constituents (*consumers* rather than citizens) and an active interest in corporations influencing the 'free market' through political donations. Previous state assets were widely privatised. Several relevant to agriculture of a much wider array in NSW include: NSW Egg Corporation; Government Insurance Office (GIO); NSW Grain Corp; a range of NSW and Sydney berths and ports; the Fish Marketing Authority; State Bank of NSW; Sydney Market Authority and numerous Irrigation Districts.[19] Since neoliberalism is geared to individual interest, it reinforced agriculture's 'get big or get out' approach that proved increasingly problematic for smaller, independent growers and cooperatives who were unable to financially compete. Neoliberalism accelerated in the 1990s and subsequently received bipartisan support, continuing to be embraced to the present day (see *Period Five* and *2021+*).

Relevant Sydney Events: 1946–1999

In 1946, there was a range of both remnant food production districts and diffuse growing operations scattered across Sydney (Figs. 6.11, and 6.12). By 1999, a significant proportion of these operations were erased by urban development, which consisted of manufacturing and light industry, commercial zones and chiefly residential suburban sprawl. In 1967, Rutherford et al. noted Sydney's "rolling wave" of unplanned development in its hinterland, whereby agricultural productivity was intentionally downplayed through strategic low-productivity agricultural land uses aiding urban land development.[20] Figure 2.7 depicts the expansion of Sydney's urban form and suggests agricultural production landscapes eliminated in the process. While some metropolitan agricultural operations relocated further afield to cheaper land, many producers sold their land, lost their leases, left the industry or retired, often without intergenerational continuance. Consequently, agricultural knowledge declined—an ongoing phenomenon in Western contexts. Local fresh food supply security continually decreased as the city's inhabitants relied more and more on industrial agricultural products transported from further afield. Ironically, this phenomenon coincided with the onset of Sydney's official and legislated planning. *Period Four* therefore reveals an intertwined narrative of planning and speculative land rezoning for urban development, routinely by building on previously agricultural production lands.

Sydney's Metropolitan Planning: 1946–1999

Sydney produced five metropolitan planning strategies over the course of the period, though the first two spanned four decades until 1988. These were the *County Cumberland Council Planning Scheme (1948)*, *Sydney Region Outline Plan (1968)*, *Sydney into its Third Century (1988)*, *Cities for the 21st Century (1995)* and *Shaping Our Cities (1998)*. The strategies are particularly significant to Sydney's food landscape history, as previously no official spatial and land use mechanisms formally existed to guide the city's trajectory. We focus on the first as it is the only plan significantly emphasising agriculture's value to the city.

Legislation was first required to create a body with the authority to devise metropolitan-scale guidance. Despite the fact that town planning advocates had been arguing for greater control and coordination of market-driven development, state planning legislation was only passed in 1945. Legislated planning was henceforth enshrined in state and local government land use statutes. This enabled local authorities to prepare zoning schemes across the state stipulating acceptable uses for given zones and land parcels, such as rural, industrial, residential and commercial, along with permitted and designated classification subsets.

A new body called the Cumberland County Council (CCC) was established in 1945 to prepare and oversee a legally binding metropolitan plan for Sydney. The plan intended to establish a vision for the city's trajectory, which local government plans would then implement in a more unified manner and at a finer grain. Sydney's first metropolitan plan covered the extent of the County of Cumberland, an area of 1650-square miles (4275-square kilometres, or 80 per cent of this book's considered area) (Fig. 2.8). This contained one quarter of the country's population, rendering the plan hugely significant at a national scale.[21] Before the CCC could develop their metropolitan strategy, they conducted considerable analysis of the city's makeup. In 1945, they published Sydney's first metropolitan-scale analysis map capturing the city's composition and land uses.[22] In 1948, the CCC then released *The Planning Scheme for the County of Cumberland New South Wales* (CCPS). This included an invaluable statistical[23] and spatial[24] analysis of Sydney's agriculture in the 1946–1947 season (Fig. 6.2).

CCC Analysis

The analysis clearly reveals that bulk of production occurred on the (then) outer rim and peri-urban fringes, unsurprisingly shrinking towards central Sydney's congested and contested urban makeup.[25] In a data table, Sydney's total horticultural area was calculated as 38,581 acres (15,613 hectares).[26] Cropping, including for fodder, was the most spatially expansive horticultural land use at 17,275 acres (6991 hectares), followed by market garden vegetable production at 8815 acres (3567-hectares), citrus fruits at 6718 acres (2719 hectares), non-citrus fruit at 3434 acres (1390 hectares), with the evidently popular passionfruit receiving its own category at 459 acres (186 hectares), grapes at 1174 acres (475 hectares), nurseries at 670 acres (271 hectares) and finally other usage totalling 36 acres (15 hectares).

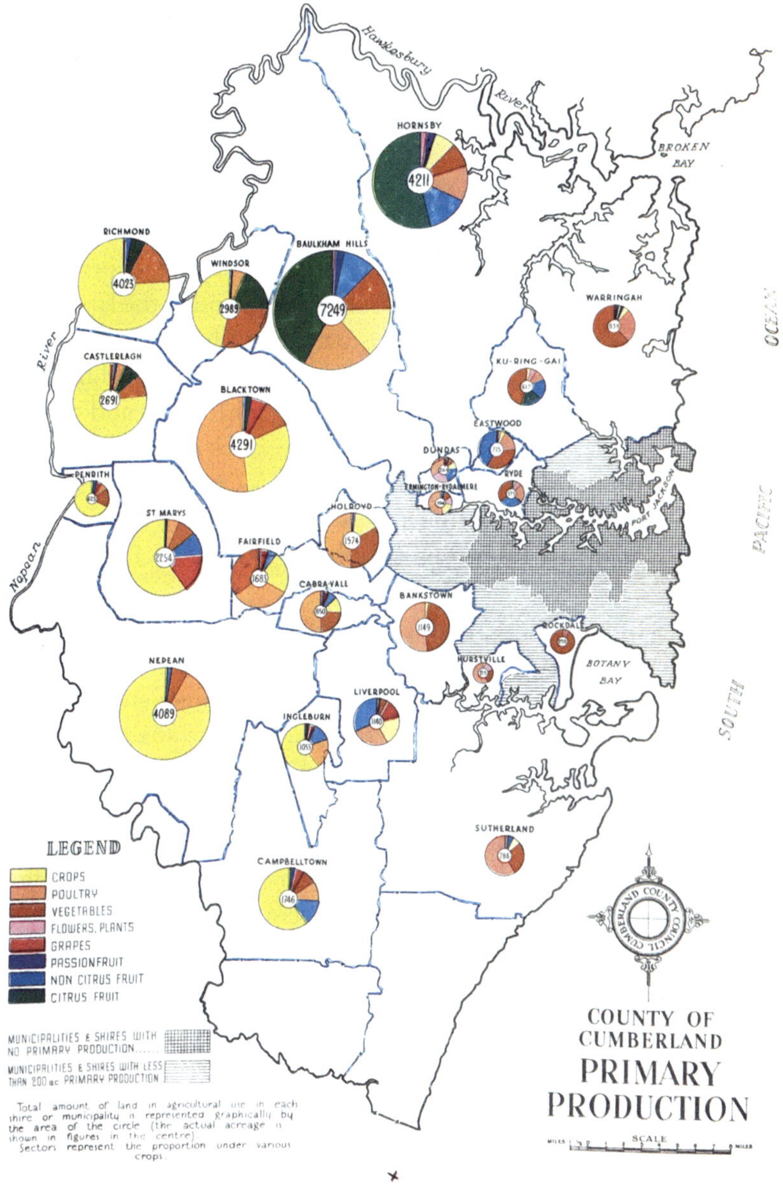

Fig. 6.2 Primary production in the County of Cumberland, 1946–1947 season. Source: Cumberland County Council, 1948: 125

We can also get a sense of the city's major production hubs. Baulkham Hills Local Government Area (LGA) had the largest area of cultivated horticultural land at just over 6000 acres (2428 hectares), followed by Richmond, Hornsby and the Nepean Shire, which all exceeded 3000 acres (1214 hectares). Castlereagh, Windsor, St Marys and Blacktown had more than 2000 acres (809 hectares) of cultivated horticultural land.[27]

The County contained over 70 per cent of New South Wales' poultry farms and was responsible for 18 per cent of Sydney's milk production.[28] Grazing livestock on unimproved pastures commanded the largest area of agricultural land use (Fig. 1.2). Livestock are excluded from map data, except for poultry,[29] and no livestock production is included in the table data.[30] Acreage for grazing cattle, sheep, goats, or for pigs and dairying is not accounted for. Fodder crops grown for livestock are captured, helping to explain why 'crops' are recorded as Sydney's largest 'primary production' land use. If livestock lands had been included, however, the County's agriculture would have been considerably larger than the 15,613-hectares specified.[31]

Despite its limitations, such a comprehensive spatial-statistical city-level overview of agricultural production had not occurred since Fletcher's appraisal for 1788–1821,[32] nor have any other subsequent district- and typology-level appraisals come close to its detail. The analysis recognised agriculture's meta-economic value, beyond short-term measures ultimately diminishing its perpetual value. It noted the significance of the 10,300 County of Cumberland agricultural workers 1948 as "immense" and "beyond their [recorded] numbers".[33] An employment survey further reveals that over 34,000 factory workers were tied to 'food' production, the third largest employment category in manufacturing.[34] In hindsight, the CCC analysis reveals Sydney's key food production districts substantially declining.

Sydney's First Metropolitan Planning Scheme

Following their analysis, the CCC released Sydney's first metropolitan-scale planning vision, *The Planning Scheme for the County of Cumberland New South Wales* (CCPS) in 1948. The strategy contained metropolitan-scale land-use zoning recommendations and a major roads scheme. It proposed new suburban and rural district centres, inner-city renewal and aspired to acquire considerable areas of open space. Most notably for this book, it specified a definitive town-country boundary intended to prevent urban sprawl and conserve agricultural and undeveloped lands. The rural lands' inner arc was dubbed the 'green belt', which became the CCPS' most renowned and popular aspect for its strong green vision. The scheme was outlined in a substantial report,[35] containing a photographic record that included several key elements of Sydney's agriculture.[36] Some agricultural regions were specifically identified as threatened by urban expansion, such as Cabramatta and Lurnea in the south-west, or their value was highlighted, as in the case of Eastwood in the north. We have redrawn the CCPS for increased legibility and clarity (Fig. 6.3).[37]

The CCPS was announced in 1951. Minister for Local Government, and soon to be NSW Premier, John Joseph Cahill declared in State Parliament,

> The rural areas of the country will be preserved not only for vital food production and as a countryside to serve the metropolis, but also for soil conservation, irrigation, afforestation and the development of towns and villages with services, amenities and opportunities for local employment. A 'green belt' will be established around the urban development of the metropolitan area, the principal function of which will be to preserve a distinct, permanent line between town and country and to retain some portion of the county at a higher standard of rural development within reasonable reach of all. It will generally take the form of farms, forests and rural institutions.[38]

Cahill's remarks demonstrate an awareness of cities' interdependencies with local life-supporting landscapes, which subsequently declined in mainstream Australian politics and indeed within many facets of the built environment professions. Both politics and the profession supported urban growth and land development requiring long supply chains over local production. This is perhaps surprising given modern environmentalism's emergence in the 1960s, but illustrates the incredible stability

Fig. 6.3 Sydney's first metropolitan planning scheme, the Planning Scheme for the County of Cumberland New South Wales (CCPS), 1948. Source: Redrawn from Cumberland County Council (1948) by Josh Gowers, Stephanie Stankiewicz and Joshua Zeunert

and economic growth that followed World War II's scarcity. This led to a disregard for Sydney's food landscapes and the importance of healthy ecosystem functions vital for sustaining ongoing productivity, including soil conservation and afforestation. Local food production was instead overridden by economic primacy and industrial-scale agribusiness in distant shadow landscapes. The CCPS devoted considerable space to the importance of conserving rural lands. Unfortunately, this would prove an anomaly in all Sydney's eight ensuing metropolitan plans.

Winston's account of the CCPS reiterates the importance of supporting metropolitan agriculture through regional planning. He believed that all Sydney's agricultural "activities need protection against the advance of urban development".[39] He emphasised metropolitan agriculture's role in supplying the city with nutritious fresh food, as well as its economic value to the state. One of his most useful inclusions is the metropolitan map summary (re-drawn in Fig. 1.2). The map classified three agricultural land uses: 'orcharding', 'pastoral' and 'intensive farming (poultry, vegetables, glasshouses)', although he did not quantify total land areas for each.[40] Based on Winston's map, we have calculated these areas as approximately: 70 square kilometres for orcharding, 240 square kilometres for pastoral and 260 square kilometres for intensive farming. This correlates with grazing constituting Sydney's most spatially extensive agricultural land use. It also provides a (very approximate) figure of 570 square kilometres for the city's overall agricultural area, mid-century. This is considerably larger (nearly 3.7 times) than the CCPS' horticultural-only measure totalling 156 square kilometres.[41]

Winston's study reveals that our current food system's 'just-in-time' logistics and food-mile-laden products were not yet feasible for many perishable and fresh foods in the 1940s and 1950s.[42] Local cultivation was advantageous and often necessary for fresh vegetables, herbs and soft fruit supply. Post-harvest techniques such as cold storage, refrigerated trucking, shipping and aviation and long food chain retail were either less available, affordable or developed than today.

The CCPS nominated rural areas to be "primarily devoted to the production of the County's fresh food" and to be declared "as essential to the County as its industry and its port".[43] The green belt was a more accessible, high-quality inner rural zone and a front-line defence against "promiscuous urbanization"[44] that was already a phenomenon only three years following the War. The CCC further recognised the need for the "stabilisation of rural land use and tenure", particularly for improving land management practices.[45] It proposed a "rural district system" that would be "defined mainly by farming activity" and focused on re-planned townships as employment hubs and to define resilient communities of interest.[46] The primary development control was a proposed minimum subdivision size of two and a half acres, considered to be the "smallest area from which a livelihood could be obtained by full agricultural development" to "discourage ribbon development along arterial roads".[47]

CCPS Shortcomings

While the CCPS' aspirations were progressive and welcome, the regulations set in place were flawed and ultimately ineffectual. A defeatist tone for agricultural land planning is palpable,

> The removal of market gardens from inner areas [...] and the threat to the existence of gardens [...] mean the progressive abandonment of the most fertile soils to other use [...] most of the suitable farming lands eventually will be given over to other uses unless the trend is checked.[48]

The CCPS did not outline a strategy to check the loss of pre-existing agriculture outside of their green belt. Thus, their pessimistic forecast came to pass, which both was and continues reveal developers' power in the face of a legislated metropolitan planning body.

The CCC subdivision control was recognised early on as manifestly inadequate. The minimum rural landholding was doubled in area when the scheme became law in 1951 on the recommendation of senior state government planner Norman Weekes, reportedly based on the standards used by the Water Conservation and Irrigation Commission in the Murrumbidgee Irrigation Area in southern NSW.[49] While the CCC recognised metropolitan agriculture overall as well as various existing areas under threat of development, it made no nuanced provisions outside of green belt and rural area zonings to zone or protect high-value agricultural lands from loss to urban development. There were no specific initiatives targeting rural tenure beyond zoning, which worked to some degree in the rural lands.[50] The CCPS' rural district system seeking the reinvigoration of rural centres for employment activities to counter commuting to the central city was stillborn. The green belt, the scheme's iconic planning proposal, was progressively eroded by ministerially approved land releases under pressure from both public and private developers seeking sizeable broadacre land for 'affordable' housing.

The CCPS was sprinkled with references to the idealistic garden city. Its richly illustrated format channelled Patrick Abercrombie's *Greater London Plan* (1944), with its green belt and prospect of new satellite towns against a rural backdrop to absorb long-term urban growth. Constraining growth, recentralising activity and utilising regional green space to articulate a spatial framework reflected the conventional wisdom in Australian planning circles at this time,[51] for example, A.J. Brown's 1937 plan of *Period Three*. The CCPS was thus characteristic of its era, firmly entrenched in the British town-and-country planning tradition. In World War II's aftermath, early planning was profoundly shaped by British methodologies that prioritised containing large urban centres within greenbelts and satellite towns.[52] Containment policies' formulas were not only motivated by cultural heritage and aesthetics of the English countryside but also influenced by the War's rationing and need to preserve agricultural land.[53]

Broader colonial cultural influences also undermined the CCPS' agricultural elements. Urban containment to conserve farmland was less a driving policy force in Australia than in Britain. This can be partly attributed to the UK's significantly larger population and comparatively small land area. Additionally, factors such as food rationing, pressures on food supply lines, dependence on food imports, constraints on spatial availability of per capita agricultural land and widespread participation in Victory Gardening initiatives were all more pronounced in the UK than in Australia. Despite having a small proportion of arable land, Australia continued to perceive available agricultural land as being in extensive supply. British planning notions of containment and countryside protection were less compelling in Australia and agricultural land conservation in Sydney became fringe and sporadic. Cultural aspirations were redirected from a British colonial preference for inland pastoral land ownership to Australia's coastward orientation exemplified by Sydney's prestigious harbour and metropolitan beaches.[54]

From a planning perspective, the sprawling city represented much more than a threat to agricultural land. The CCPS' proposed green belt was built on good intentions and a sound precedent in England[55] as a panacea for controlling urban growth, providing green space for recreation and peri-urban agriculture. It was nevertheless overly simplistic[56] and allowed the "aesthetic argument" to "prevail over the agricultural".[57] Metropolitan agriculture was relegated to amenity traits in the manner of 'countryside' or 'rural character' rather than essential food sources and food supply security. The green belt disregarded many existing and established agricultural operations in Sydney's interstitial spaces, such as in Botany, Mascot, Alexandria, St George and Brookvale, which would be replaced by industrial and residential uses.[58] Sydney's pre-existing agricultural patchwork had painstaking evolved since 1788 through considerable trial and error in response to suitability to local conditions and microclimates. The CCPS green belt's over-simplified arcing-shape did not functionally address the inherent land and soil capacity for agriculture and its dendritic and patchwork nature that followed alluvial creek and river valleys and pockets of superior soils (Figs. 2.9 and 2.12, bottom). Furthermore, many agricultural areas were associated with particular cultural groups, especially with the influx of refugees and immigrants immediately after World War II, and planning policy did not recognise the significance and strength of their social ecosystems.[59]

The CCPS' shortcomings have been more widely replicated. The scheme represented problematic 'heroic' gestures rife in twentieth-century planning and architecture. Celebrated Swiss-French architect Le Corbusier's influential urban visions, including Ville Radieuse, were a telling example, as adopted by Sydney's influential architect Harry Seidler. Seidler, for instance, proposed razing McMahons Point to erect repetitive arrays of brutalist towers over the undulating topography—mercifully only one was realised. Such post-war dense apartment buildings have become synonymous with Redfern's 'The Block', Melbourne's inner-north towers, the UK's 'council housing estates' and North America's 'projects'. Metropolitan-scale planning is a delicate process requiring a clear and tangible strategy yet needs to concurrently account for nuances such as topography, mixed land-uses, heritage and character and a multitude of other values. The CCPS, like others in its wake, was not aided by a singular drawing at a metropolitan scale conveying an entire city's land-use strategy, nor by the overuse of single land-use zonings.

Not surprisingly, then, the CCPS proved a more satisfactory urban rather than rural plan. The CCC Chairman in the late 1950s declared that "development of the Rural Area Zone should be as much a matter for positive planning as the development of living and industrial areas".[60] His survey of conditions in outer Sydney, however, identified various problems. These included large grazing properties "somewhat neglected and not used to anything like their potential", effectively demonstrating characteristics of speculative land ownership awaiting higher commercial development.[61] Similarly, many smaller agricultural lots were "only partly occupied with little or no rural use and with roads and services in a rudimentary condition". The response was to progressively increase the minimum property holding in rural areas to 40 hectares by the 1970s. By then, the damage had been done. Rural land was fragmented by speculative pressures, the green belt was no more and indeed the CCC itself was long gone.

When it came time for a serious review in the late 1960s ahead of a new metropolitan strategy in 1968, preserving "much of [Sydney's] rural character" was finally acknowledged.[62] Focusing on character, however, failed to consider economic viability for Sydney's primary producers, at a time when economic pressures from agriculture's Green Revolution were in full swing. This was a tipping point for the city's metropolitan planning, whereby agriculture's image took precedence over its operational continuance (see *Period Five, MF*).[63] At the same time, the review was troubled by the lack of robust and varied employment in Sydney's outer region and stressed the need for future planning to bear in mind the "necessity for encroachments to meet the needs of urban expansion".[64] The twin concerns of growing environmental awareness and development constraints became major strategic planning themes over the next four decades. In the 1968 and subsequent strategies, agricultural considerations became virtually non-existent (Table 6.1).

Table 6.1 Sydney's metropolitan plans and their references to agriculture 1946–1999 (With thanks to Robert Freestone. See Zeunert & Freestone, 2022)

P4 Metro Strategy	Year	Core strategy	Rural territory	Agricultural and rural land elements
County Cumberland Planning Scheme	1948	Metropolitan-scale land-use zoning, major roads scheme, new suburban and rural district centres, inner city renewal, open space acquisitions, town-country boundary, green belt.	Rural lands, green belt	Rural lands zone. Green belt zone. Overall strategic recognition of the value of preserving metropolitan agricultural lands and activities; few statutory means proposed to achieve it.
Sydney Region Outline Plan	1968	Linear growth corridor and green wedge strategy to replace green belt and satellite towns	'Non-urban' zone	Agricultural industry marginalised as of low economic importance to the regional metropolitan economy. References only fleetingly made in relation to rural regions and their status relating to development and land-use change.
Sydney into its Third Century	1988	Growth corridors, urban infill and environmental quality	Non-urban areas	High-quality agricultural lands to be kept in productive use for as long as possible for the economic benefit of the Region but pressure from greenfield land releases on the urban fringe.
Cities for the Twenty-First Century	1995	Integrated urban management	Non-urban areas	As for 1988 but with large western Sydney land tracts mapped as 'potential future urban'.
Shaping our Cities	1998	A 'peak plan' combining government directions and directives	Recognises protected agriculture and other rural production	A companion 'Shaping Western Sydney' strategy supported implementation of the 'Strategic Plan for Sustainable Agriculture—Sydney Region (1998) from NSW Agriculture.

Best of the Rest

Despite the CCPS' shortcomings, it offered the most serious consideration for metropolitan agriculture when compared to subsequent plans. The 1948 scheme gave strategic but ineffective statutory consideration, but by the end of *Period Four* metropolitan agriculture had neither strategic nor statutory teeth. At best, later plans would pay lip service or at worst, completely disregard agriculture. Most plans typically included a few token statements ultimately proving hollow (Table 6.1, *Then and Now* and *2021+*).

Table 6.1 summarises the period's five metropolitan planning schemes and their inclusion of metropolitan agriculture and rural land elements. As evidenced in the table, agriculture's relevance within Sydney's metropolitan planning declined in significance and perceived value from the 1948 scheme onwards.

Paradoxically, given its founding aim to control 'market-driven development', in its inaugural period legislated planning simply facilitated it, through its own channels, and in its own image. Post the CCC and CCPS, metropolitan planning routinely acted as a mere 'rubber stamp' for urban development. The absurdity of Sydney's urban planning in 1969 and 1970 saw projects worth hundreds of millions approved in two to three minutes by a City Planning and Improvement Committee.[65] Historian Peter Spearritt wrote at the turn of the new millennium that a "generation of town planners produced the magnificent County of Cumberland Plan, but since the 1980s most planning documents are badly written public relations exercises".[66]

Rezoning Agricultural Land

Agriculture further descended Sydney's planning hierarchy as a metropolitan land use. Lands usurped from their original custodians, cleared of tough vegetation by convict labour smoothed into fields and pastures facilitated an easy transition to built form. Sydney's agriculture was, and remains, commonly viewed as "land awaiting higher economic development"[67] by planners as well as agricultural and real estate speculators.

This brings us to the phenomenon of rezoning agricultural lands for urban development, which usually takes place on the peri-urban fringe. Rezoning agricultural land is a two-way process. It usually requires the landowner to be willing to sell their land or, in rare cases, to be compulsorily acquired, such as for state-level infrastructure projects.

Metropolitan agriculturalists continue to utilise this as a speculative opportunity for urban development. Post World War II farmers became particularly savvy to potential gains,[68] as highlighted by Lebanese-Australian market gardener Joe Elbustani,

> We think the government should aim to keep a zone for agriculture but I wouldn't want my farm to be zoned that way [...] This area, it will be 10 or 20 years maybe but it is a matter of time before it becomes a suburb [...] that's part of our investment actually.[69]

As such, Sydney's metropolitan farmers at times undermined their own industry's viability, and in some cases, transitioned to become land speculators and developers. It is unsurprising that metropolitan agricultural land continues to decline given an influential component of the industry supports its decline and transition. Unsurprisingly, then, the NSW Department of Agriculture observed in 1984 that there was a diminishing quantity of high-quality agricultural land remaining within Greater Sydney.[70]

Other Factors

There are several important meta-trends relevant to metropolitan agriculture outside of planning. These include Sydney's green movements, the composition of its labour market and city infrastructures, especially for motorised transport.

In *Period Four*, Sydney's green movements possessed various strings to their bow, several of which indirectly related to metropolitan agriculture. The first concerned conservation, namely, demanding greater protection for remnant ecosystems in park and reserve systems. This movement was an evolution of activities dating back to *Period Two*[71] but gaining noteworthy strength in *Period Four*.[72] The second saw environmentalism and environmental activism, such as protest movements, decrying pollution and degradative activities. In the 1960s, significant environmental fallout existed from a bevy of polluting industries and practices.[73] By the 1970s, environmental action against polluting companies was ramping up. Part of this stemmed from Sydney's exploding population and its westward advance into industrial areas, resulting in conflicting now adjacent land uses. Rhodes' train line, for example, was once a sharp division between an industrial zone on one side and booming residential land use on the other.[74]

Urban-planning environmental activism related to the 'green bans' of the 1970s. During the Bans, NSW Builders Labourers' Federation workers selectively refused to work on projects they considered environmentally, socially or aesthetically undesirable.[75] They took a stance against poor-quality urban development projects and, collectively, against Sydney's out-of-control and corruption-prone planning system.[76] The green bans would be one of the last significant 'mainstream' Sydney hallmarks demonstrating civic values against normalisation of capitalist-cum-neoliberal processes in urban development (see *Period Five* and *2021+*). Green bans helped to shape Sydney's urban environmental movement, as well as legislation such as the 1979 Environmental Planning and Assessment Act.[77] Sydney's green movements of the period demonstrate parallels with metropolitan agriculture in the sense that unchecked urban development and industry were adversely impacting and eroding elements of greater value to the wider community.

Period Four also saw the peak and decline of Sydney's substantial industrial sector, mirroring a national trend. In 1947, more Australians worked in manufacturing than in farms or mines.[78] Less than three decades later, a notable decline had occurred in Sydney's manufacturing and industrial production from the 1970s, including within metropolitan food production.[79] Despite this, industrial land uses displaced pre-existing agriculture, such as in greater Botany, Brookvale and numerous areas adjoining Prospect Reservoir (Fig. 5.8, bottom).[80] By the end of the period, knowledge- and service-based industries supplanted manufacturing and factory employment.

Rapid urban growth is routinely accompanied by a lack of infrastructure. In the 1950s and 1960s, Sydney was bursting at its suburban seams, with an infrastructure deficit despite its booming economy.[81] Motorised vehicles proliferated, shaping Period Four as a phase where suburban expansion defined the city's dominant form. Expanding road networks spun weblike across the city, leading to congestion and a financial burden of regular roadworks and maintenance resulting from decentralising urban form. As described by Peter Spearritt, the private car made much of architect Robyn Boyd's dystopian post-war suburbia possible: "now development could proceed well beyond railway stations, tram and bus routes".[82] Car ownership skyrocketed, enabling new suburbs to roll out untrammelled across Sydney's greenfields. The car also led to the decline and erasure of Sydney's tram network, with recent piecemeal attempts to revive it in inner Sydney coming at a significant cost and challenge. High Street precincts on tram-serviced arterial roads declined, as retail epicentres in the form of 'big-box' malls and chain retail outlets sprung up on cheap lands, alongside expanses of carparks. These became the new norm, as independent businesses were swallowed by generic chains and franchises.

Manifold roadways and suburbs therefore consumed prior agricultural landscapes. It is no exaggeration to say that cars and suburbia catastrophically impacted Sydney's food landscapes. High-resolution aerial imagery taken over much of Sydney reveals extensive loss of productive areas over time (Fig. 2.6 and *Then and Now*).[83] The decentralisation of Sydney's urban landscape came at the expense of local agricultural production. It was accompanied by increasing connections with and reliance on external markets and supply chains. A notable example was the construction of a standard-gauge railway running parallel to the broad-gauge track from Albury to Melbourne, which facilitated a seamless through-service link between Sydney and Melbourne for both freight and passengers. It is perhaps surprising this was only realised in 1962 but suggests greater prior reliance on local supplies of perishable fresh foods susceptible to spoilage. Over time, the transportation of delicate fresh foods requiring rapid delivery became increasingly feasible across longer distances, complementing prior transportation of durable foodstuffs well suited to long-distance conveyance, such as potatoes, root vegetables, citrus fruits and bananas.

The 11 Typologies: 1946–1999

Metropolitan aerial photographic coverage from 1943 was invaluable for populating the Map Database, considerably advancing spatial-geographic literacy to better understand the city's form, makeup and flux.[84] Aerial photography captured varying proportions of Sydney at relatively regular intervals after 1943 (Fig. 2.6). Numerous key extracts gleaned from aerial imagery are included throughout this chapter and in *Then and Now*. One of the most significant findings is the considerable extent of land supporting agriculture around the mid-twentieth century, with Sydney's population then under 2 million. Market gardening remained active and a noteworthy amount of dairying was still taking place. Orchards, although clearly declining, were also spatially significant in regions such as the north-west. Surveying aerial photography, however, only captures a few specific points in time across *Period Four*'s 53-year span and does not cover the full extent of this book's boundary.[85] Entries in the Map Database therefore represent a sample and snapshot of far more prevalent agricultural activities taking place in Sydney at the time.

Agricultural Facilities and Education (AF): A Last State-Based Hurrah?

Agriculture's increasingly technical nature in *Period Four* lent itself to longer, more rigorous and diversified streams of study, as well as to increased specialisation. There were several notable agricultural facilities and educational institutions established in Sydney between 1946 and 1999. Wider changes in the education sector, as well significant global shifts in agricultural practices, would see institutions substantially alter their structural makeup over the course of the period.

P.A.Yeomans, for instance, undertook internationally significant applied agricultural research and experimentation in 'keyline farming', using land contouring to increase water infiltration and retention. He conducted this on his two North Richmond properties Yobarnie and

Nevallan from 1943 to 1964 (Fig. 8.11),[86] with Yobarnie sold to Peel Dairy in the mid-1960s.[87] Yeomans documented this work in his 1965 book, *Water for Every Farm* and his principles have been internationally applied in water management for sloping landscapes and in contour farming. Although his experiments were privately initiated and geared towards dairy activities, his approach has been widely shared and applied, including within grazing, agroforestry, permaculture and food forest farming.

As Long Bay Gaol's built footprint grew, food production was scaled back but continued nonetheless. Farming activities were mentioned in a 1949 report and a wide range of vegetables were being grown in the 1960s, irrigated from a stone-lined canal.[88] Aerial photography in 1953 and 1970 (Fig. 6.4, bottom) capture verdant and sizeable plots. A 1986 image shows a large new prison facility constructed on the former farm site, with two considerably smaller gardens visible to the west and south-east.[89] Subsequent imagery reveals the western garden disappearing and the south-western garden increasing in size between 1991 and 1994.[90] Aerial imagery also shows abundant agriculture in Parramatta Gaol's farm in 1965. A 1970 image reveals that it had fallen fallow, and a 1975 photograph shows new building constructed on the former farming plots, correlating with a written historical survey.[91]

Research institutes of note during the period included the Council for Scientific and Industrial Research (CSIR, now the CSIRO) establishing a 48-hectare agricultural animal research facility in Prospect in 1946, opening in 1953. Focusing on sheep/wool but also incorporating poultry, the station operated for around half a century before the selling the land to developers in 2000 to become the suburb of Pemulwuy.[92] The 1600-hectare Elizabeth Macarthur Agricultural Institute opened in 1990 near Camden on lands part of the original land granted to John and Elizabeth Macarthur in 1805 in the Cowpastures (see *Period One*). It remains a plant and animal science biosecurity and agricultural development facility run by the NSW Department of Primary Industries.[93]

Within education facilities, James Ruse Agricultural High School in Carlingford took its first intake of students in 1957 and became a fully

(AF) Agricultural Facilities and Education

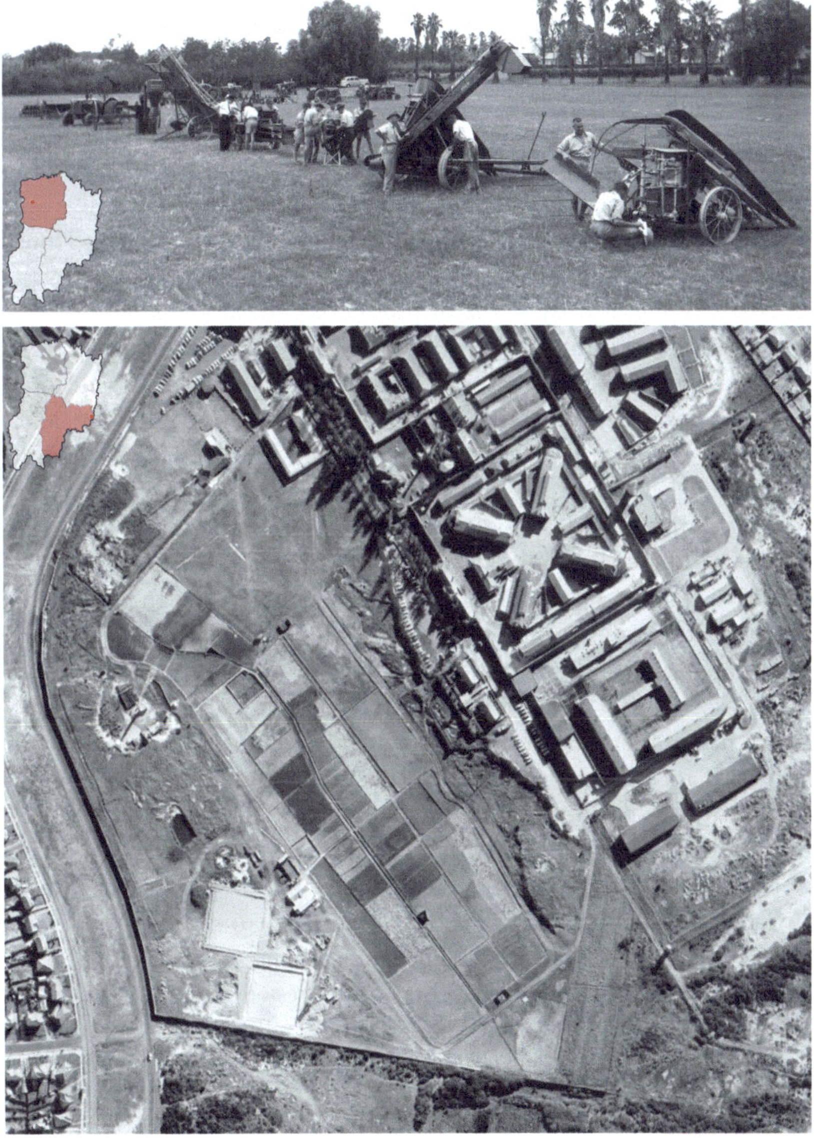

Fig. 6.4 A range of farm machinery at Hawkesbury Agricultural College c. 1950s (top), aerial image of Long Bay Gaol farm in 1970 (bottom). Sources: Western Sydney University Archives Ref. P1504, and Spatial Services NSW

functional agricultural high school in 1959. Its namesake was taken from Australia's first successful farmer (see *Period One*) and the school included a five-hectare farm.[94]

Ryde Campus was purchased for £7000 in 1946–1948 by Sydney Technical College (which would later become TAFE) to become the first specialist horticulture college campus in the state. The site was previously a successful fruit and vegetable farm called 'Longford'. Sydney's full-time horticultural teaching position commenced in 1947.[95] In 1955, the site was described as 23 acres and consisting of a three-acre orchard, five acres of vegetable and cut flower gardens, three acres of nursery, glass and 'bush' houses, three acres of 'special purpose turf' and five acres of landscaped gardens.[96] While initially a mix of productive and ornamental horticulture, the former diminished over time. By 1993, the campus was believed to be the "largest ornamental horticulture teaching establishment in the southern hemisphere [at] five hectares", with popular courses including bush regeneration, tree surgery and landscape design and construction.[97]

The University of Sydney continued to educate specialised agricultural scientists amidst significant national and global changes in agricultural research and practice. Nationally, intensification of livestock production and increasing broadacre cropping yields was mirrored in the university's research and teaching centred on bolstering spatially extensive rural, regional, national and international farming contexts. These changes can be glimpsed in the 1966 subdivision of the University's Department of Agriculture into six distinct departments: Agricultural Botany (encompassing Plant Pathology and Biometry), Agricultural Chemistry, Agricultural Economics, Agricultural Microbiology, Agronomy (including Crop Botany) and Soil Science.[98] Additionally, a new Plant Pathology chair was established, followed by the first Horticultural chair at an Australian university in 1971. Over time, numerous departments and disciplines underwent shifts but a consistent association between Agriculture and Animal and Veterinary Sciences endured.

The University of Sydney's agricultural facilities also significantly changed. A small agricultural research station and plant-breeding institute operated at Castle Hill from approximately 1952 to 1991. By 1995, the University boasted 15 farms within a 65 kilometres radius of Sydney, totalling over 1500 hectares. These farms supported teaching and research activities of both the Faculties of Agriculture and Veterinary Science, with three main centres. The 717-hectare Corstorphine Centre housed most teaching and research facilities. The 466-hectare Bringelly Centre was comprised of three complementary farms located 10 kilometres apart (Wolverton, Coates Park and John Bruce Pye Farm), while the Badgerys Creek centre incorporated the McGarvie Smith Farm and Fleurs. In 1995, Sydney University's farms contained over 400 milking cows, 1200 beef and dairy cattle, 2000 sheep, 30 horses, 170 deer, 3000 hens, 3500 broilers, 100 goats and 400 pigs.[99] While primarily dedicated to teaching and research, livestock also generated supplementary commercial income to cover basic farm expenses and assist development, research and teaching activities.[100]

Hawkesbury Agricultural College (Fig. 6.4, top) in Richmond finally admitted its first two female students to its cohort of 285 in 1971. The College's trajectory reflects a wider trend in the period's education sector, which saw many of Australia's agricultural colleges come under universities' operating umbrellas. Through this transition, agricultural courses became part of a larger suite of increasingly diversified educational offerings also offering greater specialisation.[101] The Agricultural College became the College of Advanced Education from 1972,[102] although it continued to be operated by the Department of Agriculture until 1976 when it became autonomous with its own governing council. In 1989, the College merged with the Nepean College of Advanced Education to form the University of Western Sydney, Hawkesbury campus.

Fluctuating university fees over the course of *Period Four* affected agricultural enrolments. University study was bolstered by greater government support, culminating in the Whitlam Labor Government abolishing university fees on 1 January 1974. By 1989, however, fees were reintroduced under the guise of the current HECS/HELP systems and have only increased over time under higher education's increasing neoliberalisation.

Mixed Farming (MF): Specialisation

Sydney's overall agricultural activities were mixed throughout the period. Drawing heavily on the 1948 CCC Planning Scheme, Winston noted in 1957 that,

> [C]attle, sheep and stud horses are now raised and dairying and poultry farming carried on, and in recent years there has been some pasture improvement which, with modern soil conservation and water farming methods, has greatly increased the productivity of the area [...] [T]he County [of Cumberland] is still a big producer of vegetables, fruit, eggs and dairy products, with important stud farms as well.[103]

Mixed farming on individual commercial farms, however, continued to decline as agricultural economies of scale increased. Most commercial metropolitan farms became more specialised in their approach and outputs. Combined citrus and poultry operations, for example, had or were dissolving in the early stages of *Period Four*, as poultry production became industrialised (see *IL*). As such, there was a diminishing number of individual farms commercially operating across multiple topologies in Sydney.

If we expand the lens from individual properties to localised agglomerations, certain locations contained a range of activities such as poultry, orchards, vineyards and market gardens in relative close proximity (Fig. 6.5). Sydney's regions with mixed farming and hobby farms regularly saw landscapes transition and land use patterns shift alongside changing agricultural practices. The 1948 CCPS illuminated agriculture lost during *Period Four*. It emphasised Eastwood as "invaluable" and included images of district mixed farming including poultry pens, orchards and market gardens.[104] Eastwood succumbed to urbanisation between 1943 and 1986.[105] Similarly, along the Georges River in Milperra, with its slender land parcels and mixed-farming practices, a dynamic interplay between agriculture and impending urbanisation took place. This manifested in the gradual disappearance of orchards and poultry runs and the advancement of suburbs, light industry, golf courses and ecological alterations to the river.[106]

Change was further evidenced in locales like Baulkham Hills and Northmead, where 1953 aerial imagery[107] signalled suburban blocks encroaching on agricultural plots. Imagery from 1970 imagery reveals that the suburbanisation process was then complete. Similarly, 1953 imagery shows a ranged of farming activities taking place in West Pennant Hills, including remnant orchards, market gardens, cropping, poultry, grazing, dairying and occasional vineyards, all of which had disappeared under suburbs by the 1980s.

Greater Ryde was historically one of the city's most important metropolitan agricultural production districts. 1955 imagery conveys warning signs of suburban encroachment upon remnant orchards, cropping, grazing, the former North Ryde Common and livestock operations, including poultry, piggeries and dairies. Winston emphasised that the green belt near Ryde and Eastwood produced "great quantities of fresh fruit, milk, eggs and vegetables" (Fig. 6.5).[108] Historical imagery of Macquarie Park before Macquarie University was developed shows mixed-farming activities previously widespread in the greater Ryde district with its fertile soils.[109] Finally, Nepean River floodplain at Ellis Lane and Cobbitty Paddocks, remnants of the region's agricultural heritage as part of the Macarthur empire, persisted amidst encroaching suburbanisation.[110] Although dairy farming waned, the area sustained mixed-farming pursuits encompassing irrigated pastures, market gardening, viticulture, orcharding and turf growing.

Market Gardening (MG): Mushrooming Housing

There are over 300 market garden entries in the Google Maps database for *Period Four*, hinting at the once-extensive network that was particularly noteworthy at the beginning of the timeframe. The actual number of individual market gardens was far greater. Entries in the database often represent clusters of multiple market gardens within a single district, and aerial photography offers only limited snapshots of temporal and spatial coverage. New market gardens did emerge across various districts in the period. These were chiefly situated on the peri-urban fringes, which were always in flux. Overwhelmingly, though, the period's trend was one of

Fig. 6.5 Images of greater Ryde c. 1948 (top) and 1955 (bottom) show a diverse agricultural district including orchards, market gardens, poultry and pig farms. The snaking road on the top right was Epping Highway. Sources: Mitchell Library, State Library of New South Wales and Spatial Services NSW

absorption of market gardens into the built landscape, as the accumulating pool of suburbs continued to flood outwards from the city and key nodes.

Immigrants continued to be crucial to market gardens' ongoing success, with the post-World War II influx proving fundamental to Sydney's productivity. Chain migration saw the growth of substantial Maltese and Italian communities in outer western Sydney, of which numerous early members were market gardeners or farm labourers. Sarah James estimates that Maltese market gardeners were the "largest single cultural group in vegetable farming in the Sydney Basin".[111] Serbian and Croatian market gardeners also gathered in western Sydney. Chinese market gardening was reinvigorated from increased immigration from the mid-1970s, also reintroducing vegetables such as bok choy and choy sum and bringing "new blood into the gardening workforce, which had been ageing and diminishing particularly since the 1950s".[112] Around 90 per cent of Sydney's market gardeners came from migrant backgrounds, including Vietnamese and Cambodians arriving from the early 1980s.[113] Writing in 1988, Michael Christie determined Sydney market gardeners to predominantly hail from Malta and other southern European countries and China, with a smaller number from Lebanon, Laos, Vietnam and Democratic Kampuchea (formerly part of Cambodia).[114]

In the 1948 CCPS, the CCC quantified market gardens as the second largest horticultural spatial entity in the county, spanning 8815 acres (3567 hectares).[115] Fruit production surpassed this when aggregated.[116] Furthermore, the CCPS revealed that 16.8 per cent (equivalent to 36,800 tons)[117] of New South Wales' vegetable output in the 1945/46 fiscal year originated from the county area, with significant contributions including three-quarters of the state's lettuce, half of its spinach, one-third of cabbages and one-fourth of beans. This data underscores market gardens' substantial role in the regional agricultural landscape, particularly with respect to vegetable production.[118] Further production also took place within Sydney's topographic basin (i.e. within this book's boundary), but outside the County's statistical capture, such as west of the Hawkesbury-Nepean.

As noted by Winston in 1957,

> In spite of all our canning and deep freezing, there is still no real substitute for fresh fruit and vegetables, gathered when they are "just right" and eaten soon afterwards with the minimum of handling in between: but fruit and vegetables of this kind can only be had in the city if they are grown within a reasonably short distance away. Perhaps some day we shall be able to fly them in cheaply but we can't do that yet.[119]

The long-distance produce transport and post-harvest technologies progressively improved during the period, influencing metropolitan market gardens' decline.

Winston emphatically argued that the "preservation of the farms and market gardens is therefore of considerable importance for the well being of Sydney as well as for the economy of the State", highlighting key areas and hinting at some of Sydney's metropolitan planning's shortcomings.[120] He recognised that Botany's swamps were formerly Sydney's primary source of vegetable production due to their reliable water supply. In the 1950s, many people still remembered Rose Bay's Chinese market gardens and remnants still existed in South Randwick, Mascot, Rockdale and Hurstville at the end of the decade. During the formulation of the CCPS, market gardens in Sutherland, Eastwood, St Ives and Warriewood came under threat from new subdivisions. Additionally, ongoing developments in Liverpool and Blacktown continually converted more acres of land away from agricultural use,

> Much of this development was of course necessary, but a good deal of it could quite as well have taken place on land which was not of such great agricultural importance. The [CCC] Planning Scheme therefore zoned all land not specifically required for residential, industrial or special uses as Green Belt or Rural areas, within which primary production has first consideration and subdivisions are generally limited to agricultural holdings of not less than five acres.[121]

James, however, highlights the paradox of competing agendas for limiting the city's expansion to preserve agriculture and the "discourse of global economic competitiveness [that] ran concurrent to Winston's argument about the need to protect agricultural land".[122]

Similar sentiments are voiced in a 1970 document outlining the once-extensive Chinese market gardens network in the St George district,

> It is of interest to recall the location of these fertile Chinese market gardens, as most of them, under the lack of wise planning, have been given over to factory sites, or have lain fallow because their leases have expired. The production of essential green foodstuffs for an ever-growing city community is obviously regarded as being of secondary importance in municipal circles.[123]

Market gardens (and poultry production) initially expanded in early *Period Four* in Mona Vale-Warriewood on the northern beaches.[124] Warriewood Valley's glasshouses, for example, were so considerable that the area came to be known as 'Glass City' (visible in a 1961 aerial image, see Fig. 8.13, top left), effectively acting as a precursor to contemporary mega-scale greenhouse districts such as in Almeria, Spain and Westland in the Netherlands.[125] The once-thriving district had largely disappeared under a mix of urban forms by the late 1960s due to cheaper produce shipped in from other regions and states, although some glasshouses remained in the 1970s and a few producers continued to operate in late *Period Four* (Fig. 6.6) and into *Period Five.*

Historical aerial imagery shows many regions containing clusters of market gardens, collectively playing a significant role in Sydney's food production across the period. Gardens flourished in Sans Souci adjacent to the Goomun Creek drainage channel in 1951 imagery (Fig. 6.6, bottom left). However, a gap in image coverage until 1971 and 1975 reveals abandonment or loss under development. Similarly, market gardens once thrived in what is now Todd Park in Carss Bush Park, as visible in 1943 and 1955 imagery (Fig. 6.6, top) but had disappeared by 1971. The Dee Why Valley witnessed a surge in agricultural activity, particularly market gardening, between 1943 and 1955 as evidenced by cropping, poultry farms and small orchards.[126] Activity declined sharply by 1975 with only remnants visible in 1986 and 1991. In Brookvale, fertile alluvial fans supported numerous market gardens and orchards.[127] By the 1960s, urbanisation was encroaching, with 1975 imagery illustrating the loss of agricultural land to industrial and residential development.

Fig. 6.6 Market Gardens pre-1955 in Carss Bush Park (now Todd Park) (top); A 1951 aerial image shows extensive market gardens adjoining the Goomun Creek in

The greater Botany region, once abundant in market gardens, experienced decline following its industrialisation. While this had commenced earlier, *Period Four* saw to the effective obliteration of 'Sydney's backyard vegetable garden'. This was compounded by the ongoing White Australia Policy impacting the region's Chinese gardeners. Aerial images from 1951 and 1953 reveal extensive remnants of market gardens in Botany, Hillsdale, La Perouse, Port Botany, Phillip Bay, Little Bay, Matraville[128] and Eastgardens. At the time of writing, only two small areas remain in Matraville (Fig. 7.7) and La Perouse (Fig. 7.8).

A volcanic crater in Lugarno was a novel market garden site. Aerial imagery from 1943, 1951 and 1955 shows immaculately cultivated plots in the eight-hectare crater,[129] which is now the site of H.V. Evatt Memorial Park. Tended by the Lum family,[130] the garden thrived from the mid-1950s to around 1963. Nearby production pockets were surrounded by bushland in Illawong, visible in 1953 aerial images.[131] The site of Macquarie University was also an area of abundant agricultural activity.[132] Macquarie Park, Marsfield and West Pymble were part of the greater Ryde district that were all once agriculturally active, especially in market gardening and fruit production in their previously rich soils.[133] The post-war urbanisation boom led to the loss of many agricultural areas, particularly market gardens closer to the city as near Arncliffe railway station. Lastly, aerial imagery of Marsfield in 1955 reveals a recently completed gas booster amidst market gardens, remnant orchards, poultry and cropping that were under threat from encroaching waves of detached suburban housing.[134] By 1985, only a fraction of productivity remained in the once-highly-fertile district.[135]

Despite the considerable loss of market gardens under Sydney's inexorable growth, by 1998, NSW Agriculture recorded that 85 per cent mushrooms, 70 per cent tomatoes and 95 per cent of spring onions

Fig. 6.6 (continued) Sans Souci (bottom left); A Sicilians' thriving market garden near Sydney 1972; On Garden and Macpherson Streets in Warriewood in 1995, an outdoor tomato crop is flanked by rusting glasshouses; Chinese market garden at La Perouse 1999 (bottom right). Sources: Georges River Libraries Local Studies Collection; Spatial Services NSW; National Archives of Australia (Courtesy National Archives of Australia. NAA: A12111, 1/1972/16/134); Northern Beaches Council Library Service; and Stuart Read produced in the state were nonetheless grown in the Sydney region.[136] ABS data for Sydney's agriculture is not, however, necessarily reliable, as discussed in *Period Five*.

Fruit (FR): Decimation of a Thriving Industry

Overall, fruit and citrus production declined in Sydney in the 1950s, with some growers transitioning to poultry (see *IL*).[137] A significant flood in 1956 destroyed many orchards along the Hawkesbury's riverbanks.[138] Aerial photography from early *Period Four* reveals many abandoned or transitioning orchards, particularly in the 1940s and 1950s. The CCPS separates Sydney's 1946–1947 fruit production into citrus (2719-hectares), other fruits (1390-hectares),[139] passionfruit (186-hectares) and grapes (see *VW*).[140] If we combine the first three at 4295-hectares, their spatial area exceeds that of market gardens, at 3567-hectares. Fruit production was therefore the third largest agricultural activity after cropping (6991-hectares—most was for livestock feed) and livestock grazing (Fig. 1.2). Significantly, the CCPS also records that the county produced 17.4 per cent (453,319-bushels) of citrus fruits and 7 per cent of non-citrus fruits (219,177-bushels) of New South Wales' total in 1945/46.[141]

Although significant at the start of the period, Sydney's orchards showed signs of imminent decline. Increasing dedicated regional irrigation districts and their large fruit growers, improved transportation capabilities and supermarkets' growing dominance created formidable challenges for Sydney's comparatively smaller fruit growers. Many struggled to compete, leading to what one Sydney producer termed a "crash in the citrus markets in the late 1980s".[142] Consequently, metropolitan fruit production became increasingly rarer towards the end of *Period Four.*

A 1953 aerial image of West Pennant Hills and Carlingford reveals orcharding to be the area's dominant agricultural activity (Fig. 6.7, top), although abandoned and removed orchards in the wider image indicates the typology's overall decline.[143] Orchards were also once extensively present across the Dural/Middle Dural/Galston/Glenorie districts, which cultivated a diverse array of stone and citrus fruits including peaches, plums, nectarines, lemons and oranges (Figs. 6.1 and 6.7). The Shore family orchard's meandered across the striking terrain on Cattai Ridge Road, as captured in an aerial image from 1953 (Fig. 6.7, middle). While fruit production persisted in these regions, many commercial orchards and agricultural lands would be lost, some becoming wholesale and retail plant nurseries specialising in more lucrative ornamental species. Sand mining operations in Maroota replaced orchards, evidencing agriculture's perceived lack of value and susceptibility to displacement from more immediately lucrative endeavours.[144] By the turn of the new millennium, fruit growing in metropolitan Sydney had significantly diminished, chiefly paved over by suburban expansion. Once extensive and commonplace, commercial-scale orchards were an increasingly rare metropolitan occurrence by the end of *Period Four*.

Cropping (CR): Crops, for Animals

As previously noted, the 1948 CCPS recorded cropping as the largest cultivation-based land use in Sydney at nearly 7000-hectares.[145] As a scale comparison, this is larger than today's 5280-hectare Western Sydney Parklands but comparatively a very small scale of crop production to commercial broadacre activities beyond Sydney. Although cropping is a form of agricultural cultivation, most Sydney cropping was for animal feed rather than plants for direct human consumption—mirroring a national phenomenon.[146] In this sense, cropping is integrally linked with the livestock industry.[147] The proportion of Sydney's cropping for direct human consumption rather than as fodder is unknown from the CCPS statistics but it was likely minor. Significant areas of cropped lands were lost to urban expansion. The Georges River's banks and floodplains, for instance, once supported a range of agricultural pursuits. A 1953 aerial photograph of Warwick Farm shows remnant (and some active) croplands, orchards, vineyards, market gardens and poultry farms benefitting from alluvial fans and water supply for irrigation.[148] We can also observe cropping activity taking place in the land parcels near the sinuous Prospect Creek in 1953.[149]

Fig. 6.7 A 1953 West Pennant Hills & Carlingford aerial image shows orcharding as the dominant typology. Also in 1953, the Shore family's Glenorie orchard on

Intensive Livestock (IL): From Scratchings to Billionaires

An image of dairy farmer Joseph Homer kissing his cow Annabela goodbye epitomises the forces reshaping almost all of Sydney's agricultural landscapes over the course of *Period Four*.[150] In 1950, Homer was compelled to 'sell' his dairy in Bassett Street, Winnererremy Bay, Mona Vale, following its forced resumption by the Warringah Shire Council and developer 'Pacific Paradise'.[151] The period's timeframe witnessed a continued decline in metropolitan dairying, however, saw unparalleled agricultural defiance through the exponential growth of industrial poultry operations. In both cases, concentrated livestock and intensive agricultural activities became increasingly prevalent, driven by high land prices and a thirst for profit. Sydney's city region housed animal breeding operations and stud farms, which Winston notes as still 'important' in 1957,[152] albeit many were skewed towards gambling, entertainment and recreation.

Dairy

The CCPS notes that 18 per cent of Sydney's consumed milk originated from the County of Cumberland during the 1945/46 fiscal year.[153] *Period Four*, however, saw a significant decline in Sydney's fresh dairy industry, approaching dissolution of operations with only a few noteworthy exceptions. Smaller local dairies once commonplace within or near built-up areas disappeared. Over time, the same would happen to most of Sydney's peri urban and rural dairies (Fig. 6.8, bottom). As production facilities dwindled, so too did traditional home milk deliveries and the ubiquity of suburban depots. Household milk distribution relied on bottling, a practice that became nearly universal by 1955.[154] Home deliveries became near obsolete by the end of the period. Remarkably, in 1984, home milk

Fig. 6.7 (continued) Cattai Ridge Road was later subdivided but continued fruit production. In the late 1960s or early 1970s extensive orchards still existed in the Dural/Middle Dural/Galston districts, such as Cranstons Road in Middle Dural (bottom). Sources: Spatial Services NSW, Dural and District Historical Society Inc.

Fig. 6.8 Photographs from the 1950s of a Kenthurst poultry farm typical of the smaller scale before the industry was industrialised (top left, centre); industrial shed complexes part of an Ingham's broiler 'farm' on Badgerys Creek Rd in 1975,

delivery still constituted 65 per cent of the market share, yet this figure plummeted to a mere 15 per cent by 1998 and virtually disappeared by 2003, accounting for only 1 to 2 per cent of the market.[155] This marked the transition to refrigerated, pasteurised and more processed milk products sourced from larger producers and regional production districts, offering greater shelf life but lacking the reusability of former glass bottles.

Archival imagery from the 1950s and 1960s captures dairying (and grazing) at 'St Andrews Farm' on Campbelltown Road, emblematic of the once-prevalent livestock practices across western Sydney (bottom of Figs. 6.8 and 6.9).[156] Originating from the colony's early years, the farm traces its lineage to emancipist Andrew Thompson who bequeathed a quarter of his estate to fellow Scot Governor Lachlan Macquarie on his death in 1810. Macquarie noted the farm's fine rich soils, farmhouse and paddocks stocked with sheep and cattle.[157] However, Fig. 6.8 evidences overgrazed, denuded soils and erosion and aerial imagery from 1975 unveils the onset of subdivisions. Throughout its 150-year existence, the farm bore witness to the relentless impact of hard-hoofed animals, resulting in the compaction and erosion of its once rich soils.

Poultry

Sydney's poultry industry shot to meteoric heights in *Period Four*, facilitating such social ascendancies as Tom Lewis' rise from chicken farmer to New South Wales Premier (1975–76).[158] The industry rapidly evolved from a near-backyard to industrial-scale 'factory farming'.[159] The CCC Report recorded over 70 per cent of New South Wales' poultry farms as concentrated within the County of Cumberland in 1945/46.[160] According to Sydney poultry group Baiada, the "poultry industry grew nearly sevenfold during the '50s", while the "'60s and '70s saw an increase in the supply of chicken to butchers, clubs and restaurants".[161] This

Fig. 6.8 (continued) later demolished in 2018 (top right); a 1956 image of dairying and grazing landscapes at 'St Andrews Farm' on Campbelltown Road (bottom). Sources: Malcolm Johnstone, Kingsclear Books, Spatial Services NSW, and Peter and Marie Thomson Collection, Campbelltown City Library

Fig. 6.9 Grazing lands near Campbelltown c. 1945–1948 (top); grazing lands around Eschol Park House 1960s. Sources: Mitchell Library, State Library of New South Wales, and Peter and Marie Thomson Collection, Campbelltown City Library

intensification in western and north-western Sydney corresponded with an immense rise in Australians' chicken consumption. In 1939, the average person consumed around 4.4 kilograms of chicken in a year. By 1985, this had grown to 21 kilograms per year. The staggering growth in both production and consumption spawned multi-million, and eventually billion, dollar empires.

Factory livestock farming generated unprecedented incomes, even from from modest landholdings. Families such as Ingham (Inghams Chickens, Fig. 6.8, top right)[162] and Baiada (Steggles and Lilydale)[163] joined Tony and Ron Perich's dairy farming (Leppington Pastoral Company, Fig. 7.10) as examples of intergenerational success stemming from humble farming origins in western Sydney. At one point in the period, Ingham provided Sydney with 60 per cent of its chicken.[164] Profits were reinvested in strategic property acquisitions, including greenfield land for subdivision into housing estates and commercial uses, vertically integrating agriculture-cum-land development. Recent Sydney 'rich lists' attest to Perich's and Ingham's remarkable financial success, both of whom have surpassed billionaire status, while Baiada is nearing this milestone.[165] Additionally, Cordina Farms emerged as another notable Sydney poultry family enterprise.[166]

As the poultry industry expanded, the birds' freedom contracted. Comparatively low density fenced outdoor pens were typical in *Period Three* and early *Period Four* (visible in Figs. 5.11 and 6.8). These were made economically unviable by densely stocked large sheds deployed in multiple clusters across a single site (Fig. 6.8, top right). Birds were no longer publicly visible in these 'barns' (industrial sheds), in part due to each commonly accommodating 15,000–70,000+. Ingham's former operation in Fig. 6.8's (top right) contained 56 Poultry sheds (not all are visible), which given stocking rates, illustrates how single farms contained hundreds of thousands and even millions of birds. The industry has continued to upsize, with Fig. 6.8's 100-metre-long and 12-metre-wide sheds (1200 square metres each) now at the smaller end of the commercial poultry scale.[167] Such shed arrangements were, and continue to be, used for both meat and egg production, including a 'free range' variation offering unstipulated access to small adjacent outdoor areas.

Unsurprisingly, strong ammonia odour clouds drift considerable distances from these highly concentrated sites to impact neighbouring communities. Issues arising from intensive livestock operations such as odours and noise were a key catalyst for planning controls to mitigate localised land use conflicts. This proved challenging where approved residential development encroached on pre-existing agricultural operations—a phenomenon widely encountered in Sydney's peri urban poultry industry and continuing to this day.[168] Controls gradually evolved to encompass factors including total bird populations, permitted shed concentrations and buffer zones to mitigate impacts on natural features (such as watercourses) and cultural adjacencies (like neighbouring land uses). However, due to regularly high stocking rates and total bird populations, buffer zones often proved inadequate, especially for properties situated downwind. Consequently, intensive poultry farming is incongruous with most perceptions of agricultural zoning and better suited to regional industrial zones, if considered divorced from environmental and animal welfare consequences.[169]

Despite challenges coexisting with neighbouring communities, Sydney's factory livestock industries emerged as the most influential agricultural entities in the city's highly contested metropolitan land use arena. Intensive livestock's success stories have hints of *Period One*'s bunyip aristocracy, such as the Macarthurs. Pastoral and social pretensions aside, they all amassed significant land, wealth and power despite often arriving in Sydney with very little, if any, capital. Leveraging land-based industry, development and agribusiness in-turn shaped economic dynamics and policies across multiple generations. This immigrant-to-mogul trajectory reflects the recurring theme of selling, converting and/or developing agricultural lands for resources and urbanised uses. While catalysing industries across manifold Australian locations, Sydney's overall land bank available for metropolitan agriculture was diminished.

Grazing (GZ): Sydney's Largest Agricultural Land Use

Winston's 1957 map of the County of Cumberland's metropolitan agricultural makeup (Fig. 1.2) reveals grazing as the largest food landscape.[170] While collectively significant, most Sydney graziers operated at hobby or small scales compared to remote vast pastoral estates, landholdings and leases. Winston's observations highlighted the recent pasture improvements within the County, facilitated by modern soil conservation and water farming methods augmenting productivity.[171] This potentially alludes to the keyline approach pioneered by P.A. Yeomans, discussed in *AF*—albeit, this would never be realised at scale.[172]

As evident in Fig. 1.2, there were undoubtedly far more grazing activities in Sydney than the modest number identified in the period's Map Database, with aerial photographic coverage not extending to many areas (Fig. 2.6). Despite pastoral fields often being visible, grazing we cannot assume that grazing took place unless image quality is adequate to identify livestock herds. Historic photographs and aerial imagery do clearly reveal overgrazing and land degradation. The CCPS had already noted in 1948 that "in many grazing areas, such as near Campbelltown, severe erosion of topsoil has taken place, due to overstocking and the destruction of trees" (Figs. 6.8, bottom, and 6.9).[173] Mirroring countless sites across the Cumberland Plain, historic houses, farms, grazing fields, farm dams and denuded drainage lines were further incrementally engulfed by urban sprawl of subdivisions across *Period Four*'s latter decades, continuing throughout *Period Five* to today.

Vineyard Wineries (VW): A Last Hurrah

At the beginning of the period, Sydney was still a substantial grape producer. The 1948 CCPS accounts for a significant 14.5 per cent (1129 tonnes) of NSW's grape output in 1945/46.[174] Table grapes constituted only a minor fraction of Sydney's total grape production. The overwhelming majority of Australian grapes were allocated for fortified wine in 1950, with 94 per cent used for table wine by 1995 as consumer tastes changed.[175] The industry's overall arc across the period was one of significant decline,[176] as evident in the Google Maps database[177] and historical aerial imagery. While some wineries persisted at the end of the period, they were typically at boutique scales rather than fully fledged commercial ventures.

Fig. 6.10 An area of vineyards (and orchards) on the Georges River flats at Chipping Norton, c. 1945–1948 (top). The area was extensively mined and rehabilitated

Despite Sydney having a lower proportion of wine consumption during much of *Period Four* when compared to more recent decades,[178] its latter years witnessed a transition towards growth and export, particularly in the last 15 years.[179] Notable Sydney grape-growing regions included Bonnyrigg, Cabramatta West, Wetherill Park, Orchard Hills, Cobbitty, Camden and Chipping Norton (Fig. 6.10). Camden Estate continued its ongoing legacy thanks to its deep alluvial soil and mild climate suited to Chardonnay grapes, which were planted between 1975 and 1988. Estate owner Greg Penman stated in the late 1990s that the grapes yielded between four and six tonnes per acre, depending on seasonal factors, although March rains could pose an issue from mildew or mould. Most of the grapes from the estate vineyard were sent to the Hunter Valley for processing prior to blending and bottling.[180]

Conversely, the decline of prominent institutions such as the Penfold's-owned Minchinbury revealed some of the industry's challenges. Despite Minchinbury's soil erosion and vineyard exhaustion, their sparkling wine had been "so highly regarded that it was served to Queen Elizabeth during her first Australian visit in 1954".[181] From 1962, the winery operated using grapes grown elsewhere and in 1978 it closed. Aerial imagery from 1943 to 1965 illustrates the century-long process of cumulative land degradation through an extensive monoculture's ongoing 'mining' of soil fertility, with commensurate lack of landscape-scale sustainable land management of waterways, tree canopy and groundwater. The site subsequently became a mix of land uses including residential and industrial estates and a quarry (Fig. 8.10). Some past cues remain, particularly remnants of its olive grove planted from 1912 now surrounded by houses.[182] These historical insights not only illuminate winemaking's evolving

Fig 6.10 (continued) during *Period Four* (Fig. 8.10, bottom). An aerial image (bottom) conveys the extensive size of Minchinbury's vineyards in 1947 but also suggests a century of cumulative land degradation through native vegetation clearance and vine monoculture. Minchinbury's buildings are centre right, at the southern end of the olive avenue. Sources: Mitchell Library, State Library of New South Wales and Spatial Services NSW landscape in Sydney, but also highlight the imperative for wider sustainable land management practices to mitigate environmental degradation and loss of agricultural productivity.

Agricultural Infrastructure (IN): End of a Golden Age of Metropolitan Food Processing?

One of the major changes to Sydney's food system logistics was the development and opening of the Sydney Markets in Flemington from 1975. These continue to be Sydney's central wholesale facility but also provide retail and fresh food sales.[183] Their continuance is significant given Sydney's fresh food markets' historic flux before 1975 and the extent of urban and population growth since (see *Periods Two* and *Three*).

Companies like the Green Valley Dairy Corporation in Shanes Park were engaged in dairy processing and manufacture, namely raw milk's processing rather than production. In 1955, Union Carbide acquired the Timbrol factory in Rhodes, initiating pesticide production that included DDT and found widespread agricultural use during the period. Meanwhile, poultry 'processing' plants (namely, abattoirs) were established and/or expanded to service poultry's boom, such as the Ingham plant in Tahmoor and the Baiada/Steggles production plant Pendle Hill later in the period. Wollondilly abattoir was also created, as well as rendering plants in Riverstone (AJ Bush and Sons) and Maroota (Camilleri Stockfeed). After state government consolidation of livestock slaughterhouses in *Period Three* at Homebush, the 1980s shifted to free market decentralisation and regional expansion. Once highly visible, livestock slaughter and processing became increasingly hidden due to conflicting adjacencies, refrigerated truck transport and sales benefits stemming from the concealment of processes and conditions in animal raising and slaughter.[184]

The period's transformation of Sydney's agricultural infrastructures reflected increases in land value as urban areas encroached. The private agri-food sector also trended towards mergers, agglomeration and consolidation under the period's growing neoliberalism, leading to the closure of smaller facilities, manufacturers and processors. As such, Homebush's large, centralised State Abattoir and Stockyards declined in activity for much of the period. The State Abattoir nonetheless commenced meat exports from the 1960s and $27 million was invested attempting to upgrade the complex between 1965 and 1976. Costs and ongoing maintenance impacted its profitability, with questions raised as to its management. The complex faltered, its export licences were relinquished in 1980 and by 1988 it had closed. Much of the land was redeveloped for industrial activities before becoming part of the 2000 Olympic site and Millennium Parklands.[185] Nearby, Sydney Meat Preserving and Processing Works on Haslams Creek south of Parramatta Road, as started by Sizar Elliot in *Period One*, ceased operations in 1964 and dissolved in 1973.[186] As noted by Jan O'Connell in 2000, "the rationale of 'old' Homebush, had gone".[187] As refrigerated transport and networks grew, the now urban Homebush centre gradually lost its market domination as abattoirs decentralised from urban areas to less visible regional locations.[188]

Mills and grain silos have waxed and waned across Sydney's history, and *Period Four* was no exception. Associated British Foods bought John Darling's flour mill in Rhodes in 1960 on the artificially constructed Rhodes peninsula, rebranding it as Allied Feeds.[189] As observed by Michael Wayne, "most of the product manufactured at the Allied Feeds site would end up in the stomachs of livestock sent to Homebush Abattoir, where said stomachs would then be carved up to be fed back to the populace".[190] Such processes are today referred to as local circular economies. General Mills constructed a facility at Rooty Hill in the late 1970s, while Austral Flour's (later McCorquodale Brothers') concrete grain silos looming over Parramatta Station were demolished in 1984.[191]

Aquaculture (AQ): Death of an Icon

Although aquaculture has historically received less attention than terrestrial agriculture in Sydney and wider Australia, the oyster industry stands out as one of its greatest metropolitan achievements.[192] In 1957, Winston praised the oyster leases of the Hawkesbury and Georges Rivers for providing the "the world's finest oysters".[193] 1953 aerial imagery captures extensive oyster beds in Oyster Bay's shallow waters of the lower Georges River adjoining Sylvania and Kareela, such as the Woronora River inlet near Bonnet Head.[194]

Winston argued for "greater protection" of both oyster farming and coastal fishing against the advance of urban development".[195] His warnings were disregarded, as evidenced by the fate of Australia's first oyster farm in Gwawley Bay in the Georges River region in the 1960s. Aerial imagery from the 1940s and 1950s shows the remainders of oyster beds and channels at the tidal edges being reclaimed by mangroves (Fig. 8.9).[196] These were remnants of Thomas Holt's system of kilometres of channels dug in the bay in the 1870s that enclosed the mouth with a bridge containing floodgates to control water levels.[197] Holt's extensive oyster beds were destroyed from the 1960s and the entire bay was completely remade as the artefactual suburban marina development Sylvania Waters[198] (Fig. 8.9). The most devastating blow to Sydney's oyster industry would come in the form of the 'QX virus' in 1995. This saw approximately 50 oyster operations fold, leaving only Drake's in the Georges River.[199] The industry's implosion highlights the precarity of relying on a small number or a single species, as in the case of the Sydney Rock Oyster (*Saccostrea glomerata*).

Commercial fishing in Sydney's nearby waters was already suffering from overfishing,[200] and *Period Four*'s large-scale industrial activities, regularly located directly adjacent to Sydney's bays and waterways, added significant concentrations of pollution. In the early 1950s, Auburn Council's sign at nearby Silverwater Baths warned of dangers from sharks attracted to the abattoir's wastes, which were still released into the Parramatta River.[201] Industrial activities adversely affected marine ecosystems, the quality and safety of fish and other harvested aquatic species such as prawns,[202] as well as adding unhealthy levels of pollutants into the seafood chain such as dioxins and heavy metals.[203] Perennial concerns continued about how to responsibly and sustainably manage fish stocks in Sydney Harbour against the pressures of a growing population.[204] In terms of distribution, in 1966 the second official wholesale Sydney Fish Market moved from the Haymarket to Blackwattle Bay in Pyrmont and was operated by the Fish Marketing Authority.[205] In 1997 and 1999, its monopoly was broken through deregulation.[206]

Miscellaneous (MC): Consuming Ornamentals

Three miscellanea in *Period Four* signalled significant changes on the agricultural horizon. Firstly, an impending new age of concentrated food retailing commenced that would, over time, reshape production scales to drive out many smaller growers and retailers. A 1947 advertisement for 'Buckinghams supermarket' suggests this shift.[207] Secondly, edible horticultural production lands continued to be displaced by ornamental horticulture such as cut flowers, turf and ornamental plant nurseries in a trend persisting today. This is evident in the 1948 CCPS photograph titled 'Carnation culture at St. Ives'.[208] Thirdly, development and planning continued to spiral with greater momentum, erasing former agricultural sites such as the former Iliffe Nursery family house 'Rosevale', described by the *Illustrated Sydney News* in 1884 as "the prettiest in the Colony".[209] Though its facade was preserved, the house was demolished in 1962[210] to make way for a petrol station[211]—a fitting endnote to *Period Four*'s age of the automobile.

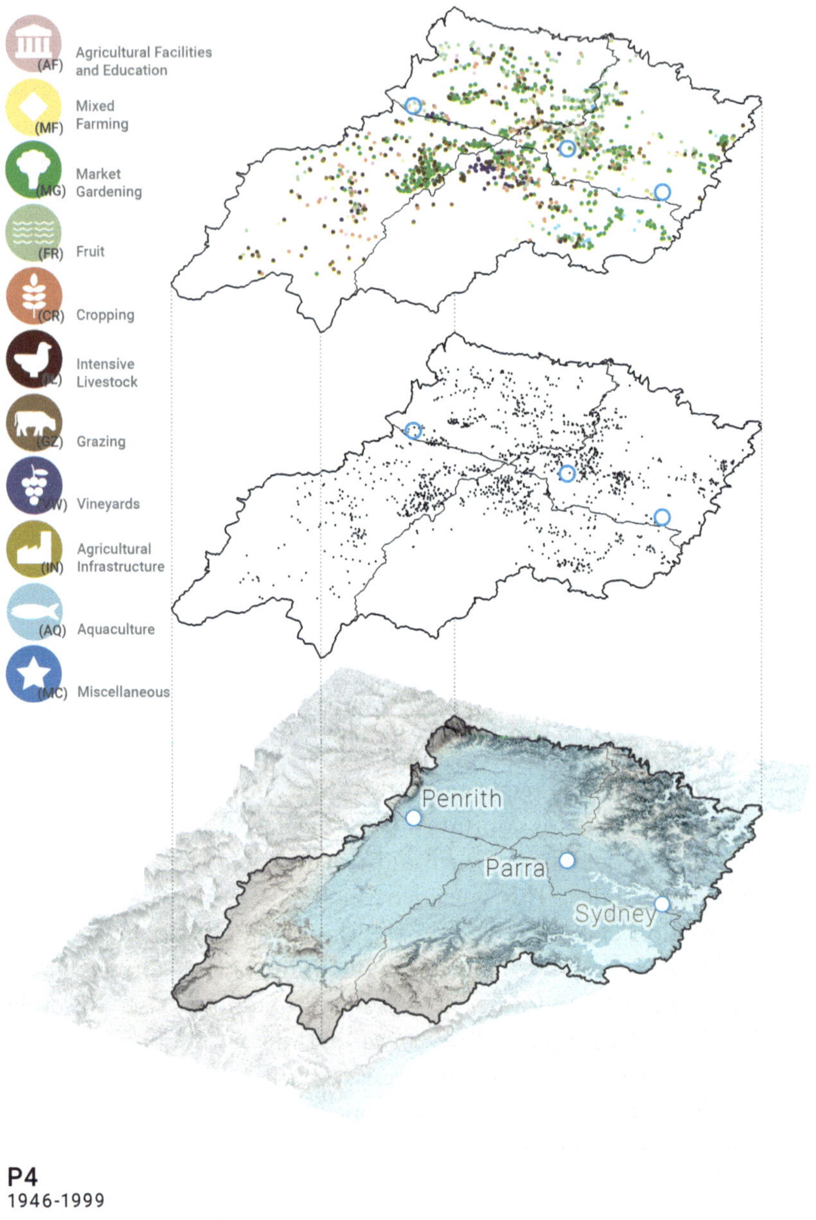

Fig. 6.11 *Period Four*'s Map Database: 3D. The period's cumulative entries paint a picture of both comprehensive (middle) and prismatic (top) metropolitan food production between 1946 and 1999. Figure 9.4, however, reveals the decline of the city's food landscapes. Source: Joshua Zeunert, Josh Gowers and Stephanie Stankiewicz

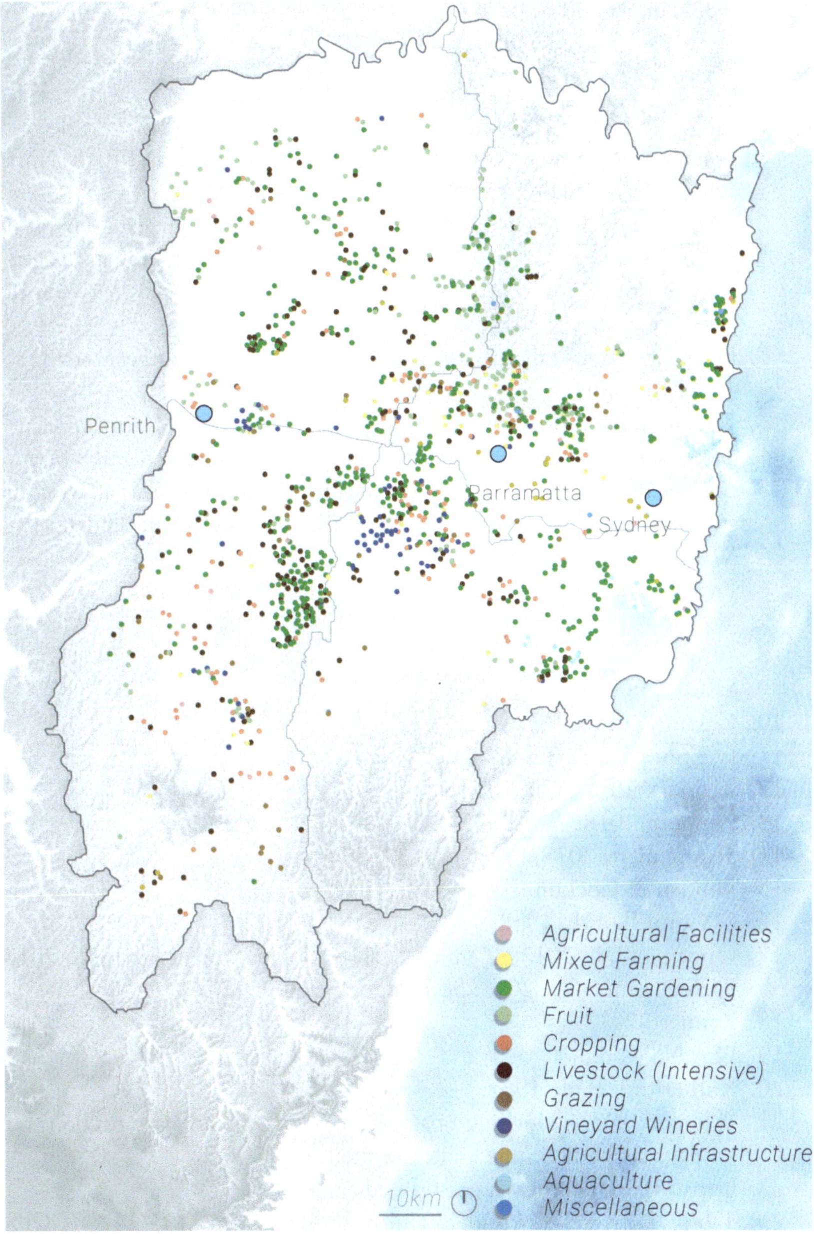

Fig. 6.12 *Period Four's* Map Database: plan. Source: Joshua Zeunert, Josh Gowers and Stephanie Stankiewicz

Notes

1. For example, in Brookvale (mainly to light industry) and Ryde/North Ryde regions (mainly to residential).
2. At this point, the ideology typically demonstrated a dualism separating culture and nature; it was focused on the latter (plants, animals, vegetation communities, ecosystems) rather than the symbiosis between the two. See Zeunert, 2013.
3. https://www.abs.gov.au/ausstats/abs@.nsf/PrimaryMainFeatures/3235.1?OpenDocument
4. This would culminated in 2000 with the concept of the 'Anthropocene', as termed by Paul Crutzen and Eugene Stoermer. Crutzen and Stoermer posited that ongoing accumulation of human impacts was altering the stability of the planet enough to warrant a new geological epoch.
5. Winston, 1957, 7. See SROP, 1968: 1.
6. Freestone, 2009.
7. Winston, 1957, 7. See SROP, 1968: 1.
8. Winston, 1957: 7–8.
9. SROP, 1968: 1.
10. Gaynor, 2006.
11. O'Connell, 2017; n.d.
12. Zeunert, 2017: 170; 2018.
13. Zeunert, 2018: 232–233.
14. See Zeunert, 2018; Erisman et al., 2008.
15. Dibden & Cocklin, 2005.
16. O'Connell, n.d., 1938.
17. Oppenlander, 2013; Zeunert, 2018; Poore and Nemecek, 2018; Springmann et al., 2018; Scarborough et al., 2023.
18. Zeunert, 2013.
19. PSA NSW, n.d.
20. Rutherford, et al., 1967.
21. Winston, 1957: vii–viii; on page seven he says "nearly two million" but on page six he says "one and three quarter".
22. https://dictionaryofsydney.org/media/1318
23. Table 2, Acreages under Cultivation, Season 1946-47 (CCPS, 1948: 133).
24. CCPS, 1948: 125.
25. These inner regions <200 acres were not captured on the map but shown in the data table. E.g. Randwick (138 acres), Botany (132 acres)

and Mascot (109 acres) did not individually meet this threshold (CCPS: 1948: 125) but are captured in the table (CCPS: 1948: 133).
26. CCPS, 1948: 133.
27. An inconsistency between the map and the table is visible where poultry appears to be the second largest spatial activity on the map, but does not appear in the table.
28. CCPS, 1948: 125.
29. CCPS, 1948: 125.
30. CCPS, 1948: 133.
31. CCPS, 1948: 133.
32. Fletcher, 1976.
33. CCPS, 1948: 125.
34. CCPS, 1948: 51. Although the extent to which local production fuelled Sydney's overall food processing sector is unclear.
35. CCPS, 1948.
36. CCPS, 1948: 125, 133.
37. We used Winston's (1957: 38) map and correlated this with the CCPS (1948) drawing. See: https://collection.sl.nsw.gov.au/record/74Vv72zNpl3l
38. Cahill, 1951: 9.
39. Winston, 1957: 10.
40. Winston, 1957: 10.
41. While we used precise Geographic Information Systems (GIS) for these calculations, the base map—Winston's hand-drawing of metropolitan Sydney's agricultural land use (Winston, 1957: 10)—was scanned from a book's page and should be seen as a rough guide only. E.g. fruit/orcharding is a common classification allowing direct comparison between the CCPS statistics at 43 square kilometres, but being 70 square kilometres on Winston's map measured by GIS. This is a 39 per cent discrepancy.
42. See Gaballa & Abraham, 2008; Zeunert, 2018.
43. CCPS, 1948: 65.
44. CCPS, 1948: 129.
45. CCPS, 1948: 129.
46. CCPS, 1948: 130.
47. CCPS 1948: 129.
48. CCPS, 1948: 125–126.
49. Ashton, 1983: 150.
50. Golledge, 1960.
51. Freestone, 1989.
52. Bunker & Houston, 2003: 304.

53. Bunker & Houston, 2003: 304.
54. Race et al., 2010.
55. Zeunert, 2017: 52.
56. See Peggy James' discussion, 2013: 96–97.
57. Freestone, 1992: 72.
58. See: https://collection.sl.nsw.gov.au/record/74Vv72zNpl3l
59. James, 2016.
60. Scutts, 1958.
61. Budge, 2007; Buxton et al., 2006.
62. SPA, 1967: 75.
63. Kevin Lynch's 1960 book, *The Image of the City* was a particularly influential text at the time contributing to this phenomenon.
64. Ibid.
65. Birmingham, 1999: 374–375.
66. Spearritt, 2000: IX.
67. Mason & Knowd, 2010: 63–64; Budge 2007; Buxton et al. 2006; Sandercock, 1975.
68. Mason & Knowd, 2010: 63.
69. Accessed 18 October 2019, https://www.smh.com.au/national/nsw/rampant-growth-threatens-market-gardens-20100303-pj4e.html
70. CCPCoC, 2009: 3.
71. Chiefly proclamation of Royal National Park in 1879.
72. See James, 2013.
73. James, 2013: 233–238.
74. See: https://pastlivesofthenearfuture.com/2013/07/09/pastlives-flashback-1-union-carbide-rhodes-nsw/
75. Mundey, 1981.
76. See Mundey, 1981; Burgmann & Burgmann, 2000.
77. James, 2016: 56, 58.
78. City Plan, 2014: 4.
79. Spearritt, 2000: 127–128.
80. Such as Girraween, Pendle Hill, Greystanes, Pemulwuy and Arndell Park.
81. Spearritt, 2000: XI.
82. Spearritt, 2000: 106.
83. See Historical Imagery Viewer.
84. Accessing the aerial imagery during *Period Four*, however, was slow, expensive and tedious, requiring requesting physical printouts from government departments, map shops or sometimes dedicated commissions, taking days, weeks or months and considerable expense to exe-

cute. Thus, it was not broadly feasible to access an entire suite of metropolitan imagery. Historic aerial imagery for Sydney has only relatively recently become publicly accessible online.

85. The initial and the most exhaustively surveyed resource was 1951/1953 imagery (captured for the military, utilities and government departments). All individual photographs in the suite are publicly available here: https://www.flickr.com/photos/angeljim46/10434808874/in/album-72157632418208341/. The release of the NSW Historical Imagery Viewer website occurred in the latter stages of the research process. It also expanded its available years of coverage very late in the research process and/or after aerial image analysis for this project had finished. The primary years used for P4 are: 1955, 1965, 1970, 1975, 1978, 1986 and 1991. See Fig. 2.6. See: Historical Imagery Viewer.
86. Also known as Redbank and Belmont Park.
87. And be controversially developed following its reported $40 million sale in 2007. See Ihlein, 2014.
88. CLS&P, 2004: 22–23 and Pers. Comm. Colleen Morris 27 February, 16 March 2020.
89. Search -33.97194, 151.24514 at Historical Imagery Viewer.
90. See the Historical Imagery Viewer.
91. TKDA, 2017: 38.
92. See Elias & Coppins, 2013: 104–05.
93. See here.
94. See: http://jamesrusepioneers.homestead.com
95. https://www.smh.com.au/national/plant-lover-broke-new-ground-in-teaching-20110915-1kbm6.html
96. Statistics are taken from signage on the Ryde Campus (Pers. Comm. Anna Richards 25 May 2021).
97. Ibid.
98. Copeland, 2020: 61.
99. Jeffs & Rose, 1995: 5.
100. Ibid.
101. O'Connell, n.d., 1883.
102. These were a form of education positioned between universities and TAFE's that existed in Australia from 1967 to the 1990s.
103. Winston, 1957: 5, 10.
104. CCPS, 1948: Figures 33 and 88.
105. Visible at: Historical Imagery Viewer.
106. See: https://www.flickr.com/photos/angeljim46/8310451662

107. See: https://www.flickr.com/photos/angeljim46/9464861622/in/album-72157632418208341/
108. Winston, 1957.
109. See: https://www.flickr.com/photos/angeljim46/11198651506
110. Willis, 2015: 59.
111. James, 2016: 12.
112. Fitzgerald, 2008: 94.
113. 2010, accessed 17 December 2019, https://www.smh.com.au/national/nsw/rampant-growth-threatens-market-gardens-20100303-pj4e.html
114. Christie, 1988.
115. CCPS, 1948: 133.
116. See *FR* section directly following.
117. Ton was then an imperial measurement equivalent to approximately 1016 kg.
118. CCPS, 1948: 125.
119. Winston, 1957: 48–49.
120. Winston, 1957: 49.
121. Ibid.
122. James further noted that such "competing agendas" continue to the time of her writing. See James, 2016: 44.
123. Gifford & Gifford, 1970: 78.
124. Burnley, 2000: 254.
125. See Court & Zeunert, 2023.
126. Some are visible at the top of this image: https://www.flickr.com/photos/angeljim46/11267157883
127. Greendale Creek flows into Harbord Lagoon which flows into nearby Curl Curl Beach, with Brookvale Creek, flowing into Manly Creek, Manly Lagoon and Queenscliff Beach. These regions supported market gardening as visible in aerial imagery: https://www.flickr.com/photos/angeljim46/11267157883
128. Also see: http://www.migrationheritage.nsw.gov.au/exhibition/atthebeach/happy-valley/index.html
129. See: https://www.flickr.com/photos/angeljim46/13934202419 and https://www.facebook.com/Igrewupinmortdale2223/posts/2087500858146957
130. https://www.facebook.com/Igrewupinmortdale2223/posts/2087084788188564
131. https://www.flickr.com/photos/angeljim46/9629173816/

132. See Historical Imagery Viewer, Ashton & Waterson, 2000: 65, and Cox, 2014: 10.
133. See: https://www.flickr.com/photos/angeljim46/11198651506
134. See Historical Imagery Viewer.
135. As visible: https://www.flickr.com/photos/angeljim46/8408558839/in/album-72157632418208341/
136. NSW Agriculture, 1998: 14.
137. McClymont, 2013: 20.
138. Nichols, 2015.
139. This is likely to predominantly consist of stone and pome fruits.
140. CCPS, 1948: 133.
141. CCPS, 1948: 125.
142. https://melandapark.com.au/about/
143. https://www.flickr.com/photos/angeljim46/9347842519/
144. McClymont, 2013: 105.
145. Albeit as noted in *GZ*, grazing was larger.
146. Zeunert, 2024: 110, 116–117, 128–129.
147. Also resulting in afar lower conversion of calories than if crops are grown for direct human consumption (see Poore & Nemechek, 2018).
148. See: https://www.flickr.com/photos/angeljim46/9198050113
149. See: https://www.flickr.com/photos/angeljim46/9310060829
150. See 1950 image here: http://www.pittwateronlinenews.com/Dorothy-Hawkins-a-new-film-by-John-Illingsworth.php
151. http://www.pittwateronlinenews.com/Dorothy-Hawkins-a-new-film-by-John-Illingsworth.php
152. Winston, 1957: 10.
153. CCPS, 1948: 125.
154. Lewis, 1980: 203
155. For NSW, overall home deliveries of milk dropped from around 80 per cent in the mid-1980s to just 5 per cent of all milk sold by 2003: https://www.smh.com.au/national/cream-goes-out-of-a-home-delivery-milkos-life-20030714-gdh3e6.html
156. See 49 images by searching "St Andrews' Farm" at: https://campbelltown.spydus.com/cgi-bin/spydus.exe/MSGTRN/OPAC/BSEARCH_ALL
157. https://standrews-p.schools.nsw.gov.au/about-our-school/history.html
158. Lewis was key in pioneering the open barn or free range method of farming chickens, which although greatly improved compared to cage

production, still contains bird densities very much reflecting factory farming of livestock. See Zeunert, 2024.

159. Harper & Tait, 2019: 3.
160. CCPS, 1948: 125.
161. https://www.baiada.com.au/our-journey/a-charlie-of-all-trades. National statistics estimated 3 million chickens were produced for meat in Australia in 1950–1951 and this would grow considerably decade upon decade: around 653 million in 2016–2017, see https://www.chicken.org.au/history-of-the-industry-in-australia/ and https://santrev.com.au/history-of-the-poultry-industry-in-australia/
162. https://www.afr.com/wealth/people/bob-ingham-empire-builder-20200923-p55yjc
163. Comprising 40 per cent of the Australian market, see Elias & Coppins, 2013: 74–75 and: https://www.afr.com/property/richest-families-make-second-fortunes-from-property-20151006-gk2600
164. Liston, 2009.
165. See https://en.wikipedia.org/wiki/Financial_Review_Rich_List_2020
166. See Elias & Coppins, 2013: 75.
167. See Scott et al. 2017 and pages 4 and 6 here: https://www.dpi.nsw.gov.au/__data/assets/pdf_file/0011/841088/buffer-zones-to-reduce-land-use-conflict with-agriculture.pdf
168. Taylor et al., 2017; https://theconversation.com/done-like-a-chicken-dinner-city-fringes-locked-in-battles-over-broiler-farms-54886
169. Zeunert, 2024.
170. Map adapted from Winston, 1957: 10.
171. Winston, 1957: 5.
172. Historic aerial photography reveals contour profiling for water management and harvesting on numerous properties across western Sydney.
173. CCPS, 1948: 128.
174. CCPS, 1948: 125.
175. https://www.wineaustralia.com/whats-happening/stories-of-australian-wine/trends/the-evolution-of-australian-wine
176. This number is also higher than the previous period, but given P3's limited aerial photographic coverage, it reflects the far greater availability of aerial imagery in P4.
177. To accurately distinguish between vines grown for table or wine grapes needs additional sources beyond aerial photography alone.
178. Table wine increased in popularity in Australia from the 1960s, see: https://theconversation.com/a-brief-history-of-alcohol-consumption-in-australia-10580

179. https://www.wineaustralia.com/whats-happening/stories-of-australian-wine/trends/the-evolution-of-australian-wine
180. https://dictionaryofsydney.org/entry/spring_farm
181. O'Connell: https://australianfoodtimeline.com.au/minchinbury-champagne/
182. See http://www.visitsydneyaustralia.com.au/minchinbury.html
183. Christie, 1988.
184. Zeunert, 2024.
185. Ibid.
186. https://archivescollection.anu.edu.au/index.php/sydney-meat-preserving-company-limited
187. Connell, 2000: 7.
188. Zeunert, 2024.
189. See: https://pastlivesofthenearfuture.files.wordpress.com/2013/07/000174.jpg
190. Wayne, 2013.
191. Pers. Comm. Emma Stockburn, 26 June 2020.
192. See for example an article and its imagery here: https://www.theleader.com.au/story/5511690/oyster-industry-collapses/#slide=0
193. Winston, 1957: 10.
194. See: https://www.flickr.com/photos/angeljim46/10753964324
195. Winston, 1957: 10.
196. See 1951 Miranda aerial image: https://www.flickr.com/photos/angeljim46/10643191264 and Sylvania & Taren Point in 1953: https://www.flickr.com/photos/angeljim46/10721451666
197. The aerial imagery from this time also reveals highly cultivated market gardens to the west of the oyster farm (see Historical Imagery Viewer).
198. And an offshoot for 'Sylvania Waters', Australia's first reality TV show (1992).
199. https://www.georgesrivertales.com/
200. Clark, 2017.
201. https://www.sopa.nsw.gov.au/About-Us/History-and-Heritage/Industrial-History
202. See for example: https://www.theleader.com.au/story/5137583/flashback-friday-photos-fishermen-get-raw-prawn-in-botany-bay/
203. See graph, https://pesticides.australianmap.net/1949-1986-union-carbide-owned-rhodes-homebush-bay-pesticides-included-245-t-24-d/

204. Anna Clark, cited at: https://www.abc.net.au/news/2018-02-18/a-short-history-of-fishing-in-australia/9449514
205. At the time of writing, the fish market is undergoing redevelopment and enlargement on the same site.
206. See: https://www.sydneyfishmarket.com.au/Corporate/History/Market-History
207. O'Connell, n.d., 1938.
208. CCPS, 1948: Figure 84.
209. Lawrence, 2011: 41.
210. Lawrence, 2011: 41.
211. Rathbone, 1986: 12.

References

Ashton, N. (1983). *New South Wales: Notes on the evolution of planning from 1945 to 1974 with particular reference to Governmental Involvement.* Unpublished. NSW Department of Environment and Planning.

Ashton, P., & Waterson, D. (2000). *Sydney takes shape: A history in maps.* Hema.

Birmingham, J. (1999). *Leviathan: The unauthorised biography of Sydney.* Knopf.

Budge, T. (2007). *Farmland preservation in Australia: Emerging issues—fragmented responses.* In W. Caldwell, S. Hilts, & B. Wilton (Eds.), *Farmland preservation around the globe: Challenges and opportunities.* University of Guelph.

Bunker, R., & Houston, P. (2003). Prospects for the rural-urban fringe in Australia: Observations from a brief history of the landscapes around Sydney and Adelaide. *Australian Geographical Studies, 41*, 303–323.

Burgmann, M., & Burgmann, V. (2000). *Green bans, red union: Environmentalism and the New South Wales Builders Labourers' Federation.* University of New South Wales Press.

Burnley, I. (2000). *Diversity and difference.* In J. Connell (Ed.), *Sydney: The emergence of a world city* (pp. 244–272). Oxford University Press.

Buxton, M., Tieman, G., Bekessy, S., Budge, T., Mercer, D., Coote, M., & Morcombe, J. (2006). *Change and continuity in Peri-urban Australia, State of the Peri-urban regions: A review of the literature.* RMIT University.

Cahill, J. J. (1951). 2nd reading, Speech delivered by the Hon JJ Cahill MLA in the Legislative Assembly on the County of Cumberland Planning Scheme. Local Government (Amendment) Bill. From Parliamentary debates, 29 May 1951. NSW Parliamentary Debates.

CCPCoC. (2009). *Economic profile of agriculture on the Central Coast Plateau, NSW*. Central Coast Plateau Chamber of Commerce. NSW Government Industry and Investment. Accessed November 3, 2020, from https://s3.ap-southeast-2.amazonaws.com/dpe-files-production/s3fs-public/dpp/213581/Final%20Farming%20Economic%20Profile%20Report%20Dec%2009.pdf

CCPS. (1948). *The planning scheme for the County of Cumberland*. Cumberland County Council.

Christie, M. (1988). *The Sydney markets 1788–1988*. Sydney Market Authority.

Clark, A. (2017). *The catch: The story of fishing in Australia*. NLA Publishing.

CLS&P. (2004). *Long bay correctional complex conservation management plan*. Clive Lucas Stapleton & Partners. https://csa.intersearch.com.au/csajspui/bitstream/10627/895/1/LongBayCMP04105%20-%20289p.pdf

Connell, J. (Ed.). (2000). *Sydney: The emergence of a world city*. Oxford University Press.

Copeland, L. (2020). A century of agriculture in the University of Sydney. *Agricultural Science, 31*, 58–65.

Court, T., & Zeunert, J. (2023). Novel horticultural adaptations to climate and their unique landscapes. In Y. Yang & A. Taufen (Eds.), *Routledge handbook of sustainable cities and landscapes in the Pacific Rim* (pp. 164–179). Routledge. https://doi.org/10.4324/9781003033530-16

Cox. (2014, June 12). *Macquarie University Campus Master Plan*. Cox and Macquarie University.

Dibden, J., & Cocklin, C. (2005). *Sustainability and change in rural Australia*. University of New South Wales Press.

Elias, J., & Coppins, S. (2013). *Pictorial history holroyd*. Kingsclear Books.

Erisman, J., Sutton, M., Galloway, J., Klimont, Z., & Winiwarter, W. (2008). How a century of ammonia synthesis changed the world. *Nature Geoscience, October*(1), 637.

Fitzgerald, S. (2008). *Red tape gold scissors: The story of Sydney's Chinese*. Halstead Press.

Fletcher, B. (1976). *Landed enterprise and penal society: A history of farming and grazing in New South Wales Before 1821*. Sydney University Press.

Freestone, R. (1989). *Model communities: The garden city movement in Australia*. Thomas Nelson.

Freestone, R. (1992). Sydney's Green Belt 1945-1960. *Australian Planner, 30*, 70–77.

Freestone, R. (2009). The first professor: Denis Winston comes to Sydney, 1949. *Australian Planner, 46*(1), 36–37. https://doi.org/10.1080/07293682.2009.9995288

Gaballa, S., & Abraham, A. (2008). *Food miles in Australia: A preliminary study of Melbourne, Victoria*. Centre for Education and Research in Environmental Strategies.

Gaynor, A. (2006). *Harvest of the suburbs: An environmental history of growing food in Australian cities*. UWA Press.

Gifford, E., & Gifford, E. (1970). The Chinese market gardeners of St George district. *St George Historical Society Bulletin, December*, 75–81.

Golledge, R. G. (1960). Sydney's metropolitan fringe: A study in urban-rural relations. *Australian Geographer, 7*, 243–255.

Harper, K., & Tait, A. (2019). Livestock industries in Australia: Production systems and management. In V. R. Squires & W. L. Bryden (Eds.), *Livestock: Production, management strategies and challenges* (pp. 79–136). Nova Science Publishers.

Historic Image Viewer. Spatial Services NSW. https://portal.spatial.nsw.gov.au/portal/apps/webappviewer/index.html?id=f7c215b873864d44bccddda8075238cb

Ihlein, L. (2014). *A heritage vision for sustainable housing goes to ICAC*. The Conversation. https://theconversation.com/a-heritage-vision-for-sustainable-housing-goes-to-icac-21616

James, P. (2013). *Cosmopolitan conversationalists: Greening modern Sydney*. Australian Scholarly.

James, S. (2016). *Farming on the fringe: Peri-urban agriculture, cultural diversity and sustainability in Sydney*. Springer International Publishing.

Jeffs, R., & Rose, H. (1995). *Faculty of agriculture handbook*. The University of Sydney. https://ses.library.usyd.edu.au/bitstream/handle/2123/1883/agriculture_1995.pdf?sequence=1&isAllowed=y

Lawrence, J. (2011). *St George pictorial memories: Rockdale, Kogarah, Hurstville*. Kingsclear Books.

Lewis, M. (1980). Milk, mothers, and infant welfare. In J. Roe (Ed.), *Twentieth century Sydney: Studies in urban and social history* (pp. 193–207). Southwood Press.

Liston, C. (2009). *Pictorial history: Liverpool & district*. Kingsclear Books.

Lynch, K. (1960). *Image of the city*. MIT Press.

Mason, D., & Knowd, I. (2010). The emergence of urban agriculture: Sydney, Australia. *International Journal of Agricultural Sustainability, 8*(1–2), 62–71.

McClymont, J. (2013). *Pictorial history: The hills district*. Kingsclear Books.

Mundey, J. (1981). *Green bans and beyond*. Angus & Robertson.

Nichols, M. (2015). *Fruit growing in the Hawkesbury*.s http://hawkesburyheritage.blogspot.com/2015/

NSW Agriculture. (1998). *Strategic plan for sustainable agriculture: Sydney region*. Department of Agriculture.
O'Connell, J. (n.d.). *Australian food history timeline*. Accessed May 12, 2019, from https://australianfoodtimeline.com.au/
O'Connell, J. (n.d.) 1883: *Australia's first agricultural college*. https://australianfoodtimeline.com.au/first-agricultural-college/
O'Connell, J. (n.d.) 1908: *Minchinbury Champagne first produced*. https://australianfoodtimeline.com.au/minchinbury-champagne/
O'Connell, J. (n.d.) 1938: *First Australian supermarket?* https://australianfoodtimeline.com.au/first-australian-supermarket/
O'Connell, J. (2017). *A timeline of Australian food: From mutton to MasterChef*. New South Publishing.
Oppenlander, R. A. (2013). *Food choice and sustainability: Why buying local eating less meat and taking baby steps won't work*. Langdon Street Press.
Pers. Comm. Anna Richards 25 May 2021.
Pers. Comm. Colleen Morris 27 February, 16 March 2020.
Pers. Comm. Emma Stockburn 26 June 2020.
Poore, J., & Nemecek, T. (2018). Reducing food's environmental impacts through producers and consumers. *Science, 360*, 987–992.
PSA NSW. (n.d.). *NSW privatised assets*. The Public Service Association of New South Wales and Community and Public Sector Union. Accessed March 25, 2024, from https://psa.asn.au/privatisation-hurts-everyone/nsw-privatised-assets/
Race, D., Luck, G., & Black, R. (2010). Patterns, drivers and implications of demographic change in rural landscapes. In G. W. Luck et al. (Eds.), *Demographic change in Australia's rural landscapes* (1 Landscape series) (Vol. 12, pp. 1–22). https://doi.org/10.1007/978-90-481-9654-8_1
Rathbone, R. (1986). *From White Gum flat to the suburb of Rockdale*. Rockdale City Council.
Rutherford, J., Logan, M., & Missen, G. (1967). *New viewpoints in economic geography*. Martindale Press.
Sandercock, L. (1975). *Cities for sale: Property, politics and urban planning in Australia*. Melbourne University Press.
Scarborough, P., Clark, M., Cobiac, L., et al. (2023). Vegans, vegetarians, fish-eaters and meat-eaters in the UK show discrepant environmental impacts. *Nature Food*. https://doi.org/10.1038/s43016-023-00795-w
Scott, A., Singh, M., Toribio, J.-A., Hernandez-Jover, M., Barnes, B., & Glass, K. (2017). Comparisons of management practices and farm design on

Australian commercial layer and meat chicken farms: Cage,barn and free range. *PLoS One, 12*(11). https://doi.org/10.1371/journal.pone.0188505
Scutts, L. (1958). *Subdivisions in the rural areas of the County of Cumberland.* Chairman of the Cumberland County Council. Department of Environment and Planning records, EPA, TR333.76. NSW State Archives.
SPA. (1967). *Prelude to a plan.* State Planning Authority.
Spearritt, P. (2000). *Sydney's century: A history.* UNSW Press.
Springmann, M., Clark, M., Mason-D'Croz, D., et al. (2018). Options for keeping the food system within environmental limits. *Nature, 562*, 519–525. https://doi.org/10.1038/s41586-018-0594-0
SROP. (1968, March). *Sydney region outline plan.* State Planning Authority of New South Wales. https://archives.cityofsydney.nsw.gov.au/nodes/view/1754543?keywords=1968%20sydney%20region%20outline%20plan&highlights=WyJzeWRuZXkiLCJyZWdpb24iLCJvdXRsaW5lIiwicGxhbiJd
Taylor, E., Butt, A., & Amati, M. (2017). Making the blood broil: Conflicts over imagined rurality in peri-urban Australia. *Planning Practice & Research, 32*(1), 85–102. https://doi.org/10.1080/02697459.2015.1028252
TKDA. (2017). *Parramatta north historic sites consolidated conservation management plan.* Tanner Kibble Denton Architects report for UrbanGrowth NSW.
Wayne, M. (2013). *Past/lives flashback #1: Union Carbide—Rhodes, NSW.* Accessed January 25, 2020, from https://pastlivesofthenearfuture.com/2013/07/09/pastlives-flashback-1-union-carbide-rhodes-nsw/
Willis, I. (2015). *Pictorial history: Camden & District.* Kingsclear Books.
Winston, D. (1957). *Sydney's great experiment: The progress of the Cumberland County Plan.* Halstead Press.
Zeunert, J. (2013). Challenging assumptions in urban restoration ecology. *Landscape Journal, 32*(2), 231–242.
Zeunert, J. (2017). *Landscape architecture and environmental sustainability: Creating positive change through design.* Bloomsbury.
Zeunert, J. (2018). Challenges in agricultural sustainability and resilience: Towards regenerative practice. In J. Zeunert & T. Waterman (Eds.), *Routledge handbook of landscape and food* (pp. 231–252). Routledge.
Zeunert, J. (2024). Bridging rural and urban disconnections: Spatial graphic explorations of Australia's livestock landscapes. In R. Pernice & B. Chen (Eds.), *Urban regeneration and rural revitalization in Australia and China* (pp. 110–143). Routledge.
Zeunert, J., & Freestone, R. (2022). From rural lands to agribusiness precincts: Agriculture in metropolitan Sydney 1948-2018. In C. Brisotto & F. Lemes de Oliveira (Eds.), *Re-imaging resilient food landscapes - perspectives from planning history* (pp. 247–272). Springer.

7

Period Five: 2000–2020

Box 7.1 Population 2000 and 2020

- *Population 2000:* 4+ million
- *Population 2020:* 5 million
- Increase = ~25 per cent / ~1 million increase in two decades.

J. Zeunert, A. Daroy, *Sydney's Food Landscapes*,
https://doi.org/10.1007/978-981-96-0710-5_7

Fig. 7.1 Horsley Park, 2019. Source: Nearmap and Joshua Zeunert

Box 7.2 Key Events 2000–2020

- Significant population growth rates in the city and nationwide, predominantly from immigration;[1]
- Sociocultural trends towards individualism and consumptive lifestyles;[2]
- Growing gap between the rich and poor;
- Increased neoliberal governance, prioritising global economic markets and trade;
- Democratic erosion from 'open' to 'narrowed';[3]
- Competitive productivism and agricultural intensification and industrialisation;
- Technological innovation and digital growth including the internet's considerable advancement;
- Continued development of long-chain food systems through plant breeding, cold storage, postharvest technologies and transportation logistics;
- Increasing genetic intervention in animal breeding (such as featherless chickens with more breast meat and favourable cuts);[4]
- Increasing use of computational technologies in agriculture such as remote sensing, GPS, drones, robotics, automation and sensors;
- Continued departure from seasonally available fresh produce to a 'permanent global summertime'[5] by major retailers;
- Mass species extinction including ecosystem damage heavily influenced by agricultural land clearance, particularly for grazing;[6]
- Decreasing pollinators and biodiversity through ongoing agricultural chemical use;
- Emergence of the Anthropocene concept, namely, the notion that human activities have become the dominant influence on the planet's geological stability.

Box 7.3 Key Events in Sydney 2000–2020

- The 2000 Sydney Olympics, marketed as the 'Green Games', seeking landscape rehabilitation and utilising native and indigenous planting;
- The 2005 Metropolitan Strategy committing to significant urban expansion, especially in north-west and south-west growth centres;
- The 2018 Metropolitan Strategy proposing to a 'third' city region in Sydney, the 'Western Parklands City';
- Significantly more high-rise residential developments, particularly around public transportation nodes;
- Continued emphasis on private car ownership as the dominant transportation mode, with some additional rail;

(*continued*)

Box 7.3 (continued)

- Realisation of the world's most extensive and expensive private toll road system;[7]
- The start of construction for the 24/7 Western Sydney Nancy-Bird Walton Airport ~55 kilometres from the CBD and its 'aerotropolis' precinct;[8]
- The catastrophic 'Black Summer' bushfires of 2019–2020, resulting in the destruction of 5.5 million hectares of land across NSW and 3 billion burnt or displaced (vertebrate) animals across south-east Australia;[9]
- The Covid-19 pandemic from 2019, resulting in supply chain interruptions, lockdowns, border closures and panic buying of food, essentials, agricultural animals and edible plants.

Box 7.4 Sydney's Metropolitan Agriculture Trends 2000–2020

- The overall decline of metropolitan agriculture across the 11 typologies and consequently, a significant reduction of local food security;
- Further loss of existing metropolitan agriculture, especially due to greenfield developments in urban growth regions;
- Policy and planning's overall failure to protect agricultural land loss against rezonings;
- Animal agriculture's continuing intensification and upscaling (poultry, feedlots, dairying) resulting in conflicts, relocations and production decline in peri-urban regions;[10]
- Increasing horticultural intensification through greenhouses, crop protection and the use of artificial (hydroponic) growing mediums;
- The continued shift from food production to amenity outputs, such as turf, cut flowers and ornamental-plant nurseries, on agricultural land;
- Agriculture officially included within the Western Sydney Parklands;
- A proposed agribusiness export precinct to accompany the new Western Sydney Airport;
- Agglomeration of agricultural institutions;
- Neoliberal and corporate-influenced educational shifts to agri-business, agri-science, agri-tech and agri-food, evident in tertiary institutions and academic research trends;
- An upsurge in the demand for local food industries during the Covid-19 pandemic;[11]
- Increased interest and participation in public forms of urban agriculture;
- Strong ongoing local mushroom production.[12]

While only two decades, the scale and pace of *Period Five*'s changes significantly diminished the city's food bowl. In the year 2000, Sydney's population was just over 4 million. By 2020 it and had reached 5 million—adding the combined populations of Canberra, Hobart, Darwin, Cairns and Bendigo. More homes were built than at any time in Sydney's history.[13] Not content with growth levels, the 2018 Metropolitan Strategy proposed to add more new dwellings in the following 20 years than were built, for instance, in Adelaide since it was founded in 1836.[14] It is perhaps unsurprising, then, that Greater Sydney was one of the top 10 fastest-growing regions in the Western world.[15] Remaining agricultural lands were subsumed in a frenzy of urban expansion accompanied by a lack of effective policy, legislation or practice measures to halt metropolitan agriculture's evisceration.

Relevant Global and National Events: 2000–2020

Period Five marked a key moment in human history. In addition to the commencement of a new millennium, in 2000 Paul Crutzen and Eugene Stoermer proposed that exponentially heightened human activities since the industrial revolution have altered the planet's geological and atmospheric stability to the extent of ushering in a new geological epoch—the 'Anthropocene'.[16] These historically unprecedented and ongoing ramifications stem from "unsustainable energy use, land use and land-use change, lifestyle and patterns of consumption and production across regions".[17] Their severity and extent will ensure that "[hu]mankind will remain a major geological force for many millennia, maybe millions of years, to come".[18]

The period therefore saw increased impacts accumulating from agricultural land-use, food production and dietary consumption patterns[19] that risk ecosystemic collapse.[20] Anthropogenic (human-caused) activities increasingly exerted, and continue to exert, adverse impacts on ecosystems and in-turn on food and water security.[21] Unsustainable expansion and intensification propelled by economic imperatives and consumer choices led to, for example, extensive deforestation for beef production.[22]

An increasingly unstable wider planetary climate correlated with record-breaking droughts, heatwaves, bushfires and intensified rainfall and flooding events across Australia during the period. The complex mechanisms and processes eroding planetary carrying capacity were underpinned by the relentless pursuit of infinite economic growth, paradoxically supported by finite planetary resources. Accordingly, competition for land and water resources continued to increase.[23] Despite planetary, national and metropolitan warning events, Sydney's appetite for, and manifestation of, perpetual growth continued its unrelenting advance.

Fossil fuels, particularly oil, served as the lifeblood for transport, including in shipping, aviation, freight trains, trucks and cars. Coal, meanwhile, undergirded Australia's and New South Wales' electricity grids, powering essential services such as passenger trains, lighting and battery-operated vehicles and devices. Natural gas was extensively used in heating, cooking and the industrial production of agricultural fertilisers vital to national and state food production systems. The (unholy) trio of oil, coal and natural gas therefore served as the cornerstone of Australian industries and lifestyles despite their environmental implications, contributing significantly to greenhouse gas emissions and driving planetary climatic instability.[24]

Awareness of our precarious dependency on finite resources experienced fluctuations in *Period Five*. Only baby steps were taken to transition from acute fossil-fuel dependency.[25] 'Peak oil' was initially a concern, although this dwindled with widespread and problematic fracking to extract oil and gas from low-grade sources. Prolonged droughts precipitated the construction of saltwater desalination plants, creating a consumer illusion of freshwater abundance. These energy-intensive behemoths transforming saline water to fresh, however, only exacerbated climate change by drawing from the coal- and gas-fired Australian electricity grid.[26] While Sydney's desalination plant incorporated a dedicated wind farm, the state's grid was dominated by coal power.

Sydney's agricultural practices were significantly determined by economic policy and neoliberal governance across *Period Five*. As discussed in *Period Four*, neoliberal governance is marked by short-term gearing, privatisation, declining social responsibility, reduced political and corporate accountability and erosion of democratic processes.[27]At a federal level, former Prime Minister John Howard (1996–2004) ushered in the period with an "investor state" ethos.[28] Tim Battin notes that the subsequent Australian Labor Party leadership's (2007–2013) "unwillingness or inability to mark out a clear alternative to neoliberalism was fundamental" to Australian political directions.[29] Within New South Wales itself, *Period Five*'s first decade was held by the Australian Labor Party (ALP), and the second, by the Liberal Party.[30] While individual premiers sought to differentiate party policies and practices, bipartisan trajectories reflected the overall subscription to neoliberalism with minor nuanced differences in execution.

Neoliberalism continued to reshape political, legal and decision-making processes. While privatisation rose, civil liberties fell. Activism and non-violent direct action were increasingly muzzled and criminalised by chiefly Liberal Party Federal and State governments. This saw Australian democracy be downgraded from 'open' to 'limited',[31] moving it "in line with the United States, Ghana and Botswana in terms of civil freedoms".[32] The insidious erosion of egalitarian values met little resistance, aided by increased criminalisation of physical protests. Australia's, and Sydney's, social order hence increasingly facilitated a 'survival of the richest' trajectory, as noted by the World Economic Forum's *Future of Global Food Systems* report.[33] The gap between rich and poor considerably widened.[34] Other downward macro trends in the period included declining happiness,[35] housing affordability, real wage growth and job security.[36] The neoliberal ethos fed through federal, state and most local governments and their planning regimes, becoming increasingly normalised. This was aided by concentrated mass media ownership that continually broadcast and normalised neoliberal worldviews.

How, though, did this affect agriculture? By the 2000s, Australian farmers faced minimal government financial support under the paradigm of 'competitive productivism',[37] compelling perpetually increasing outputs on larger and larger farms. This was compounded by neoliberalism in food retail, with competition diminishing amidst Coles and Woolworths' growing duopoly. Many smaller and local growers were unable to comply with the retailers' stringent production standards for cosmetic and uniform produce, equating to greater industrialisation across the food system. The free market-driven importation of vegetables exerted pressure on domestic horticulturalists and contributed to the decline of segments within the industry, prompting concerns about Australia's future food security.[38] Writing in 2015, Claire Baker notes that the community "has been deeply marked by the economic and social restructuring brought about by neoliberal and globalising trends and their impact on agricultural production".[39] She traces the shift from post-Second World War "ex-servicemen turned farmers and nation-builders" to a "deep recalibration of values to those associated with market performance, productivity and competitiveness".[40]

The neoliberal model of governance lacks the intergenerational perspective and ecocentric ethic required to address big-picture environmental and social issues in a time of ecological fragility and existential crisis. Australia's short-term growth came at a cost to ongoing ecosystem provisioning and support, planetary biocapacity and carrying capacity.[41] Consequently, many currently farmed areas were and are "undergoing salinisation, acidification and other forms of soil degradation as part of the productivist (intensive, chemically based) farming practices that have been in place since the 1960s".[42] In 2006, for instance, Australian agriculture used a staggering 70 per cent of the country's freshwater.[43]

Remnant industrial and manufacturing modes also continued to diminish. *Period Five*'s dominant sectors were service- and knowledge-based industries. Urban manufacturing, celebrated in earlier periods, further dissolved, relocated or moved offshore due to resident complaints, health concerns and cheaper labour markets comparatively unencumbered by worker rights. The John Howard-led Federal Liberal Party imposed global, increasingly free trade policies that reduced prices for many items but 'mothballed' local industries and their workforces. Many long-established factories, port infrastructures and farms closed, while others were redeveloped into commercial and residential areas, taken up by consumption-oriented white-collar migrants. Local consumption increased but demand was increasingly supplied by overseas factories and producers, whose working conditions and environmental implications became imperceptible.[44] Environmental impacts therefore escalated but were less visible. Local skies and waters became more crystalline while manufacturing sites abroad grew hazier, as glimpsed in the grey skies in the 2008 Beijing Olympics broadcasts.[45]

The period hence consolidated a production shift from local to global, and from material lifecycle efficiencies to expanded logistics and transport facilitated by crude oil and finite energy sources. Additional relevant factors during *Period Five* include concerns over foreign ownership of Australian farmlands; finance capital purchasing properties for capital gains; production of foods and biofuels for export to overseas investors; largely unregulated mining capital investments in coal and gas reducing high-quality farmland and the growth of intensified productivist farming creating significant environmental damage.

Relevant Sydney Events: 2000–2020

Sydney's appetite for urban growth significantly diminished its food bowl. In the words of former Hawkesbury farmer Gary Adams,

> I am frustrated because we were the food bowl and we no longer are. We rely on properties outside our area to supply the goods. We have navel oranges coming in from America, and we used to grow them here [...] At one stage we did have greenbelt areas brought in by the government, set aside to grow vegetables, where the land had to stay rural [...] That has all been thrown out [...] There are a lot of people [who] should look in their own backyard to see how we got here. It is not just the growers, it is politicians, local councils, town planners.[46]

Adverse environmental impacts were increasingly pushed beyond the city.[47] *Period Four*'s 'green bans' of the 1970s were inconceivable a mere three or four decades later under Sydney's unrelenting neoliberal thrust.[48] Over the years, Australia's concentrated news media ownership, coupled with hard-right politicians, industry lobbyists (some posing as think tanks), alongside a persistent strain of anti-intellectual sentiment and individual self-interest, gradually marginalised environmental, climate, interspecies and social justice discourses.[49]

By the end of the period, Sydney was "still devouring agricultural land as suburbia move[d] outwards and with motorway networks extending far to the west of the old CBD".[50] Maurie Daly offers a scorching appraisal in *Sydney Boom, Sydney Bust*, describing Sydney's suburbia as "mottled sub divisions, which pock-mark the broad acres of an expanding fringe, a result of developers' greed, atrocious planning, and a century-old passion for one's own house and land".[51] The city's addition of another one million people saw the decrease of arable land and natural resources per capita and the continued reduction of overall carrying capacity.[52] By the end of the period, Sydney's population exceeded the region's estimated total First Nations people in 1787 by 1667–2500 times,[53] and the Australian continent's total First Nations population prior to occupation by 5–17 times.[54] Given this population density in an area almost 1500 times smaller than the continent, the city's lifecycle demanded substantial resource flows from national and international sites in a complex pulsating web of supply networks.

Sydney's Metropolitan Planning: 2000–2020

Sydney's urban development was led by federally imposed, state-endorsed and metropolitan-planned population growth.[55] This in-turn unlocked a growth frenzy. The hierarchical 'pecking-order' of Australian governance frequently left Sydney's local governments burdened with accommodating and servicing urban growth. The top-down structure regularly rendered local governments relatively powerless to push back up the chain of command, albeit, many mayors and councillors welcomed growth. Four metropolitan plans were developed for the Sydney region across a mere 13-year period, more than in any prior timeframe, reflecting the pace of expansion and extent of change: *City of Cities* (2005), *Metropolitan Plan for Sydney 2036* (2010), *A Plan for Growing Sydney* (2014) and *A Metropolis of Three Cities* (2018).[56]

Official planning was characterised by reactive and facilitatory processes responding to federal government population growth intakes. These were geared to satisfy the economy's incessant demand for ongoing growth but lacked accompanying meaningful debate.[57] As a result, population growth and urban expansion encountered minimal effective resistance, both in metropolitan Sydney and across the nation. Planning efforts lacked aspirations and vision beyond maintaining the status quo and struggled to accommodate the substantial population increases set without due consideration of what figures translated to in built terms.[58] Australia's growing population was primarily funnelled into existing centres, with Sydney and Melbourne absorbing the largest influxes.[59] Apart from accommodating growth, most other planning initiatives were peripheral or reactionary and aimed to mitigate negative impacts.[60]

Planning therefore lacked the visionary scale required for Sydney to achieve anything resembling intergenerational sustainability.[61] There was a complete absence of recognition and response to issues such as decoupling from perpetual growth (planned degrowth),[62] navigating finite resource scarcity and moving to a post-carbon/greenhouse gas world (transitioning from dependence on fossil fuels—oil, gas and coal to 100 per cent renewables), mitigating mass extinctions and species loss (protecting and enhancing remnant ecosystems and containing the urban footprint), regenerating local and planetary biocapacity (ecosystem and environmental restoration),[63] offsetting greenhouse gas emissions and proactively adapting to climate change (not developing in and relocating from flood-, storm- and bushfire-prone areas). Relocalisation, namely, reorienting towards or at least enabling local systems and economies of supply, trade and exchange and restoring degraded lands through regenerative agriculture[64] and agroecology[65] were ignored and excluded. This meant that movements like the Transition Towns Network[66] and initiatives like Shadow Planning,[67] both seeking ecologically viable, socially just and biocapacity-regenerating futures, were rendered hopelessly idealistic when pitted against Sydney's Goliathan commercial development culture.

Twentieth-century notions of Western 'progress' as development, growth and intensification continued to prevail, despite their ongoing environmental costs. These manifested in Sydney through its population increase of one million, construction of a 24/7 airport, the world's most extensive private toll road network, a sea of private-vehicle-dependent and energy-inefficient detached housing estates, and a raft of unsolicited developments such as a casino.[68]

Metropolitan Planning and Agriculture

Table 7.1 lists and summarises specific mentions of agriculture within Sydney's four metropolitan plans during *Period Five.*

We note that certain commentators have highlighted a considerable variability in the extent of agricultural engagement across each strategy.[69] Yolanda Gil et al. are a rare exception to a prevailing published view that agriculture is inadequately incorporated in Sydney's metropolitan plans,[70] instead asserting that "there is a strong commitment to urban farming in metropolitan Sydney strategic plans and directives".[71] As evident in Table 7.1, there are only cursory and fleeting references made to agriculture, rather than forming an integral element in the strategies. Regardless of any stated aims, there was a marked disjunct between planning rhetoric and the tangible built environment outcomes (Figs. 8.11–8.15). While Sydney's metropolitan planning schemes theoretically acknowledge the city's limited physical capacity concerning cultivation and transportation of produce, in practice, peri-urban agriculture was perceived, treated or rendered as land awaiting development. All nine metropolitan plans failed to formally establish or actualise the preservation of agricultural lands, as evidenced by the significant and ongoing disappearance of agriculture since the release of Sydney's inaugural official metropolitan plan in 1948.[72] Plans have consistently neglected to address changes in agricultural land use and rezoning, nor prevent encroachment on existing agricultural areas.[73] Consequently, any assertions regarding the importance of agriculture to the city rang hollow.

Peri-urban land speculation consistently eroded metropolitan agriculture's productive land. This manifested in two ways: firstly, agricultural lands were often left fallow as landowners awaited lucrative opportunities to sell to developers, and secondly, incompatible adjacent land uses encroached upon agricultural areas, rendering farming operations increasingly challenging or unfeasible. The latter frequently resulted in incumbent land uses, such as residential areas, being prioritised over the rights of pre-existing agricultural operations. Notably, a critical mass of complaints from new collective ratepayers often held considerable sway with local councils or regulatory authorities, impeding farming activities.

Proximate agriculture did not necessarily hold appeal for residents, with the realities of squeals from a nearby abattoir or smells from poultry

Table 7.1 Sydney's metropolitan plans and their references to agriculture 2000–2020[a]

P5 Metro Strategy	Year	Core strategy	Rural territory	Agricultural and rural land elements
City of Cities	2005	– A polycentric global and growing city for 1.1 million more people in 25 years (to 2031), '5 cities–3 corridors–1 global city'. – 640,000 new homes; 500,000 more jobs; 7500-hectare new industrial land; 6.8 million square metre new commercial floor space; 3.7 million square metre new retail space. – Growth though north-west (NW) and south-west (SW) growth sectors, accommodating 160,000 new dwellings[b] and "provid[ing] between 30 and 40 per cent of new housing over the next 25–30 years".[c]	Rural and resource land	– Policy foci to maintain rural activities and resource lands, exclude incompatible and inappropriate uses, and encourage investment. – Conflicts identified with vegetable production and greenhouses in proposed north-west and south-west growth centres. – "Maintain rural activities and resource lands; Protect resource lands from incompatible and inappropriate uses" (p. 43).

(*continued*)

Table 7.1 (continued)

P5 Metro Strategy	Year	Core strategy	Rural territory	Agricultural and rural land elements
Metropolitan Plan for Sydney 2036	2010	– A 'sustainable, affordable, liveable, equitable and networked' vision for metropolitan Sydney. – Updated 2005 strategy and integrated with the Metropolitan Transport Plan.	Rural/resource lands	– The major strategic direction, "F: Balancing Land Uses on the City Fringe", has four main objectives: To contain Sydney's urban footprint, maintain and protect agricultural activities and resource lands, encourage investment in agriculture and resource lands, and ensure Sydney's soil health. – Accentuation of conflicts with greenfield development in the suburban growth centres with loss of agricultural land and residential complaints about agricultural activities. – "Contain Sydney's urban footprint" (p. 7,8),—"maintain and protect agricultural activities and resource lands, encourage investment in agriculture and resource lands, and maintain Sydney's soil health" (p. 161). – Budge (2013: 374) finds 18 references to the term 'food' in the 2010 plan, however, notes the absence of a food strategy for Sydney.

A Plan for Growing Sydney	2014	– A strong global city – A great place to live	Metropolitan rural area	– A "rural and bushland backdrop" to be managed to balance local growth needs, environmental protection, resource management, agriculture, tourism and culture, research activity, military uses and community safety. – Economic value acknowledged along with a need for a new "strategic framework" (p. 98).
A Metropolis of Three Cities	2018	– Three Cities (Harbour, River, Parkland). – A productive, liveable and sustainable poly-centric '30 minute' conurbation of 8 million people by 2056. – Place-based planning.	Metropolitan rural area	– Four major landscape 'types' to which the 'Metropolitan rural area' (MRA) encompasses agricultural and rural land elements. Objective 29 focuses on the MRA, where "environmental, social and economic values in rural areas are protected and enhanced" (p. 23, p. 160). – Strategy 24.3 endeavours to "protect and support agricultural production [...] by preventing inappropriately dispersed urban activities in rural areas" (p. 140).

[a]With thanks to Robert Freestone. See Zeunert & Freestone, 2022
[b]CoC, 2005: 28
[c]CoC, 2005: 57

sheds less appealing than the resultant bacon and egg breakfast.[74] Activities also frequently took place in contested, 'messy' sites. Figure 7.1 shows Horsley Park in 2019. On the left, we see part of a 43-hectare landfill and resource recovery facility licensed to take up to 430,000 tonnes per annum of general solid non-putrescible waste. Adjacent are hay fields providing feed for livestock, bisected by high-voltage power lines. A gas line and facility is just out of frame. Remnant bushland, flammable for much of the year, wraps around the hay fields and links into a regional vegetation corridor, which is important for ecological habitat connectivity. Market gardens are squeezed between the M7 toll road (a system carrying over 190,000 vehicles per day in Quarter 1 of 2020)[75] and a junior dirt bike track—although the gardens were abandoned in the following years.

Figure 7.2 illustrates typical (sub)urban expansion into agricultural areas. The rezoning of previously agricultural land in Leppington enabled residential developments to intrude upon established market gardens, as depicted in the 2018 image. Former poultry sheds and a nursery lay fallow, awaiting their inevitable transmutation into suburbia. By 2021, they were demolished and replaced predominantly by detached houses. Figure 7.3's proposed 2019 residential development in Picton and Tahmoor highlights the juxtaposition between agricultural, industrial, recreational and infrastructural elements. The aerial view reveals state and local government approved rezoning of peri-urban and former agricultural lands to accommodate residential greenfield subdivisions. These starkly contradict the objectives outlined in the state Government's metropolitan strategy for preserving rural agricultural and resource lands, as cited in Table 7.1. Marketing obscures potential residents' awareness of nearby peri urban land use impacts, such as spraying, odours, pollution, dust and noise.[76]

Planning and local councils routinely privileged urban over agricultural land use. Farmers' 'right to farm' bills responding to urban encroachment failed across peri-urban Sydney. For many farmers, it was easier to relocate further afield or sell their land than resist and fight. Planners and decision-makers regularly justified the loss of peri-urban agricultural land by arguing that it was minimally used, fallow, exhausted of fertility, or of a low agricultural land capacity classification.[77] These claims must be considered in the light that purportedly fallow or exhausted agricultural

Fig. 7.2 Rezoned agricultural land in Leppington with residences encroaching upon a pre-existing market garden. Sources: Joshua Zeunert using Nearmap image

lands facilitate rezoning and inflate their financial value. This is a routine and widespread speculative game played in peri-urban regions.[78] An example of agricultural land rezoning can be found in poultry mogul Baiada's development arm, Celestino, and their persistence to push ahead with the prolonged process of rezoning 11 agricultural properties in Glossodia. The subsequent 580-lot residential estate called 'Jacaranda' was supported by Hawkesbury Council, whose remit includes Sydney's most valuable lands for agriculture, and Australia's oldest continually operating farmland (Fig. 8.13).[79]

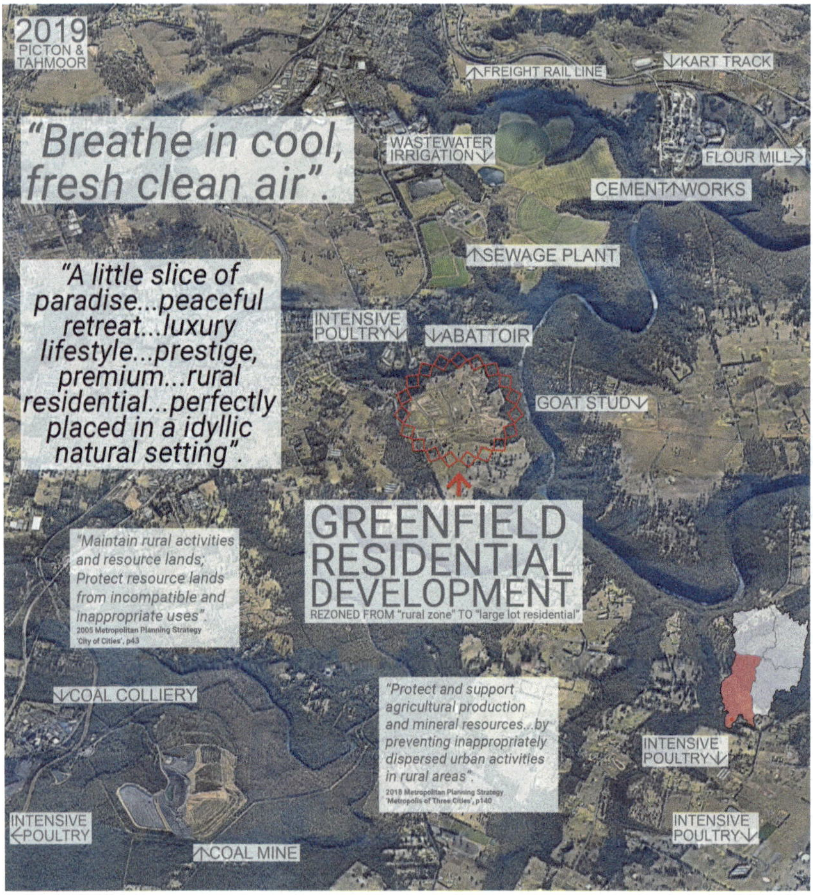

Fig. 7.3 A proposed residential development in Picton and Tahmoor, sampling marketing spin and planning rhetoric that contrast with actual land use activities. Sources: Joshua Zeunert using Nearmap image

Unfettered Growth

Sydney's metropolitan plans failed to accurately depict population growth and dwelling projections, and specifically, their translation to urban form and associated infrastructures.[80] Instead, vague 'blob-diagrams',[81] contradictory,[82] or outright misleading[83] misrepresentations obscured the true extent of Sydney's spatial growth.[84] It is difficult to pinpoint whether this demonstrates a lack of spatial ability to accurately present the actual size of planned growth, or too great of a challenge for planning's and governance's public relations to wash—or both. Regardless, Sydney's metropolitan plans' growth numbers failed to accord with their spatial graphics, just as their rhetoric did not accurately correlate with the reality of the city's urban development (Figs. 8.12–8.15).

The absence of clearly delineated spatial and visual planning efforts hindered comprehension of the immense scale of urban development required to meet targets in their condensed timeframes. Instead, the best examples of spatially accurate projections, based on potential densities and measured configurations, was executed in academic publications by landscape architects Richard Weller and Julian Bolleter.[85] These seminal works clearly and accurately conveyed the magnitude of national and state planned growth and politicians' and planners' tremendously ambitious neoliberal targets.[86]

The inception of Sydney's first Metropolitan Strategy in 2005 posed a significant challenge to the preservation of peri-urban farmland, and to farmers themselves. Specifically, the Strategy's proposed establishment of north-west and south-west growth centres was deemed highly detrimental to Sydney's ongoing capability to provide fresh produce and foodstuffs.[87] A 2009 study revealed that over 50 per cent of Sydney's vegetable-growing enterprises and 60 per cent (by area) of its greenhouses were located within these two proposed growth centres.[88] However devastating for metropolitan agriculture, the 2005 Strategy's implications were merely a precursor to subsequent plans' far-reaching consequences.

Examples of colossal growth numbers are highlighted through both the 2005 and 2018 Sydney metropolitan planning strategies. The former planned for 640,000 new dwellings by 2031[89] based on its 2005 publishing year. The latter, published in 2018 by the Greater Sydney Commission (GSC), proposed 725,000 new dwellings by 2036[90] and a "net jobs growth of 817,000 to 2036".[91] To help put this in context, 725,000 dwellings is 163,000 more than the total in Greater Adelaide over 180 years to 2016[92] and slightly less than Greater Perth's total of 818,000 over 187 years (Perth supported almost 2 million people in 2016).[93] Sydney's official plan was to pump out a comparative amount to Adelaide and Perth in under two decades.

Despite this, the 2018 metropolitan plan claimed that there would be "little change in [Sydney's] outward spread".[94] It further stated that 81 per cent[95] of the earmarked 184,500 dwellings for the GSC's 'Western Parklands City' (WPC) would be detached housing,[96] otherwise known as suburban sprawl.[97] Recent aerial photography not only evidences widespread new sprawl in the WPC, but new developments on many prior agricultural lands.[98] The GSC's 2018 Metropolitan Plan also failed to specify a proportion of 'infill' versus 'greenfield' development targets.[99] Instead, the 2018 Strategy set 'housing targets' across districts but did not specify whether these were new land releases or infill developments.[100] This absence signalled the relaxation of new land releases, translating to increased loss of agricultural land and counter-actions to their claim of little outward spread.

Manipulating and Bypassing Planning

Although one of the world's epicentres of urban growth, Sydney's Metropolitan Strategies' incredible extent of planned urban growth did not satiate the city's rampant development culture. Developers employed several key tactics to circumvent the New South Wales planning system. Firstly, they utilised Ministerial discretionary powers, which granted the State Planning Minister and the Director-General of the Department of Planning the authority to assume delegation for the direct approval of purportedly large-scale or 'state-significant' planning proposals. This discretion was typically final and indisputable, and somewhat parallels Lieutenant-Governor Grose's supplanting of Magistrates with central military figures during *Period One*. Secondly, developers frequently sought to curry favour through political donations, often through shell businesses. Under the state Labor government's administration from 2005 to 2011, these practices were conducted under 'Part 3A' of the NSW *Environmental Planning and Assessment Act*. Part 3A became notorious for its susceptibility to corruption and bias, as it provided a loophole for lobbying and donations by affluent interest groups.[101] Despite pledges by the incumbent Liberal government in 2011 to address the issue, Ministerial discretionary powers persisted, simply rebranded as 'State Significant Developments' (SSD) and 'State Significant Infrastructures' (SSI). However, as noted by James Weirick, these powers were often applied arbitrarily to projects lacking genuine state significance to manipulate desired outcomes.[102]

Sydney's escalating commercial practices of 'unsolicited urbanism' and 'uninvited developments' were in many ways more concerning than ministerial (in)discretion. In the latter stage of the period, developers gained the liberty to propose large-scale commercial private projects as 'unsolicited developments'.[103] This allowed developers to effectively dictate their own terms and proposals without adhering to established planning codes.[104] Dallas Rogers and Chris Gibson characterise unsolicited urbanism as "a neoliberalization of urban development" that "connects money and power in new ways to reshape cities" and bypass the planning system, councils and environment courts by normalising "planning-as-deal-making".[105] Consequently, top-down decision-making and prevailing commercial development culture exploited these and other vulnerabilities within the planning system. A continuous stream of cases concerning urban development, including land rezonings and contentious major project approvals,[106] have been (and continue to be) referred to the Independent Commission Against Corruption (ICAC).[107]

These high-level circumventions of the planning system exposed its inherent flaws and structural cracks. Planning in New South Wales' could offer a smooth conduit for wealthy and politically influential individuals to bypass congested ground level procedures. Bureaucratic hurdles for simple tasks such as constructing a shed, veranda or house extension could be more bureaucratically arduous than those for developer-initiated projects, such as a 1500-lot rezoned subdivision of agricultural land for greenfield housing development or an unsolicited $3 billion tunnel and tollway project. The hive of development activity required by ambitious urban growth targets pushed metropolitan agriculture to the sidelines in favour of neoliberal agendas and economic growth as Sydney frequently hovered at the top of the 'world's most expensive city' rankings.

Unreliable Statistics

Period Five invokes questions of how best to capture metropolitan agriculture's value and extent. The ABS states, for instance, that the value of agricultural production in Greater Sydney[108] in 2015–2016 was $645 million or 5 per cent of the value of New South Wales' agricultural output.[109] This may be dismissed as insignificant when compared with a single development, such as 'Sydney Science Park' at Luddenham by Celestino on formerly agricultural land, with a self-claimed value of $5 billion dollars. It should be remembered that development's high, one-off, short-term windfalls are not an equal measure of agriculture's perpetual economic value. If properly managed, agriculture can generate economic returns in perpetuity, even without considering its life sustaining nature. As both the County Cumberland Council[110] and Denis Winston noted during *Period Four*,[111] local agricultural production's value to the city is meta-economic. But unlike metropolitan elements such as biodiversity, water and air quality, scenic and rural amenity, recreation and heritage,[112] agriculture's benefit beyond its annual economic value is rarely recognised.

It is difficult to discern the true extent of Sydney's metropolitan food production in *Period Five.* Certain elements are underrepresented, suggesting that agricultural activity may have been higher. Others appear to be overinflated. As James et al. noted in 2010,

> Despite the need for clear and coherent data, however, previous studies of Sydney's agricultural land have consistently failed to create a coherent picture of Sydney's agricultural industry and its changing nature and extent [...] Characterised by small scale farms run by culturally and linguistically diverse groups, Sydney's vegetable farms have been seen to belong to a class of agriculture—peri-urban farming—which has been systematically under-recognised by an ABS data collection methodology focused on broad scale Agriculture.[113]

James et al. further appraised the official Australian Bureau of Statistics (ABS) data from at least the early 1990s until 2005–2006 as inadequately reporting and failing to capture the value of metropolitan and peri-urban agriculture in Sydney.

Our audit of the ABS 2015–2016 data revealed a significant discrepancy in which accountants' addresses rather than farm properties were included. Location data may hence be attributed to Sydney-based accountancy firms and farms may "not even be located within the Sydney Basin area".[114] Conversely, as noted by James, farms within the region may have been overlooked. Further research is needed to determine the degree of inaccuracy. Additionally, as discussed in *Site and Topic, 2021+* and demonstrated in Fig. 2.8, measurements of Sydney's agriculture are based on a greatly varying boundary area. Studies frequently fail to clearly delineate exact boundaries and specify their total square kilometre area. Often, these areas surpass the city's geographical limits yet are still referred to as 'Sydney's agriculture' (see *2021+*).[115] Inaccuracies are concerning, given that authoritative bodies such as the Department of Primary Industries[116] and Greater Sydney Commission[117] directly rely on ABS statistics to inform high-level decision-making.[118] Organisations like the National Trust[119] and journalists and media outlets also parrot this data, leading to a perplexing array of mismatched areas, statistics and economic values.

Landscape Heritage and Planning

Landscape heritage was another concern in the period, rendered quaint by Sydney's metropolitan development and planning regimes. While heritage protection mechanisms are well established in the built environment context, particularly for tangible built historic sites, landscape heritage preservation lacks agency. Colleen Morris and Geoffrey Britton's 2000 study underscored the importance of incorporating cultural landscape heritage into planning processes, with a specific emphasis on agricultural landscapes. The study recommended the establishment of a special zoning category within the environmental planning framework, such as Environmental Protection Zone 7, "especially for the conservation of important cultural landscapes as a whole and large landscape curtilages".[120] They suggested that landscape heritage items include "gardens, vineyards, orchards, paddocks, fences, cemeteries, grant areas, windbreaks and accessways"[121] and advocated for incorporating agricultural landscape elements.

Where Sydney estate houses were preserved for heritage, their landscapes often received inadequate protection, especially because annexing estate land for development can fund maintenance and improvements to built heritage fabric. Advocates such as James argued for leveraging heritage to safeguard Sydney's agriculture and particularly its migrant market gardens.[122] Although some progress has been made, planning largely continued to prioritise economic gains at significant cost to landscape heritage and its dynamic spatial, cultural and temporal patterns.

Food Planning

Although Sydney's metropolitan plans neglected or offered only tokenistic gestures towards including agriculture (Table 7.1),[123] 'food planning' internationally (re)emerged following the turn of the millennium.[124] Momentum began to build in certain city regions, especially across Europe.[125] The planning discipline in the United States also renewed its focus on the concept, as evidenced by the American Planning Association's 2007 *Policy Guide on Community and Regional Food Planning*.[126] Similarly, in the United Kingdom, The Royal Town Planning Institute issued a policy statement *Planning for Food* in 2010. The Planning Institute of Australia (PIA), on the other hand, was largely inactive with respect to food planning beyond fringe and sporadic aspirations. Despite submitting commentary in 2011 regarding the Federal Government's Food Plan, a promised "policy position statement on how to plan for a secure food supply" was notably absent from the PIA's website a decade later. Their 2011 submission astutely characterised metropolitan food production as 'contested landscapes' within the realm of urban development, however, this remained a battle they appeared unwilling to meaningfully engage in.

Figure 7.4 depicts two emblematic examples of contested land use. Situated along the Nepean River at Spring Farm (top), the site epitomises a nexus of suburban encroachment, competition between land uses and conflicting values. These include minerals and natural resources, safety and pollution, agricultural production, suburban expansion, landscape heritage, waste management and flood-prone lands.[127] Spring Farm's

Fig. 7.4 Spring Farm (top) and Tahmoor (bottom), 2019. Sources: Joshua Zeunert using Nearmap image

once agriculturally prized river flats (and northern floodplain) have been allocated to detached low-density residential housing. Part of Camden Park Estate is visible top-left, while Camden's sprawl exists just out of view to the west (top).

Tahmoor, in Sydney's far south-west, (Fig. 7.4, bottom) presents a contrasting case. Farming activities including an olive grove and market gardening within greenhouses are adjacent to poultry sheds and a car-wrecking yard, all overshadowed by the spectre of a coal colliery. Invocating the agricultural concept of *terroir* is intriguing in this production context, namely, the essence of a local context imbued in an agricultural product.[128] Both of Fig. 7.4's instances illustrate the contested, competing and conflicting characteristics of contemporary peri-urban land use relating to food production.

While the Commonwealth Association of Planners released a food planning discussion paper in 2011, it was chiefly tailored to subsistence-based food planning and metropolitan agriculture in developing-world Commonwealth contexts[129] rather than Australian cities and neoliberal governance challenges.[130] Trevor Budge highlighted food planning's invisibility within the Australian context in 2013, emphasising that the "sustainable supply of fresh nutritious food at a reasonable cost and distributed so that all consumers across a city can equitably access it is not part of the agenda in Australia's metropolitan planning strategies".[131] Budge argued for addressing food security as a key issue,

> Concerns about food production and access—related to peak oil, climate change, and population health—are contributing to a realization in some cities within the developed world that food needs to once again be incorporated into metropolitan planning strategies [...] Australian metropolitan areas need to consider adopting similar agendas. In order to ensure the future economic, social, and environmental sustainability of cities, food needs to become a key component of future metropolitan planning.[132]

In 2020, Federico Davilla similarly noted that "land at the moment is prioritised for development over agriculture [...] Farming is not seen as a useful activity in urban areas according to NSW government planning laws".[133] This reflects the significant departure from local food systems and food security as prominent concerns from Sydney's founding until early in *Period Four*.

Numerous attempts at food planning failed to significantly alter development's course during *Period Five*. Much of food planning's responsibility fell upon researchers and academic groups. The Victorian Eco-Innovation Lab (VEIL) at Melbourne University were particularly notable but their medium and long-term perspectives were disregarded by Sydney's planning and urban development communities focused on short-term profits.[134] A 2009 report highlighted the concerning trend of decreasing agricultural land in the face of growing awareness of food security.[135] A decade later, production land continued to dwindle despite the Covid-19 outbreak exposing food supply vulnerabilities in Sydney's 'just-in-time' logistics-oriented system. The city's surging population and urban footprint continued to require increasingly intricate and intensive supply systems as Sydney headed into its next phase.

At the end of *Period Five*, Megan Lehmann stated that there were "more than 2000 farms between the Blue Mountains and the coast, from the fruit orchards of Bilpin to the cabbage and cauliflower farms of the Hawkesbury river flats, from the market gardens around Penrith to poultry and livestock in the Wollondilly shire".[136] Sydney's planning trajectories paint a terminal picture for these farms' ongoing viability, with the collective metropolitan area confronting a persistent diminishment in local food production capacity and food security.

The 11 Typologies: 2000–2020

Period Five saw the considerable growth of high-resolution satellite-captured aerial imagery, marking a significant advancement in land monitoring technology. These heralded a shift in the transparency of land use activities that could no longer be readily concealed through on-ground 'visual management' screening approaches. Readily available aerial views revealed activities and temporal patterns, proving transformative for research and facilitated by online platforms like Google Earth and Google

Maps, both launched in 2005. Key platforms used to populate the period's database include 2004 and 2005 imagery in the Historical Imagery Viewer,[137] Nearmap imagery where available (usually at regular intervals from 2009 to 2020 but with varying metropolitan coverage)[138] and Google Maps satellite imagery. Along with aerial imagery, the maturation of Geographic Information Systems (GIS) and their increased precision, usability, open-source and public availability greatly increased accessibility to land use data.

Agricultural Facilities and Education (AF): Defunding and Decline

Period Five's ongoing decline in traditional agri*culture* led to an increasing shift to agri*business*. Aerial photography suggests that Long Bay Gaol abandoned its almost century-long legacy of significant food production somewhere between 2004 and 2005.[139] This reflected an overall decline in government support for social services and their programmes as these sectors became increasingly privatised across the period. Despite the remarkable expansion of Australia's tertiary education sector largely driven by international enrolments, the discipline of Agriculture did not experience commensurate growth.[140] In 2013, the ABC noted a decline in the number of campuses offering Agriculture and Agricultural Science degrees, dropping from 23 in the 1980s to just nine in 2011, along with insufficient graduates to fill approximately 4000 agriculture-related positions advertised annually.[141]

The decrease in Agriculture degrees[142] reflects neoliberalism's infiltration of higher education. Like a microcosm of agriculture itself, *Period Five* shifted towards more centralised and consolidated educational operations and institutions in major centres, with international students flowing to universities highly ranked in global metrics. Unsurprisingly, a reduction of agricultural offerings resulted in fewer completions. Jim Pratley and the Australian Council of Deans of Agriculture quantified an alarming 53 per cent decline in undergraduate completions in Agriculture between 2001 and 2010.[143] Richard Harper and Pratley's data from 2002 to 2012 also reveals diminishing graduate numbers across Agriculture, Viticulture, Horticulture and Farm Management, with Animal Sciences demonstrating growth and then decline.[144] National data recorded between 2012 and 2018, however, showed steady university intakes into these disciplines[145] and growth across New South Wales from 2012 to 2016.[146] An increased proportion of female enrolments was significant given previous low enrolments (see AF in *Periods Three* and *Four*), exceeding 50 per cent in the state.[147] Also encouragingly, an increased proportion of the Agricultural sector possessed higher education qualifications, although the total overall proportion remained low and was less than half than all other industries combined.[148]

University fees increased over the course of the period, while graduate income thresholds for debt repayment fell. At the University of Sydney in the mid-2000s, "student enrolments in the agricultural sciences degrees declined, in parallel with a worldwide pattern of decreasing student interest in Agricultural Science courses, despite a strong demand for graduates".[149] Australia's high levels of urbanisation and continued focus on its capital cities and major urban regions during the timeframe partly account for this trend. Farm consolidation and agglomerations were also a factor. Agriculture-related educational offerings did, however, continued to expand into areas such as agri-tech, agri-science, agri-food and related areas such as logistics. Academic offerings increasingly catered to large-scale rural agribusiness operations or the corporate hubs supporting them. Smaller-scale metropolitan and peri-urban agricultural education was minimal, reflecting the increased corporatisation of Australia's agri-sector.

An east-facing aerial image from 2019 (Fig. 7.5, top) shows part of the University of Sydney's Camden main campus in Brownlow Hill, including the Corstorphine Dairy Farm. Pastures adjacent to the Mount Hunter Rivulet are irrigated to support a 350-head dairy herd. Like other elements in the University's agriculture portfolio, it functions as both an educational facility and supplementary commercial enterprise.

Fig. 7.5 University of Sydney's Camden main campus 2019 (top), Cobbitty and Ellis Lane 2019 (bottom). Source: Nearmap

Mixed Farming (MF): Mixed Farming Precincts

By *Period Five*, mixed farming had significantly diminished, if not disappeared entirely from Sydney's commercial agricultural operations. Mixed Farming entries in the *Period Five* Map Database focused on localised clusters of varying agricultural activities (e.g. Figs. 7.2, 7.3 and 7.4) rather than individual farms engaged in commercial-scale production across multiple typologies. As industries upscaled to remain competitive within the globalised food system, they simultaneously became more specialised and volume-oriented.[150] Although some hobby-scale, peri-urban properties continued to produce mixed agriculture in the city's outer regions, they seldom met the classification of 'primary producers'.[151] These were also critiqued by some professional farmers due to contributing to the fragmentation of agricultural lands and malpractices such as overgrazing and poor animal welfare.[152]

While some entries document heritage properties that historically practised mixed farming, such activities were not necessarily ongoing. Remnant building(s) were more commonly found, often accompanied by landscape curtilages that had lost their agricultural functions over time. Brownlow Hill Farm is a noteworthy exception as a historic and progressive establishment continuing commercial mixed farming alongside diversified income streams such as events.[153] As depicted in a 2019 south-facing aerial image, regions like Cobbitty (Fig. 7.5, bottom right) and Ellis Lane (Fig. 7.5, bottom left) also continued localised mixed farming such as irrigated crop pastures for livestock, market gardening, viticulture, orcharding and turf growing. The region also included agricultural education and research at the University of Sydney's Animal Reproduction Unit and Plant Breeding Institute (bottom centre).

In a positive move, mixed farming was accommodated within Australia's largest urban park, the Western Sydney Parklands. The Parklands sit across a formerly significant but diminished mixed-agricultural production region, via a 27-kilometre-long public open space corridor of approximately 5280-hectares. A 2016 article noted that the Parklands Trust committed to providing 10 per cent (about 500-hectares) of land for urban farming as per the Western Sydney Parklands Plan of Management 2020.[154] The Parklands website, however, halved this in 2020 stating "approximately 5% (264 hectares) of the Parklands has been set aside for Urban Farming".[155] It is also important to note that agricultural activities were required to resemble an aesthetic idyll of farming that does not necessarily equate to production realities under agribusiness' commercial pressures.[156] The Parklands hence sought to maintain "rural character" by "allowing sustainable agriculture, horticulture or forestry"[157] and "creating a place of peri-urban agriculture that celebrates a genuine rural character".[158] Emphasising visual character is problematic in light of the significant decline and cessation of many traditional agricultural forms, including orchards, small-scale market gardening, cropping and (genuinely) free range poultry. Instead, intensified production, plastic greenhouses and poultry and mushroom sheds have increased to stay economically competitive within neoliberal agribusiness' demands.[159]

Market Gardening (MG): Usurpation

Sydney's market gardens continued to retreat from the city's rapidly expanding edges. In late *Period Three* into *Period Four*, Sydney still possessed large agglomerations of market gardens clustered across expansive production districts. Most remaining areas, and indeed Sydney's overall market gardening industry, continued to diminish and fragment during *Period Five.* Many farming operations were abandoned and left to await urban land use rezonings and dissolve into new urban development areas. As such, the overall patterning of market gardening devolved into diffuse scatterings further and further away from the city centre. This is visible in the period's food landscape maps in Figs. 7.14 and 7.15. Subsequent operational market garden sites primarily fanned across peri-urban regions of the extensive western Sydney region, branded from 2018 as the 'Western Parkland City'.[160]

Horsley Park's market gardens and agricultural production diminished in *Period Five* (Fig. 7.1) and Spring Farm's and Camden's valuable lands succumbed to urban sprawl (Fig. 7.4). The Hoxton/Austral/Kemps Creek region offers an illustrative case of fragmentation and dissolution (Fig. 7.16).[161] A zoomed-in aerial view of Austral in 2019 (Fig. 7.6, top) illuminates new suburban street grids and houses gradually engulfing agricultural allotments. Further west, Greendale (Fig. 7.6, bottom) was less impacted but its proximity to the emergent Western Sydney Airport spells a likely demise, as with other agricultural regions within the airport's orbit. Wallacia, positioned on the opposite side of the Nepean River, may have some degree of protection from urban encroachment, or at least stave it off for longer given the expense of bridge building.

Fig. 7.6 Austral 2019 (top), Wallacia and Greendale 2019 (bottom). Source: Nearmap

The Map Database's 165 market garden entries for *Period Five* include multiple local production clusters within a single entry. This is reflected in Peter Malcolm and Riad Fahd's 2009 study that identified over one-thousand vegetable growers (1052).[162] These growers collectively cultivated 2025 hectares, in theory equating to an average area of 1.9 hectares each. One quarter of outdoor-field vegetable growing enterprises, however, constituted 65 per cent of Sydney's total of 1775-hectares,[163] meaning that the average land size for 75 per cent of Sydney's vegetable growers was less than one hectare. Greenhouse vegetable production occurred on 294 properties totalling 189-hectares.[164]

Aerial images reveal Sydney's few remnant operational market gardens that remained in the well-established urban fabric at the end of the period. Areas included Arncliffe, Kyeemagh, Matraville (Fig. 7.7) and La Perouse (Fig. 7.8). Figure 7.7 shows Arncliffe and Kyeemagh's remnant market gardens adjacent to Kingsford Smith Airport and Botany Bay in 2016 (top). The middle satellite aerial image reveals Kyeemagh's extraordinary market garden patterning in April 2020. Matraville contained one of the few remaining urban market gardens on Bunnerong Creek, shown in 2015 (bottom). Figure 7.8 presents an aerial view of the historic La Perouse Market Gardens in their unique urban context. The historically and culturally significant market gardens have operated constantly for 150 years,[165] albeit at perennial threat from the expansion desires of the adjacent cemetery.[166]

Sydney's mushroom production was the key sub-typology to defy the city's overall agricultural decline, with an impressive proportion of the state's output taking place in the Hawkesbury region. As the birthplace of Australia's modern mushroom industry, the region contained the greatest concentration of mushroom growers in the country. A 2008 study quantified 78 per cent of New South Wales' mushroom production and nearly a quarter of the national figure, at 11,335 tonnes of mushrooms per year from 15 of the state's 23 growers.[167] Figure 7.8 shows mushroom growers in Londonderry (middle, 2017) and Glossodia (bottom, 2019). Australian commercial production typically takes place in indoor semi-industrial shed facilities, thus bearing closer resemblance to intensive poultry facilities than other forms of agricultural production. Consequently, an absence of the wine or fruit industries' picturesque charm is unlikely to

Fig. 7.7 Arncliffe & Kyeemagh 2016 (top), Kyeemagh 2020 (middle) Matraville 2015 (bottom). Sources: Nearmap and Joshua Zeunert

Fig. 7.8 The historic La Perouse Market Gardens in 2016 (top). Mushroom growers in Londonderry (middle, 2017) and Glossodia (bottom, 2019) typify Australian commercial production, taking place in large shed complexes. Sources: Joshua Zeunert using Nearmap image

leverage similar volumes of agri-tourism. Moreover, the spatially efficient and intensive nature of mushroom cultivation presents a viable model for achieving agricultural success in metropolitan areas. Mushroom growing may prove to be the last remaining viable agricultural production form in Sydney.

Finally, another way of conceiving 'mixed farming' is through a lens of cultural diversity. Though beyond this book's scope, it is worth noting that market gardening captures the greatest cultural array of any typology. ABS data, for instance, reveals growers from over 50 different countries across Sydney's recorded agricultural industry.[168] In 2010, Sydney's vegetable production industry comprised 80 per cent to 90 per cent culturally and linguistically diverse (CALD) growers[169] and the Sydney basin contained the highest number of culturally diverse horticulturalists for any Australian region.[170]

Fruit (FR): Shadows of the Past

As we have previously observed, Sydney's once-thriving commercial fruit production industry had already significantly declined since *Period Three*. By the end of *Period Five*, only a handful of commercial growers remained. Their persistence bucked the trend, with growers partly or wholly resisting pressure to sell their land to developers or land bankers. The mounting pressures from increasing economies of scale drove other growers further afield into regional production districts, or closure altogether. Figure 7.9 (top) shows Cedar Creek Orchard in Thirlmere, for example. During *Period Five* it was a 45-hectare orchard with 23,000 fruit trees (apples, peaches, nectarines and persimmons) also producing 700,000 litres/year of apple juice.[171] In 2018, the orchard had subsequently declined with part of the area since being subdivided. Canoelands Orchard (Fig. 7.9, bottom), shown in 2020, is in an isolated locale enclosed by National Parks but steadily encroached on by suburbanisation.

The majority of the Map Database entries for fruit (Figs. 7.14 and 7.15) represent declining orchards and small remaining remnants during rather than commercial-scale operations. Numerous hobby-scale or

Fig. 7.9 Cedar Creek Orchard in Thirlmere shown in 2018 (top) has subsequently declined. An orchard in Canoelands in 2020 (bottom). Source: Nearmap

patches of prior, large orchards are scattered across once-vast orcharding regions such as the Hills District and the Hawkesbury, as well as areas in the west and south-west.

Cropping (CR): Feeding Hobby and Race Horses

Sydney's area under crop production also declined after *Period Four.* In *Period Five,* residual cropping activities primarily revolved around smaller-scale fodder production catering to fragmented uses such as equine needs (i.e. not food production), small-scale livestock operations, educational institutions and hobby farms. Irrigated crop production supported a diminishing number of commercial dairies (see Intensive Livestock and Fig. 7.12, top). There was a reduction in cattle slaughtering and therefore decreased demand for local fodder. Furthermore, intensive livestock operations increasingly favoured a grain-centric diet due to its fattening properties, with grain-producing districts and their economies of scale situated in drier regions beyond Sydney. Figure 7.1 illustrates cropping for livestock fodder at Horsley Park in 2019. Cropping also takes place at the Picton Sewage Treatment Plant and Picton Farm, aided by distribution of nutrient-rich wastewater by centre-pivots and irrigation onto crop fields (Fig. 7.3).

Intensive Livestock (IL): Clash of the Titans

The phenomenon that was Sydney's peri-urban poultry production in *Period Four*, like all of this book's agricultural production typologies, fell from its former heights. In 2010, there were 88 broiler (chicken meat) farms located in metropolitan Sydney.[172] According to key poultry figure John Cordina, the city's production capacity was "decimated".[173] Cordina argued that,

> Any reasonable assessment could only conclude that previous planning strategies have failed badly in relation to the poultry industry and other forms of agriculture. Accordingly, the Metropolitan Strategy needs to incorporate into its planning measures, protection of existing agricultural

> activity and also make provision, areas for future expansion. Land availability for poultry farming (and intensive agriculture) in the Sydney basin has been in decline and time is literally running out to reverse this disturbing trend.[174]

Cordina's recommendations[175] to protect the metropolitan poultry industry and Sydney's agriculture have been ignored. Production downturn was reflected in only 65 broiler businesses operating in 2020, a further 26 per cent fewer after Cordina's 2010 statement.[176] Some operations moved out of Sydney due to speculative land development of former poultry lands.[177] Poultry (Baiada and Ingham) and dairy (Perich) heavyweights[178] were in fact key catalysts spearheading western Sydney's transition from agricultural to urbanised land uses. Relocating poultry operations further afield became increasingly necessary due to the city's expanding urban footprint encroaching on the odorous industry, which also continued to increase the scales of its rural operations. This in turn placed economic pressures on now smaller metropolitan producers.[179]

If residents moving to the city's fringes held idealised conceptions of bucolic rurality, they were surely regularly rudely awakened by the stench of being within intensive livestock operations' orbit.[180] Figure 7.10's images (top, second row) are indicative of peri-urban poultry production, with sheds each containing tens of thousands of birds as shown in Freemans Reach, Cobbitty and a smaller-scale producer with close neighbours in Llandilo. The images, though, cannot capture poultry production's heavy ammonia-odours. Peri-urban intensive livestock production therefore required planning buffer zones, or navigating their lack.[181] So while the poultry sheds themselves are incredibly spatially and economically 'efficient', their true (smell) impact can extend for kilometres. In the bottom half of Fig. 7.10, we can see that being a neighbour proximate to a Greendale cattle feedlot would be similarly challenging, despite it being very small by Australia's rural standards.[182]

Australia's poultry industry itself did not decline in the period, with per capita consumption increasing.[183] Rather, poultry production diminished within Sydney itself, though abattoirs continued to operate across the metropolitan area. Poultry production is more independently distributed than processing and many of Sydney's peri-urban producers

Fig. 7.10 Poultry sheds in Freemans Reach, Cobbitty 2019 and Llandilo 2017 (top). A cattle feedlot in Greendale (bottom). Sources: Joshua Zeunert and Nearmap

contract their supply to city processors. Therefore, despite its overall decline, poultry remained the city's most financially valuable agricultural industry by some degree. The combined measure of nurseries, flowers and cultivated turf followed, with egg production in third place.[184]

Australia's cattle and dairy production systems continued their trend towards large concentrated feedlot and intensive industrial-scale operations, rendering it difficult for smaller and metropolitan producers to remain viable.[185] Like most of the agricultural typologies (and similar to Sydney's fruit industry), the city's dairy industry continued a gradual downward trajectory. This is not surprising, given Sydney's escalating land prices and the increasingly large scales and operational intensity required to remain economically competitive.[186]

Despite the overall downturn, Sydney managed to retain a handful of commercial dairy operations by the end of *Period Five*. Among these were Leppington Pastoral Company (LPC) and Country Valley, representing polar industrial and artisanal scales respectively (Fig. 7.11). LPC (top, centre and inset) operates at the rural commercial scale, as reflected by its integration into Australia Fresh Milk Holdings amidst their regional dairy acquisitions and expansions in the period. LPC is a self-confessed US-style of modern dairy farming, namely, intensive factory farming producing maximum milk volumes from breeds with exaggerated udder capacities, supported by irrigated cropping to produce feed.[187] LPC milked 2000 cows three times per day in an operation that is "one of the largest found in Australia".[188] Their Leppington operation is located directly adjacent to the Western Sydney Airport. When the proposed airport development began, LPC initially objected due to impacts on their farm's productivity but subsequently saw an opportunity to part with some of their land in a lucrative and controversial sale.[189] In contrast, Country Valley, nestled into a small valley in Picton, epitomised a smaller producer reliant on niche markets rather than mass distribution channels (Fig. 7.11, bottom).[190] Their small creek catchment is prone to varying climatic patterns and in the 2019 drought, the dairy stated that the "creek's dried out, the dams dried up—we've been trucking water in for our cows".[191]

Figure 7.12 depicts Peel Pastoral's Hambledon Park dairy. The Peel family were part of Sydney's dairy fabric for around a century. They purchased Hambledon Park in North Richmond in 1972, developing it into

Fig. 7.11 Dairying in Leppington (top, centre and inset) and Picton 2018. Sources: Joshua Zeunert and Nearmap

Fig. 7.12 Peel Pastoral's Hambledon Park in North Richmond in 2019. Source: Nearmap

a 1200-cow dairy farm before replacing dairy with a cattle feedlot for a breeding herd of around 300 Black Angus cows late in *Period Five*.[192] The dairy was listed for sale at $140 million in 2018 and aimed to attract commercial developers for residential rezoning. The dairy was decommissioned and in mid-2020 it was purchased by Baaida poultry offshoot Celestino for a reputed $60 million.[193]

Grazing (GZ): Keeping the Grass Down

Grazing hard-hooved livestock continued across western Sydney, as had been the case since Governor Phillip's herd of cattle escaped in 1788. Similarly, cattle continued to remain more prevalent than sheep. Livestock grazing usually occurred at non-commercial, educational and hobby scales to keep pasture grasses mown, rather than operating as full-time commercially viable enterprises. Grazing often took place on urban fringe properties awaiting purchase or on lands impending development. Grazing activities in (western) Sydney can be gauged to an extent through the total number of farm dams in paddocks, although many have been abandoned from prior grazing pursuits. Their steady disappearance often reflected the transition to peri-urban and urban land uses and chiefly, residential developments.

Vineyard Wineries (VW): A Rare Drop

Sydney's self-claimed oldest operating winery,[194] Vicary's in Luddenham, was forced to close in late 2015 to make way for western Sydney's new airport (Fig. 8.16). Like many wineries, Vicary's purchased grapes from local growers who on closure needed to find another market for their produce. As another key historic vineyard, the once-extensive Camden Park Estate significantly shrunk but managed to retain a portion in Elderslie. Despite the many challenges viticulture faced in *Period Five*, the Map Database includes boutique scales of production towards the outer reaches of the book's boundary. Nonetheless, Sydney's winemakers regularly relied on grapes imported from beyond this book's boundary. As such, most vineyards were vignettes rather than self-sufficient or commercial-scale grape production sites.

Agricultural Infrastructure (IN): A Few Remnants

Period Five witnessed nuanced changes to Sydney's agriculture infrastructures. Overall, an ongoing shift trended to large-scale supply chain and distribution systems, rather than metropolitan food production itself. Some local processing remained but required increasingly external sourcing as most metropolitan food production industries locally declined. A key example was a notable expansion in food logistics warehouses and cold-storage facilities along key transportation routes, enabling enhanced food freight movement from more distant locations. While this approach is currently economically feasible, it prompts sustainability concerns for a future of declining fossil fuels in an expansive continent. The modestly sized Central Coast Plateau, for instance, was the nearest area outside of the book's boundary accessible by road to central Sydney (over 100 kilometres). More substantial production districts included the Central-West (250–400+ kilometres), Riverina (over 600 kilometres), Richmond-Tweed (700–850+ kilometres) and Murray (550–800+ kilometres). Road-based supply chain and infrastructure investment were needed to bring 'fresh' (cold-stored) produce from production districts further afield. Such long-distance, diesel-and-AdBlue--dependent (required by trucks) food chains hence became normalised in the period.

Allied Mills was constructed in Wilton as a rare addition to the city's agricultural infrastructure and was listed in 2015 as capable of processing up to 300,000 tonnes of wheat and maize per year.[195] It was strategically located on a rail line and near the Hume Highway allowing for importation of wheat and grains from further afield through key transportation networks. Sydney Water incorporated new practices that adapted age-old use of sewage for agriculture, as commonplace in Sydney before installation of reticulated sewage infrastructure from late in *Period Two*. Investments were made to centralise the treatment and production of nutrient rich biosolids as a wastewater by-product at their St Marys Water Recycling Plant.[196] The Picton Sewage Treatment Plant also paired with Picton Farm, where the recycled water was utilised for crop and pasture irrigation to produce livestock fodder.

Returning to Sydney's poultry industry, *Period Five* saw shifting dynamics between production (usually poultry sheds) and processing (namely, abattoir) facilities, as previously discussed. In 2010, three of

New South Wales' five vertically integrated chicken meat processing companies had facilities in Sydney: Inghams in Liverpool and Ingleburn, Cordina in Girraween/Summertime in Galston and Red Lea in Blacktown.[197] Baiada/Steggles Pendle Hill processing operations were closed by 2010. Egg distribution was less centralised than poultry and included independent sellers across Sydney's peri-urban regions, while smaller poultry abattoirs also operated.

Some agricultural infrastructure managed to remain. The Flemington markets, for instance, continued to be Sydney's primary fresh produce wholesale traders in *Period Five*. Each day, 394 growers from the Sydney basin brought their produce to market for buyers to compare and trade products.[198] The variety of available fruits and vegetables had evolved over the years to mirror Australia's cultural diversity, with European produce grown from the 1950s, 1960s and 1970s and the integration of more produce from Middle-Eastern, Asian and African countries in recent decades.[199] Finally, Kellogg's manufacturing plant in Botany continued to operate, as did Arnott's Sydney-based production facilities.[200]

Aquaculture (AQ): Stuttering Respawning

Like Flemington Markets, The Sydney Fish Market consolidated its near monopoly on Sydney's seafood. In *Period Five*, the market was the largest of its kind in the southern hemisphere and third biggest worldwide, handling around 14,500 tonnes a year and generating $135 million in revenue in the 2012/13 fiscal year.[201] The market is a far cry from Sydney's first fish market where baskets of catch were dumped onto the floor amidst the din of trading, now conducting auctions via computer automated processes and becoming a polished tourist attraction, with restaurants, tours and cooking classes.[202] By the end of *Period Five* this is reflected in only half the market's seafood sourced from New South Wales, let alone from Sydney itself.[203]

In 2006, a commercial fishing ban was placed for Sydney Harbour's waters due to high levels of toxins in the marine food chain.[204] Although first touted as temporary, it remained in place at the end of *Period Five*. As fish stocks recovered during the ban, this in turn increased the appeal of recreational tourism fishing.

In the aftermath of the near-total destruction of Sydney's oyster industry in the Georges River due to the 'QX virus' in late *Period Four* (1995), *Period Five* instigated a slow and steady process of reconstruction in its first decade. At first, only Drake's managed to persist out of 50 previous operations but by 2010, 13 lease applications aimed to restore 58 hectares back to production.[205] In the same year, the Pacific Oyster Mortality Syndrome (POMS) virus struck, almost decimating Drake's as well.[206] In addition to the biological challenges of pests and viruses, the industry was increasingly compromised by the cumulative impact of human activities affecting the health of the Georges River and bay waterways. By the end of *Period Five*, Sydney's oyster industry was a faint echo of its former self.[207]

Miscellaneous (MC): Feeding the Highest Bidder

Period Five witnessed an ongoing transition from agricultural to amenity production, including turf, flowers, native and indigenous plants and other forms of non-food-based horticulture. These frequently proved more financially rewarding than peri-urban agriculture that needed to compete with regional economies of scale.

The aforementioned Western Sydney Airport at Badgerys Creek marked the inception of an 'aerotropolis'—a concept inspired by the growth-geared strategies of American aviation management expert J.D. Kasarda.[208] A significant component of this industry-focused project was its 'agribusiness precinct', as outlined in the late 2020 draft land use plans.[209] The precinct was designed to include an 'Integrated Intensive Production Hub', 'Integrated Logistics Hub', 'Food and Pharma Hub' and 'Australian Centre of Excellence in Food Innovation'.[210] The extent of food production within the precinct is, however, unclear.[211] The agribusiness precinct represents a shift from Sydney's traditional urban horticulture, which was socioeconomically and socioculturally marginalised and largely subsistence-based.[212] The new vision proposed the use of highly technological, automated food production systems, such as robotic greenhouses and warehouses.[213] These artificial growing mediums and protected greenhouses are largely independent of ground and natural climatic limitations to produce processed, value-added premium exports for international markets such as China. Despite being marketed as 'green' due to the air and water quality used in their production, these food systems rely on fossil-fuel-intensive global logistics chains, contradicting their environmental claims due to their considerable embodied energy. In essence, Sydney's modern agricultural planning vision aimed to cater to wealthy non-Sydney-based consumers in an aviation-enabled global market.

Period Five also saw a modest resurrection of interest in public urban agriculture.[214] This regularly centred on public and social forms of urban agriculture as distinct from this book's focus on commercial, private and peri-urban food production.[215] Interest in public urban agriculture was also channelled into social and intellectual discourse, policy and guideline documents, as well as inner-city, public domain and community urban agriculture initiatives.[216] This reflected increasing urbanisation, densities and challenges (e.g. lack of solar access) for many inner-city and smaller-plot-size suburban dwellings to accommodate edible gardens. Key examples include community gardens,[217] public farms, street verges,[218] rooftop gardens and farms, green walls, temporary and event gardens, among others. Rather than high-volume commercial producers,[219] these primarily served as platforms for knowledge exchange, social infrastructures and connection with living elements.[220]

Figure 7.13, for example, depicts part of a larger community revitalisation project by Canterbury-Hurlstone Park RSL Club called 'Camperdown Commons', which includes 'Pocket City Farms' (top, pictured in 2017). The overall project is part market garden, part educational facility, part cafe, part community hub and function centre. The market garden was formerly a bowling green. Sydney City Farm (bottom, 2019) is an example of public urban agriculture in St Peters. The City of Sydney initiative hosts volunteers, conducts educational programmes and donates harvests to charity.[221] The project reflects the City of Sydney's progressive social and environmental agenda under the long-term leadership of Mayor Clover Moore, in stark contrast to successive New South Wales state governments in the period.

There were more community gardens across Sydney than those captured in the Map Database, whose activities waxed and waned over time. Private edible gardens were another component of Sydney's food landscapes[222] outside of the scope of this book due to the distinction between

Fig. 7.13 Pocket City Farms Camperdown, 2017 (top); Sydney City Farm St Peters, 2019 (bottom). Source: Nearmap

commercial and hobbyist scales. Both indicated a positive, albeit niche, desire for greater connection with and engagement in local food systems.

By the end of the period in 2020, Sydney's urban development continued to subsume agricultural land (Figs. 7.14 and 7.15). Peri-urban areas with both agriculture and proximate greenfield developments included: Camden (e.g. Spring Hill); (North) Richmond; Windsor; North Kellyville; Austral; Rossmore; Leppington (Emerald Hills, Denham Court, Gledswood Hills and Catherine Field); Oran Park; Spring Farm; Horsley Park; Luddenham; Badgerys Creek; Mulgoa (north); Marsden Park; Schofields; Rouse Hill; Riverstone; Box Hill; Pitt Town; North Richmond; The Oaks; Orchard Hills and Claremont Meadows. These and numerous other agriculturally active regions are threatened by future urban growth due to their proximity to existing greenfield developments in western Sydney and the metropolitan plan's notion of the Western Parkland City. Additional vulnerable areas are not limited to: Catherine Field; Cecil Park; Kemps Creek; Mount Vernon; Bringelly; Greendale; Wallacia; Mulgoa; Llandilo; Shane's Park; Vineyard; Oakville; Scheyville; Maraylya; Yarramundi; Grose Wold; Camden; Camden South; Grasmere; Bickley Vale; Cawdor; Gilead; Wedderburn; Menangle; Menangle Park; Appin; Wilton; Picton; Tahmoor and Thirlmere.

Figure 7.16 conveys a pattern widely replicated across Sydney's peri-urban fringes in Austral—waves of suburbia spilling across West Hoxton, Horningsea Park and Middleton Grange. Agricultural allotments reveal their impending doom through signs of being eaten up by new street grids and houses under construction appear amongst the market gardens and farms.

In mid-2020 near the end of *Period Five*, Lehmann posed,

> Farmers charged with filling Sydney's food bowl have long warned of the threat that rampant development poses to the resilience of the city's food systems. Now, in the shadow of a global pandemic, could their decades-long battle against urban sprawl have taken on a new imperative? In the absence of the white noise of overconsumption and the clamour of an over-heated housing market, will their voices ring louder?[223]

The next chapter, *Then and Now*, presents a visual essay contrasting early and recent Sydney before turning to the final period *2021+* and exploring the city's likely future food landscapes.

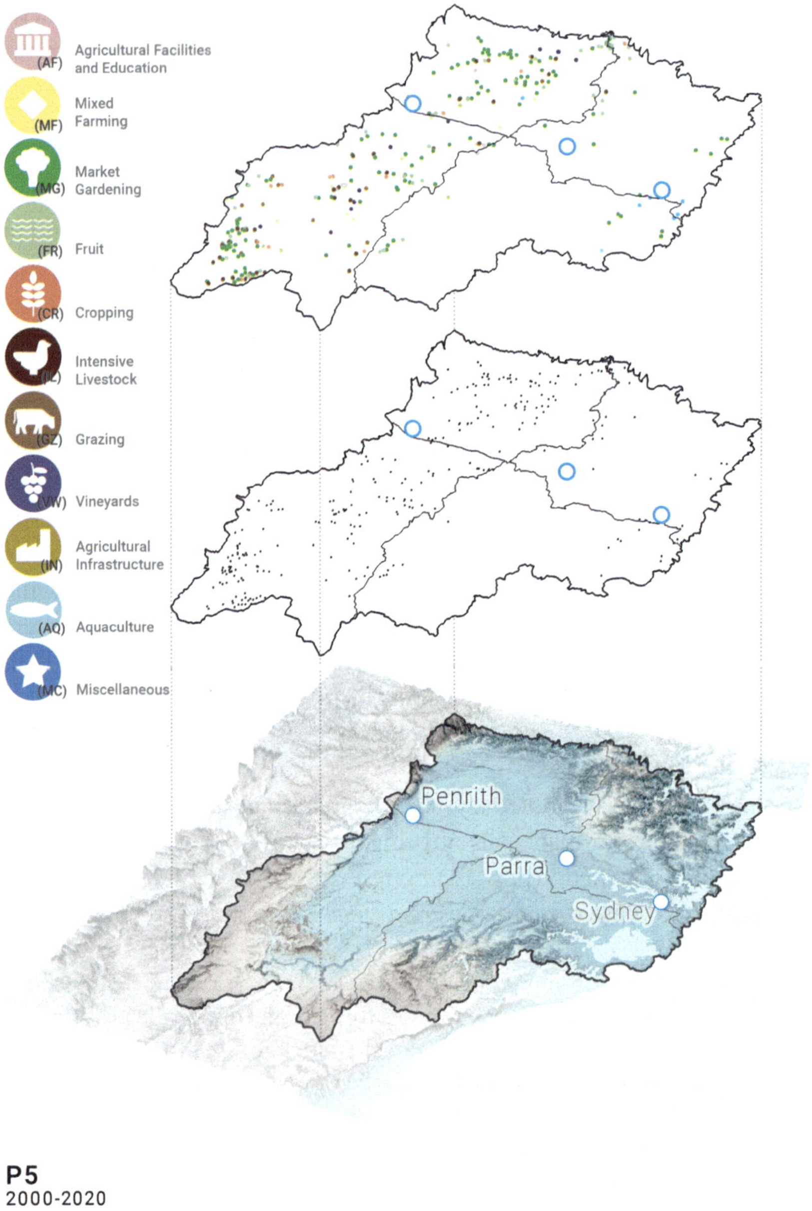

Fig. 7.14 *Period Five*'s Map Database: 3D. Source: Joshua Zeunert, Josh Gowers and Stephanie Stankiewicz

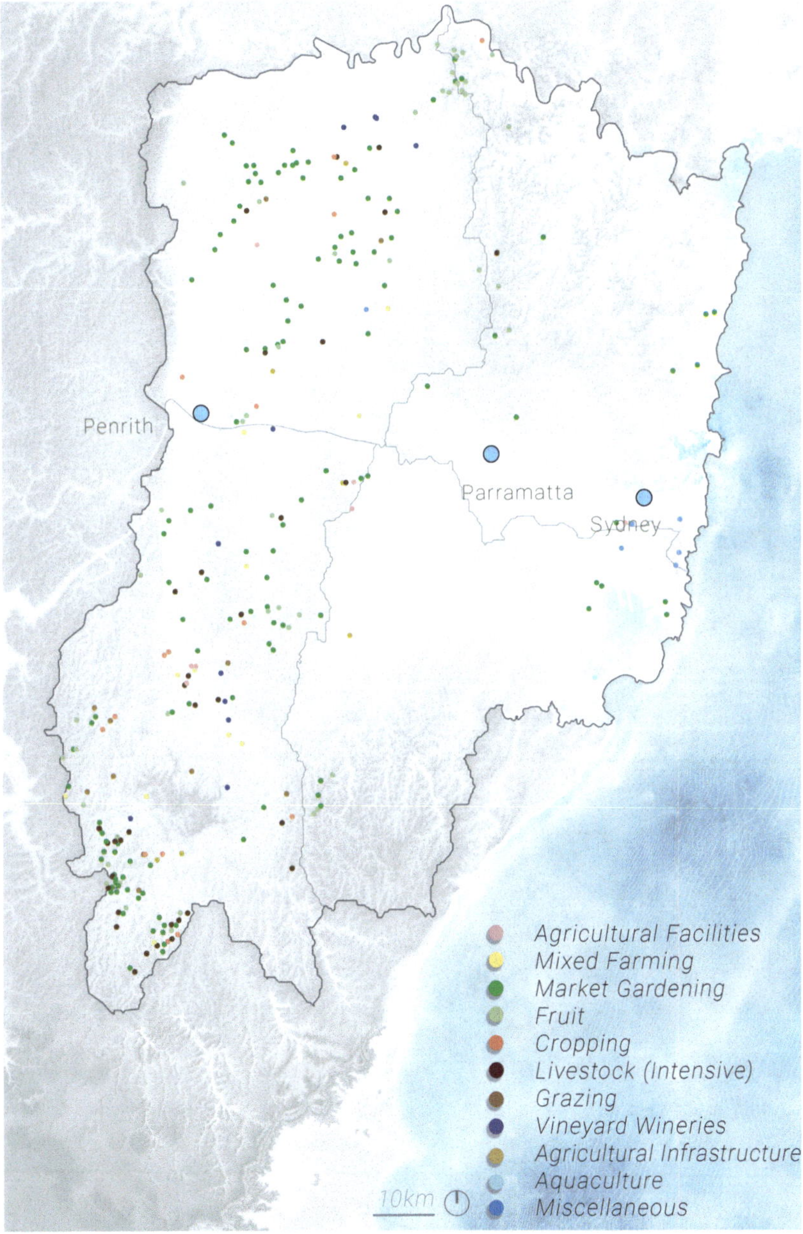

Fig. 7.15 *Period Five's* database maps reveals a significantly reduced number of entries from *Period Four*, as well as entries located considerably further out of the city. Source: Joshua Zeunert, Josh Gowers and Stephanie Stankiewicz

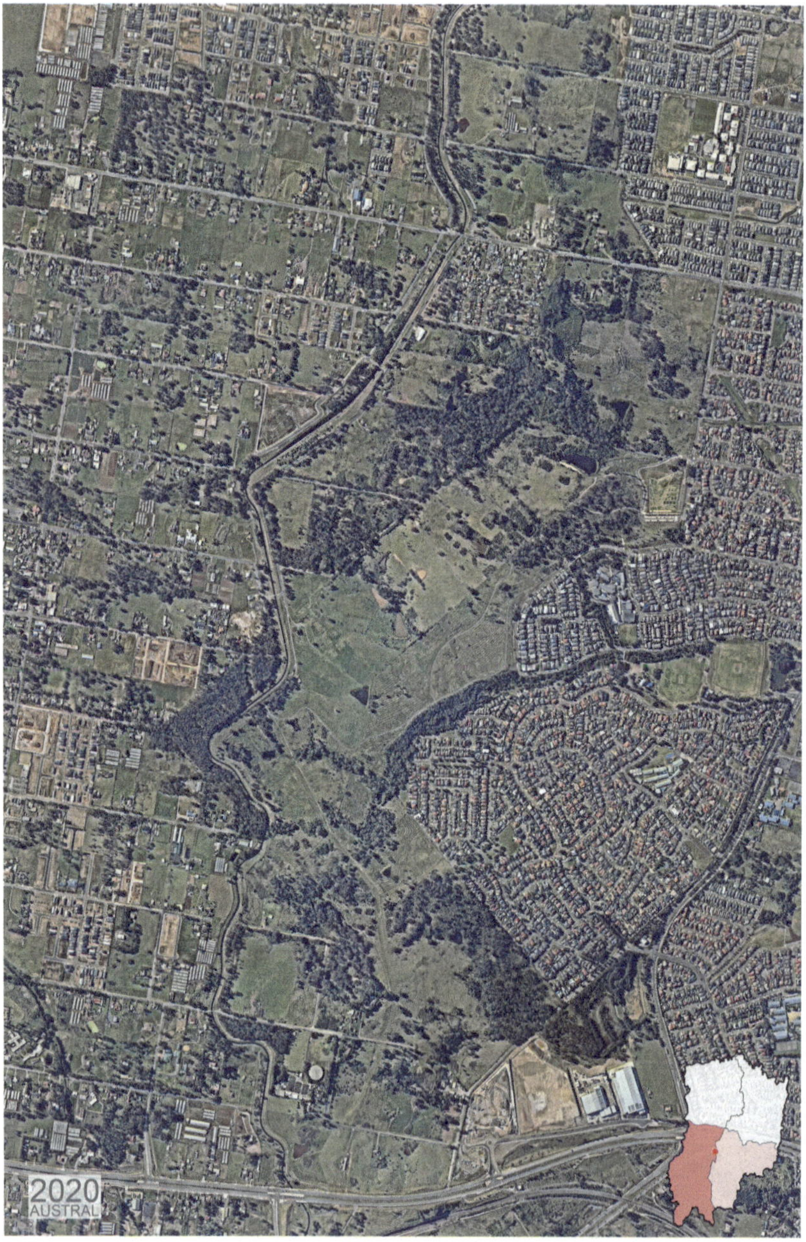

Fig. 7.16 West Hoxton, Horningsea Park and Middleton Grange (right) in October 2020, reflecting a much wider tsunami of encroachment on agriculture. Source: Nearmap, October 2020

Notes

1. The Covid pandemic shifted this pattern, which had previously been in place post WWII. See https://www.aihw.gov.au/reports/australias-welfare/profile-of-australias-population
2. Wiedmann et al., 2020.
3. See https://www.sbs.com.au/news/the-feed/australia-s-democracy-has-been-downgraded-from-open-to-narrowed and https://civicus.contentfiles.net/media/assets/file/GlobalReport2019.pdf accessed 23 February 2020.
4. See e.g. Hadad et al., 2014.
5. See Blythman, 2004.
6. Poore & Nemecek, 2018; Zeunert, 2024a.
7. https://www.abc.net.au/news/2019-02-26/sydney-has-most-toll-roads-in-the-world/10845944?nw=0
8. Zeunert & Freestone, 2022; https://www.planning.nsw.gov.au/Plans-for-your-area/Priority-Growth-Areas-and-Precincts/Western-Sydney-Aerotropolis
9. https://www.soe.epa.nsw.gov.au/all-themes/land/fire
10. Taylor et al., 2017; Zeunert, 2024a.
11. Lehmann, 2020.
12. This is also reflected nationally, with mushrooms being the fourth most financially valuable vegetable crop after potatoes, tomatoes and leafy salad vegetables. See: https://www.nswfarmers.org.au/NSWFA/Posts/The_Farmer/Innovation/Australian_mushroom_growers_the_crop_taking_off.aspx
13. GSC, 2018a: 58.
14. GSC, 2018a.
15. GSC, 2018a: 28. This is a considerable given growth levels in China's numerous urban regions during this period.
16. See Crutzen, & Stoermer, 2000; Steffen et al., 2011.
17. IPCC, 2023: 42.
18. Crutzen, & Stoermer, 2000: 18.
19. The IPCC define balanced diets as those that "feature plant-based foods, such as those based on coarse grains, legumes fruits and vegetables, nuts and seeds, and animal-source foods produced in resilient, sustainable and low-GHG emissions systems, as described in SRCCL" (2023: 70).

20. Campbell et al., 2017.
21. Ibid.
22. Cox, 2019.
23. Zeunert, 2024a.
24. See numerous reports from the Intergovernmental Panel on Climate Change during this period: https://www.ipcc.ch/reports/
25. Holmgren, 2009.
26. Zeunert, 2017: 32–33, 101–102, 118–119.
27. See for example Harvey, 2005; Leitner et al., 2007; Springer et al., 2016; Goodall, 2019.
28. Redden, 2019. Commitment to neoliberalism is set out in the Liberal Party's Federal Platform as "an economy based on private property, free enterprise and competitive markets will produce the wealth and jobs Australians want". Liberal Party Federal Platform, 2002: 9. https://cdn.liberal.org.au/pdf/FederalPlatform.pdf (accessed 27 January 2024).
29. Battin, 2017: 146.
30. Robert Carr (1995–2005, A.L.P), Morris Iemma (2005–2008, A.L.P), Nathan Rees (2008–2009, A.L.P), Kristina Keneally (2009–2011, A.L.P), Barry O'Farrell (2011–2014, Liberal), Michael Baird (2014–2017, Liberal), and Gladys Berejiklian (2017–2021, Liberal).
31. See https://www.sbs.com.au/news/the-feed/australia-s-democracy-has-been-downgraded-from-open-to-narrowed and https://civicus.contentfiles.net/media/assets/file/GlobalReport2019.pdf accessed 23 February 2020. https://civicus.contentfiles.net/media/assets/file/GlobalReport2019.pdf
32. See https://www.sbs.com.au/news/the-feed/australia-s-democracy-has-been-downgraded-from-open-to-narrowed
33. 'Survival of the Richest' is one of four scenarios presented by the World Economic Forum's *Future of Global Food Systems* report. See WEF, 2017.
34. *Period Five's* Occupy Movement, for example, although taken up with greater enthusiasm in the USA than Australia, called out the "1 percent" of the super rich for skewing democracy. See https://en.wikipedia.org/wiki/Occupy_movement and https://www.theguardian.com/world/occupy
35. https://theconversation.com/why-happiness-is-becoming-more-expensive-and-out-of-reach-for-many-australians-170877 and Morris et al., 2021.
36. Noting that the proportion of the Australian workforce as union members continued to decline during this period.
37. Dibden & Cocklin, 2005.

38. Lawrence et al. 2013: 31.
39. Baker, 2015: 195.
40. Ibid: 195–196.
41. Zeunert, 2018b.
42. Ibid.
43. ABS 2006 in Lawrence et al., 31.
44. Plumwood, 2008.
45. Despite a shut down of much manufacturing in the lead-up to these Olympics in an attempt to improve air quality.
46. https://www.hawkesburygazette.com.au/story/5333704/too-hard-to-grow-food-in-the-hawkesbury-says-longtime-farmer/
47. Plumwood, 2008.
48. See for example, overseas discourse such as Leitner et al., 2007.
49. See for example, Goodall, 2019.
50. Connell & Thom, 2000: 330.
51. Cited in Connell & Thom, 2000: 330–31.
52. Click on Australia at: https://data.footprintnetwork.org/?_ga=2.205770195.1149517287.1620266401-2013417519.1619960320#/
53. Hunwick, 2018: 7 cites a Sydney First Nations population estimate of 2000–3000.
54. There are no definitive numbers, but the pre-European Australian landscape has been estimated to have supported an Aboriginal population in the range of 300,000–1,000,000, see ABS, 2002.
55. See Weirick, 2021.
56. Zeunert & Freestone, 2022.
57. See Meadows et al., 2004.
58. See Weller, 2009 for variations on how Perth's urban growth could be implemented.
59. See Weller and Bolleter, 2013 for alternative visions for growth.
60. See McDonough & Braungart, 2002, 2013; Zeunert, 2017.
61. See Crutzen & Stoermer, 2000; Steffen et al., 2011.
62. Trainer, 2023; Holmgren, 2009.
63. As distinct from nativist ecological restoration. See Zeunert, 2013.
64. Shepard, 2013; Zeunert, 2018b.
65. See Austin, 2018.
66. See: https://transitionnetwork.org/
67. Paul Downton, an architect critical of official Australian planning, articulated a broad-scale and temporal sustainable planning vision for Adelaide with his students in 1996. This utilised Richard Register's

'shadow planning' process to project intergenerationally sustainable futures to 2076 and 2136, from a baseline of 1836. See Downton, 2009: 314–315, 481.

68. See a list of unsolicited developments in Rogers & Gibson, 2021: 542–543.
69. Wilkinson, 2011; Wynne et al., 2020.
70. See e.g. Sinclair, 2001, 2015a, b; Bunker & Houston, 2003; Sinclair et al., 2004; Houston, 2005; Budge, 2007, 2013; Cordina, 2010; James, 2014, 2016a, b, James & O'Niell, 2016; Carey & James, 2018; Wynne et al., 2020; and https://www.hawkesburygazette.com.au/story/5295478/urban-sprawl-risks-more-than-hawkesburys-country-lifestyle-says-uts-researcher/
71. Gil et al. 2016: 187.
72. See the following Chap. 8: *Then and Now.*
73. Zeunert & Freestone, 2022.
74. A similar pattern has occurred in Sydney Harbour, where residential developments have usurped the primary function of the harbour for shipping and commercial marine activities—consumers want their imported goods and chattels, but not proximity to the container ships carrying and unloading them, nor the trucks distributing them.
75. https://chartingtransport.com/2012/03/03/traffic-volumes-on-australian-toll-roads/ and https://public.tableau.com/app/profile/chris.loader/viz/TollRoadsv2/DBvolumestrend
76. Houston, 2005.
77. Clarke, 2017.
78. From Lehmann, 2020: "It's not just dairy farms being sold to property developers", Moore says. "It's beef and orchards too. We've seen cashed-up overseas investors land-banking in the hope that in the next 20 to 25 years they'll be able to move in and unlock 500-acre parcels of land."
79. See e.g.: https://www.hawkesburypost.com.au/post/glossodia-s-jacaranda-ponds-major-housing-development-inches-closer
80. Let alone the embodied resources and provisioning ecosystems required to support such growth. See for example Rees, 1992; Wackernagel, 1994; Mostafavi and Doherty, 2010: 216–217; Weller, 2009: 440–441.
81. While diagrams are a fundamental means of communicating ideas clearly, when they are spatial (e.g. plan diagrams), they should also be accurate. See vague examples here: GSC, 2018a: 7, 10–11, 14–15, 17, 19, 21.

82. The extent of GSC's Western Parklands City, for example, is difficult to ascertain when studying either of the GSC's reports, 2018a and 2018b.
83. The 2018 metropolitan strategy claims that there will be "little change in [Sydney's] outward spread", despite its intent to realise 725,000 new dwellings to 2036. See the following paragraphs and GSC, 2018a: 31.
84. The original 1948 CCC Planning scheme was likely the best in this regard. See Zeunert & Freestone, 2022.
85. See Weller, 2009 and Weller & Bolleter, 2013. Albeit, the perspectives presented in these works demonstrate alignment with neoliberal ideology that perpetual growth is inherently beneficial or necessary.
86. As discussed in Zeunert, 2018c: 177. In their 2013 book 'Made in Australia', for example, Weller and Bolleter dismiss the argument that ongoing population growth and its implications are unsustainable, describing Paul Ehrlich's and others' views that Australia's carrying capacity can ultimately support less than 20 million people as "inflammatory" (p. 17). Weller and Bolleter thus place the economy above the environment, as captured in their statement "those who wish to stop the growth are obliged to explain how an economy such as Australia's will function in relative stasis" (p. 17).
87. CoC, 2005: 8.
88. Malcolm & Fahd, 2009: 3.
89. CoC, 2005: 28.
90. GSC, 2018a: 10, 48, 58.
91. GSC, 2018a: 80.
92. At ~562,000 dwellings in the ABS 2016 Census.
93. https://quickstats.censusdata.abs.gov.au/census_services/getproduct/census/2016/quickstat/5GPER?opendocument
94. GSC, 2018a: 31.
95. GSC, 2018b: 9.
96. GSC, 2018a: 62.
97. Pejorative for large mass-produced suburban housing. See for example: https://mcmansionhell.com/post/157157014086/mcmansion-hell-australia-edition and Tim Ross' 'Streets of Your Town' ABC TV series.
98. See Chap. 8: *Then and Now*.
99. The 2010 Metropolitan Plan for Sydney stipulated 70 per cent infill versus 30 per cent renewal. It stated on page 42, "The NSW Government will target at least 70 per cent of Sydney's housing growth in established areas and focus renewal where transport capacity and improvements can be closely linked to the amount of development.

100. See GSC, 2018a, specifically Table 3 and Figure 16.
101. And a contributing factor leading to the landslide state election defeat of Labor in 2011.
102. See Weirick, 2021.
103. Rogers & Gibson, 2021.
104. Weirick (2021) characterised these as "a backdoor mechanism for unsolicited proposals".
105. Rogers & Gibson, 2021: 542, 525.
106. See an agriculturally relevant example, albeit at a relatively small scale, but regarding a very significant agricultural site in Sydney: https://www.smh.com.au/national/nsw/surprise-witness-says-suspended-liberal-mp-garry-edwards-planned-to-lie-to-icac-20140903-10boha.html
107. In 2021, thenNSW Premier Gladys Berejikilian resigned from such a scenario. ICAC was established in 1989 in NSW following a 1988 Act. See for example: https://www.smh.com.au/topic/independent-commission-against-corruption-1lwm
108. It is assumed that 'Greater Sydney' correlates with the GSC's notion of Sydney's boundary and specifically the metropolitan rural area—see Fig. 2.8 and GSC, 2018a: 14–15, 144, 157.
109. GSC, 2018a: 140, citing Australian Bureau of Statistics 2017, Value of Agricultural Commodities Produced, Australia, 2015–2016, Catalogue no: 7503.0, ABS, Canberra, Australia.
110. CCPS, 1948: 125.
111. Winston, 1957: 7–8.
112. Clarke, 2017.
113. James et al., 2010: 6.
114. Pers. Comm., Jenny Spencer, ABS, 21 April 2021. For example, production is reported for the northern beaches and other metropolitan LGAs where farming businesses of the financial value reported do not exist.
115. See *2021+* and specifically, *Sydney Food Futures.*
116. E.g. DPI, 2020, see: https://www.dpi.nsw.gov.au/__data/assets/pdf_file/0011/1260488/Greater-Sydney-Snapshot.pdf
117. GSC, 2018a: 140.
118. DPI, 2020.
119. Who stated that "After the Murray Basin and Murrumbidgee, Sydney Basin is the third most important area for vegetable production in NSW". See National Trust, 2010: 2.
120. Morris and Britton, 2000: 4.

121. Morris and Britton, 2000: 4.
122. James, 2016a: Chapter 5, 'The Things We Want to Keep: Migrant Market Gardens as Sydney's History and Heritage'.
123. Zeunert & Freestone, 2022.
124. Waldheim, 2010.
125. See for example APA, 2007; Raja et al., 2008; Viljoen & Wiskerke, 2012; Miazzo & Minkjan, 2013; Ilieva, 2016; Roggema, 2016.
126. See APA, 2007.
127. See e.g: https://dictionaryofsydney.org/entry/spring_farm; http://www.mcollins.com.au/divisions/quarry/; https://www.suez.com.au/en-au/who-we-are/suez-in-australia-and-new-zealand/our-locations/waste-management-spring-farm?utm_source=Google&utm_medium=Business&utm_campaign=BG-Listings
128. See Nowak, 2018.
129. Caldwell et al., 2011.
130. See Zeunert, 2018a.
131. Budge, 2013: 367.
132. Quoting Farmar-Bowers et al., overviewing his chapter. See Farmar-Bowers et al., 2013: 13.
133. Lehmann, 2020.
134. See, for example, works by Kirsten Larsen and Rachel Carey. Generic info here: https://msd.unimelb.edu.au/research/projects/completed/VEIL/home/about
135. CCPCoC, 2009: 3.
136. Lehmann, 2020.
137. Historical Imagery Viewer.
138. The quality of this imagery improved over time.
139. Search -33.97194, 151.24514 at Historical Imagery Viewer
140. Albeit a decline at the end of P5 following the Covid pandemics impact on international student enrolments
141. O'Connell, n.d., 1883.
142. P30: https://acda.edu.au/resources/Agricultural-education-rebooted2.pdf
143. P49: https://acda.edu.au/resources/Agricultural-education-rebooted2.pdf
144. Harper & Pratley, 2015: 2.
145. P72: https://acda.edu.au/resources/Agricultural-education-rebooted2.pdf
146. P73: https://acda.edu.au/resources/Agricultural-education-rebooted2.pdf
147. P75-77: https://acda.edu.au/resources/Agricultural-education-rebooted2.pdf

148. P89: https://acda.edu.au/resources/Agricultural-education-rebooted2.pdf
149. Copeland, 2020: 64.
150. See e.g. Cochrane, 1958; Ward, 1993; Willis, 2007.
151. https://www.ato.gov.au/Forms/Information-for-primary-producers-2020/
152. Ruoso, 2018: 38, 79, 106, 148.
153. Albeit supplementing income through non-agricultural activities. See the Map Database: -34.02516, 150.64527.
154. Gil et al., 2016: 185.
155. https://www.westernsydneyparklands.com.au/places-to-go/urban-farms/
156. Zeunert, 2024a, b.
157. Gil et al. 2016: 186–187.
158. https://www.westernsydneyparklands.com.au/places-to-go/urban-farms/. This reflects Kevin Lynch's influential book, 'Image of the City' (1960).
159. Zeunert, 2024a.
160. GSC, 2018a, b.
161. See also: https://www.sydneyimages.com.au/-/aerial-photography-gallery/abstract-aerial-photography/-/medias/9ef22838-dc76-11e1-b561-6165a9bd3245
162. Malcolm & Fahd, 2009.
163. Malcolm & Fahd, 2009: 3.
164. Ibid.
165. Humphreys et al., 2001: 143, in James, 2016a: 114.
166. The rare adjoining ecosystem fragment sand dunes to the market gardens' south also face pressures from all adjacencies.
167. See P1.5: https://majorprojects.planningportal.nsw.gov.au/prweb/PRRestService/mp/01/getContent?AttachRef=MP08_0255%2120190802T055943.939%20GMT
168. Parker 2004, cited in James, 2016a: 11.
169. James et al., 2010: 29.
170. Ibid.
171. https://cedarcreekorchards.com.au/about/
172. See p5: https://www.pc.gov.au/__data/assets/pdf_file/0011/196364/sub064-agriculture-attachment2.pdf
173. P3: https://www.pc.gov.au/__data/assets/pdf_file/0011/196364/sub064-agriculture-attachment2.pdf
174. Cordina, 2010: p6: https://www.pc.gov.au/__data/assets/pdf_file/0011/196364/sub064-agriculture-attachment2.pdf

175. https://www.pc.gov.au/__data/assets/pdf_file/0011/196364/sub064-agriculture-attachment2.pdf
176. DPI, 2020. See p1: https://www.dpi.nsw.gov.au/__data/assets/pdf_file/0011/1260488/Greater-Sydney-Snapshot.pdf
177. John's daughter and subsequent CEO Louise Cordina sold Cordina Farms for $300 million to an overseas private equity firm in 2022. See: https://hamiltonlocke.com.au/hamilton-locke-advises-sale-cordina-farms-pag/ https://www.afr.com/street-talk/pag-processes-deal-for-chicken-group-cordina-farms-20220315-p5a4p8
178. Such as their 'Greenfield Developments Company'. See e.g. https://www.smh.com.au/national/nsw/billionaire-perich-family-wins-again-as-judge-cancels-27-million-stamp-duty-bill-20230504-p5d5na.html
179. Zeunert, 2024a. During *Period Five* Sydney's poultry production was more independently distributed than processing, which was comparatively concentrated. Many of Sydney's peri-urban poultry producers contract their supply to one of Sydney's processors.
180. Taylor et al., 2017.
181. Taylor et al., 2017.
182. Zeunert, 2024a; Condon & Coombs, 2014.
183. https://chicken.org.au/our-product/facts-and-figures/
184. DPI, 2020. See p1: https://www.dpi.nsw.gov.au/__data/assets/pdf_file/0011/1260488/Greater-Sydney-Snapshot.pdf
185. Condon & Coombs, 2014.
186. Zeunert, 2024a.
187. See the image of the dairy cow at http://www.lpcmilk.com/
188. http://www.lpcmilk.com
189. See: https://www.theguardian.com/australia-news/2021/mar/10/minister-described-30m-western-sydney-airport-land-purchase-as-perfectly-sensible; https://www.abc.net.au/news/2021-05-25/leppington-triangle-report-into-western-sydney-aiport-tabled/100162580; https://www.accessnews.com.au/people-news/item/1792-one-time-airport-objector-now-a-fan
190. See e.g.: https://www.sbs.com.au/news/small-business-secrets/article/2018/03/20/highs-and-lows-boutique-dairy-production
191. https://www.sbs.com.au/news/small-business-secrets/article/2018/03/20/highs-and-lows-boutique-dairy-production
192. https://www.afr.com/property/richmond-dairy-farm-for-new-housing-at-140m-20180917-h15glj

193. See: https://www.realestate.com.au/news/richmond-north-farm-hambledon-park-sells-for-60m/ and https://hawkesburypost.com.au/new-village-homes-for-seniors-ag-college-funding-towards-bridge-roads-the-hambledon-park-plan/
194. Tizzana Winery in Ebenezer was still operating at the end of P5, having had its first vines planted in 1882, however the continuance of operation is unknown, along with whether it was excluded from a capture of 'Sydney'.
195. P7: https://alliedpinnacle.com/wp-content/uploads/2017/09/Picton-Mill-OEMP-2015.pdf
196. See for example https://www.sydneywater.com.au/web/groups/publicwebcontent/documents/document/zgrf/mdc5/~edisp/dd_079028.pdf
197. Albeit Summertime and Red Lea appear under Cordina ownership, see P1: https://www.pc.gov.au/__data/assets/pdf_file/0011/196364/sub064-agriculture-attachment2.pdf
198. See: https://www.sydneymarkets.com.au/index.php?id=33
199. This number was listed from late 2020 to early 2022, see: https://www.sydneymarkets.com.au/index.php?id=33
200. Arnott's do not provide information on these on their website.
201. O'Connell, n.d., 1872.
202. Ibid.
203. Sydney Fish Market Annual Report, 2022: 7.
204. See: https://www.dpi.nsw.gov.au/fishing/recreational/fishing-skills/fishing-in-sydney-harbour
205. https://www.youtube.com/watch?time_continue=353&v=jCQlxgG903Y&feature=emb_logo
206. https://www.theleader.com.au/story/5511690/oyster-industry-collapses/#slide=3
207. Exacerbated by floods in 2021 and 2022. Even in comparatively pristine waters, the industrial concentration of contemporary oyster production can nonetheless create challenges for the industry. Coffin Bay's cobalt waters in South Australia, for example, saw a Vibrio parahaemolyticus outbreak causing gastroenteritis in late 2021 shut down production for four weeks. See e.g.: https://www.theguardian.com/australia-news/2021/nov/17/coffin-bay-oysters-off-menu-as-south-australia-authorities and https://theleadsouthaustralia.com.au/industries/business/oyster-company-bounces-back-from-coffin-bay-shutdown/-investigate-climates-role-in-vibrio-outbreak

208. Kasarda & Lindsay, 2012.
209. NSW Government, 2020.
210. NSW Government, 2020; Zeunert & Freestone, 2022: 268.
211. Ibid.
212. Christie, 1988; James, 2016a.
213. Thus, bypassing the intrinsic soil capability. This approach is copied from the successful Dutch model—see Viviano, 2017.
214. Zeunert, 2018a.
215. E.g. James, 2016a.
216. Zeunert, 2018a.
217. Nettle and Crouch, 2018.
218. For example see Mobbs, 2012 and his edible street in Chippendale as one example.
219. Zeunert, 2018a.
220. Downton et al., 2017.
221. https://www.cityofsydney.nsw.gov.au/community-gardens/sydney-city-farm
222. See e.g. Naidoo, 2011 for an example stemming from Potts Point in Sydney.
223. Lehmann, 2020.

References

ABS. (2002). *Year book Australia*. Commonwealth of Australia: Australian Bureau of Statistics. Accessed July 27, 2011, from http://www.abs.gov.au/ausstats/abs@.nsf/94713ad445ff1425ca25682000192af2/bfc28642d31c215cca256b350010b3f4

APA. (2007). *Policy guide on community and regional food planning*. American Planning Association. Accessed November 20, 2013, from https://www.planning.org/policy/guides/adopted/food.htm

Austin, G. (2018). Conservation and ecology. In J. Zeunert & T. Waterman (Eds.), *Routledge handbook of landscape and food* (pp. 253–270). Routledge.

Baker, C. (2015). From nation-building to neoliberalism: Agriculture, change and the social imaginary in Australia. In *Refereed Proceedings of TASA's 2015 Conference. Neoliberalism and Contemporary Challenges for the Asia-Pacific* (pp. 191–197). James Cook University.

Battin, T. (2017). Labouring under neoliberalism: The Australian Labor government's ideological constraint, 2007–2013. *The Economic and Labour Relations Review, 28*(1), 146–163. https://doi.org/10.1177/1035304616687951

Blythman, J. (2004). Permanent global summertime. *The Ecologist.* Accessed May 15, 2017, from www.theecologist.org/investigations/food_and_farming/268630/permanent_global_summertime.html

Budge, T. (2007). Farmland preservation in Australia: Emerging issues—fragmented responses. In W. Caldwell, S. Hilts, & B. Wilton (Eds.), *Farmland preservation around the globe: Challenges and opportunities.* University of Guelph.

Budge, T. (2013). Is food a missing ingredient in Australia's metropolitan planning strategies? In Q. Farmar-Bowers, V. Higgins, & J. Millar (Eds.), *Food security in Australia: Challenges and prospects for the future* (pp. 367–380). Springer.

Bunker, R., & Houston, P. (2003). Prospects for the rural-urban fringe in Australia: Observations from a brief history of the landscapes around Sydney and Adelaide. *Australian Geographical Studies, 41*, 303–323.

Caldwell, W., Collett, A., Ludlow, T., Sinclair, I., & Whitehead, J. (2011, February). *Planning and food security within the commonwealth: Discussion paper.* Commonwealth Association of Planners.

Campbell, B. M., Beare, D. J., Bennett, E. M., Hall-Spencer, J. M., Ingram, J. S. I., Jaramillo, F., Ortiz, R., Ramankutty, N., Sayer, J. A., & Shindell, D. (2017). Agriculture production as a major driver of the Earth system exceeding planetary boundaries. *Ecology and Society, 22*(4) https://www.jstor.org/stable/26798991

Carey, R., & James, S. (2018). Peri-urban agriculture in Australia: Pressure on the urban fringe. In J. Zeunert & T. Waterman (Eds.), *Routledge handbook of landscape and food* (pp. 213–227). Routledge.

CCPCoC. (2009). *Economic profile of agriculture on the Central Coast Plateau, NSW.* Central Coast Plateau Chamber of Commerce. NSW Government Industry and Investment. Accessed November 3, 2020, from https://s3.ap-southeast-2.amazonaws.com/dpe-files-production/s3fs-public/dpp/213581/Final%20Farming%20Economic%20Profile%20Report%20Dec%2009.pdf

CCPS. (1948). *The planning scheme for the County of Cumberland.* Cumberland County Council.

Christie, M. (1988). *The Sydney markets 1788–1988.* Sydney Market Authority.

Clarke, M. (2017). *Values of the metropolitan rural area of the greater Sydney region*. AgEconPlus Consulting. Accessed December 10, 2018, from https://gsc-public-1.s3.amazonaws.com/s3fspublic/Values_of_the_Metropolitan_Rural_Area_of_the_Greater_Sydney_Region_(Ag_Econ_Plus).pdf

CoC. (2005). *City of cities: A plan for Sydney's future*. Department of Planning: NSW Government.

Cochrane, W. (1958). *Farm prices: Myth and reality*. University of Minnesota Press.

Condon, J., & Coombs, B. (2014). *Grain fed: The history of the Australian lot feeding industry*. The Australian Lot Feeders' Association.

Connell, J., & Thom, B. (2000). Beyond 2000: The post-olympic city. In J. Connell (Ed.), *Sydney: The emergence of a world city* (pp. 319–343). Oxford University Press.

Copeland, L. (2020). A century of agriculture in the University of Sydney. *Agricultural Science, 31*, 58–65.

Cordina, J. (2010). *Poultry industry submission: Metropolitan strategy review*. Submission 64. https://www.pc.gov.au/__data/assets/pdf_file/0011/196364/sub064-agriculture-attachment2.pdf

Cox, L. (2019, August 8). *Beef industry linked to 94% of land clearing in Great Barrier Reef catchments*. Accessed May 5, 2022, from https://www.theguardian.com/australia-news/2019/aug/08/beef-industry-linked-to-94-of-land-clearing-in-great-barrier-reef-catchments

Crutzen, P. J., & Stoermer, E. F. (2000). The "Anthropocene". *Global Change Newsletter, 41*, 17.

Dibden, J., & Cocklin, C. (2005). *Sustainability and change in rural Australia*. University of New South Wales Press.

Downton, P. (2009). *Ecopolis: Architecture and cities for a changing climate*. CSIRO Publishing.

Downton, P., Jones, D., Zeunert, J., & Roös, P. (2017). Biophilic design applications: Putting theory and patterns into built environment practice. *KnE Engineering, 2*(1), 59–65. https://doi.org/10.18502/keg.v2i2.596

DPI. (2020). *Agriculture industry snapshot for planning: Greater Sydney Region*. NSW Department of Primary Industries. https://www.dpi.nsw.gov.au/__data/assets/pdf_file/0011/1260488/Greater-Sydney-Snapshot.pdf

Farmar-Bowers, Q., Higgins, V., & Millar, J. (2013). *Food security in Australia: Challenges and prospects for the future*. Springer.

Gil, Y., Kirkland, D., & Sergi, R. (2016). Implementing the urban farming master plan in Horsley Park, Western Sydney: From planning to reality. In B. Maheshwari, V. Singh, & B. Thoradeniya (Eds.), *Balanced urban develop-*

ment: Options and strategies for liveable cities. Springer International Publishing.

Goodall, J. (2019). *The politics of the common good: Dispossession in Australia.* UNSW Press.

GSC. (2018a). *Greater Sydney Region plan: A metropolis of three cities.* Greater Sydney Commission: NSW Government. https://www.greater.sydney/publications

GSC. (2018b). *Western city district plan.* Greater Sydney Commission: NSW Government. https://www.greater.sydney/publications

Hadad, Y., Halevy, O., & Cahaner, A. (2014). Featherless and feathered broilers under control versus hot conditions. 1. Breast meat yield and quality. *Poultry Science, 93*(5), 1067–1075.

Harper, R. J., & Pratley, J. E. (2015). *Recent trends in Australian tertiary agricultural science education.* Australian Council of Deans of Agriculture.

Harvey, D. (2005). *A brief history of neoliberalism.* Oxford University Press.

Historical Imagery Viewer. Spatial Services NSW. https://portal.spatial.nsw.gov.au/portal/apps/webappviewer/index.html?id=f7c215b873864d44bccddda8075238cb

Holmgren, D. (2009). *Future scenarios: How communities can adapt to peak oil and climate change.* Chelsea Green.

Houston, P. (2005). Re-valuing the fringe: Some findings on the value of agricultural production in Australia's Peri-urban regions. *Geographical Research, 43*(2), 209–223.

Hunwick, H. (2018). *The food and drink of Sydney: A history.* Rowman & Littlefield.

Ilieva, R. (2016). *Urban food planning: Seeds of transition in the global north.* Routledge.

IPCC. (2023). *Climate Change 2023 synthesis report.* Accessed January 27, 2024, from https://www.ipcc.ch/report/ar6/syr/

James, S., Dimeski, B., Crabtree, L., & O'Neill, P. (2010). *Sydney's agricultural lands: An analysis.* University of Western Sydney. https://researchdirect.westernsydney.edu.au/islandora/object/uws:11823/datastream/PDF/view

James, S. (2014). Protecting Sydney's Peri-urban agriculture: Moving beyond a housing/farming dichotomy. *Geographical Research, 52*, 377–386.

James, S. (2016a). *Farming on the fringe: Peri-urban agriculture, cultural diversity and sustainability in Sydney.* Springer International Publishing.

James, S. (2016b). Beyond 'Local' food: How supermarkets and consumer choice affect the economic viability of small-scale family farms in Sydney, Australia. *Area, 48*, 103–110.

James, S., & O'Neill, P. (2016). Planning for Peri-urban agriculture – A geographically-specific, evidence based approach from Sydney. *Australian Geographer, 47*, 173–194.

Kasarda, J., & Lindsay, G. (2012). *Aerotropolis: The way we'll live next*. Farrar, Straus and Giroux.

Lawrence, G., Richards, C., & Lyons, K. (2013). Food security in Australia in an era of neoliberalism, productivism and climate change. *Journal of Rural Studies, 29*, 30–39.

Lehmann, M. (2020, June 13). Farming on Sydney's fringe; can it survive urban sprawl? *The Weekend Australian Magazine*. https://www.theaustralian.com.au/weekend-australian-magazine/farming-on-sydneys-fringe-can-it-survive-urban-sprawl/news-story/048b857628c86161f482319238dc9bad

Leitner, H., Peck, J., & Sheppard, E. (2007). *Contesting neoliberalism: Urban frontiers*. The Guilford Press.

Liberal Party Federal Platform. (2002). Accessed January 27, 2024, from https://cdn.liberal.org.au/pdf/FederalPlatform.pdf

Lynch, K. (1960). *Image of the city*. MIT Press.

Malcolm, P., & Fahd, R. (2009). *Ground truthing of Sydney vegetable industry in 2008*. Horticulture Australia Limited, NSW Department of Primary Industries. https://www.futuredirections.org.au/wp-content/uploads/2015/05/VG07073_Groundtruthingsurveyofvegindustrystats.pdf

McDonough, W., & Braungart, M. (2002). *Cradle to cradle*. North Point Press.

McDonough, W., & Braungart, M. (2013). *The upcycle: Beyond sustainability – Designing for abundance*. North Point Press.

Meadows, D. H., Randers, J., & Meadows, D. L. (2004). *The limits to growth: The 30-year update*. Chelsea Green.

Miazzo, F., & Minkjan, M. (2013). *Farming the city: Food as a tool for today's urbanisation*. Transcity Valiz.

Mobbs, M. (2012). *Sustainable food*. Choice Books.

Morris, C., & Britton, G. (2000). *Colonial landscapes of the Cumberland Plain and Camden, NSW: A survey of selected pre-1860 cultural landscapes from Wollondilly to Hawkesbury LGA*. National Trust of Australia.

Morris, R. W., Kettlewell, N., & Glozier, N. (2021). The increasing cost of happiness. *SSM-Population Health, 16*, 100949.

Mostafavi, M., & Doherty, G. (2010). *Ecological urbanism* (pp. 216–217). Lars Müller.

Naidoo, I. (2011). *The edible balcony: How to grow fresh food in a small space plus 60 inspiring recipes*. Penguin Books Australia.

National Trust. (2010). *Classification report: La Perouse Chinese market gardens*. National Trust of Australia.

Nettle, C., & Crouch, D. (2018). Allotments and community gardens: History, culture, and landscape in Britain, North America, and Australia. In J. Zeunert & T. Waterman (Eds.), *The Routledge handbook of landscape and food* (pp. 462–474). Routledge.

Nowak, Z. (2018). Terroir: A socially constructed subterranean landscape gone global. In J. Zeunert & T. Waterman (Eds.), *Routledge handbook of landscape and food* (pp. 543–551). Routledge.

NSW Government. (2020). *Western Sydney Aerotropolis precinct plans*. https://www.planningportal.nsw.gov.au/WSAPP

O'Connell, J. (n.d.). *Australian food history timeline*. Accessed May 12, 2019, from https://australianfoodtimeline.com.au/

O'Connell, J. (n.d.) *1872 First Sydney fish market*. Accessed May 12, 2019, from https://australianfoodtimeline.com.au/first-sydney-fish-market/

O'Connell, J. (n.d.) *1883 Australia's first agricultural college*. Accessed May 12, 2019, from https://australianfoodtimeline.com.au/first-agricultural-college/

PIA. (2011). *Submission to the national food plan issues paper*. Planning Institute of Australia.

Plumwood, V. (2008). Shadow places and the politics of dwelling. *Australian Humanities Review, 44*. https://australianhumanitiesreview.org/2008/03/01/shadow-places-and-the-politics-of-dwelling/

Poore, J., & Nemecek, T. (2018, June 1). Reducing food's environmental impacts through producers and consumers. *Science, 360*(6392), 987–992. https://doi.org/10.1126/science.aaq0216

Raja, S., Born, B., & Russell, J. (2008). *A planners guide to community and regional food planning: Transforming food environments, facilitating healthy eating*. American Planning Association.

Redden, G. (2019). John Howard's investor state: Neoliberalism and the rise of inequality in Australia. *Critical Sociology, 45*(4–5), 713–728.

Rees, W. (1992). Ecological footprints and appropriated carrying capacity: What urban economics leaves out. *Environment and Urbanization, 4*, 121–130.

Roggema, R. (Ed.). (2016). *sustainable urban agriculture and food planning*. Routledge/Taylor & Francis.

Rogers, D., & Gibson, C. (2021). Unsolicited urbanism: Development monopolies, regulatory-technical fixes and planning-as-deal-making. *Environment and Planning A: Economy and Space, 53*, 525–547.

Ruoso, L. (2018). *The politics of place identity in peri-urban environments. What role for productive farming landscapes? A case study of Wollondilly Shire, NSW, Australia*. PhD Thesis, UTS Sydney.

Shepard, M. (2013). *Restoration agriculture: Real world permaculture for farmers*. Acres USA.

Sinclair, I. (2001). Lifestyle living. In A view from the edge: Issues in rural and metropolitan fringe planning. *New Planner, 49*. Planning Institute of Australia, NSW Division.

Sinclair, I., Docking, A., Jarecki, S., Parker, F., & Saville, L. (2004). *From the outside looking in: The future of Sydney's rural land: Background issues and workshop outcomes*. University of Western Sydney.

Sinclair, I. (2015a). *Growing food in a residential landscape*. Paper presented at the State of Australian Cities Conference 2015, 9–11 December 2015, The Gold Coast, Queensland.

Sinclair, I. (2015b). *Sydney Peri-urban network issues paper*. Edge Land Planning.

Springer, S., Birch, K., & MacLeavy, J. (2016). *The handbook of neoliberalism*. Routledge.

Steffen, W., Grinevald, J., Crutzen, P., & McNeill, J. (2011). The anthropocene: Conceptual and historical perspectives. *Philosophical Transactions of the Royal Society A: Mathematical, Physical and Engineering Sciences, 369*, 842–867.

Sydney Fish Market Annual Report 2022. (2022). https://www.sydneyfishmarket.com.au/Portals/0/adam/Content/IMGVO4gvbkefxTaur6o9aA/ButtonLink/Annual%20Report%202022%20-%20FOR%20WEB-1.pdf

Taylor, E., Butt, A., & Amati, M. (2017). Making the blood broil: Conflicts over imagined rurality in peri-urban Australia. *Planning Practice & Research, 32*(1), 85–102. https://doi.org/10.1080/02697459.2015.1028252

Trainer, T. (2023). On degrowth strategy: The simpler way perspective. *Environmental Values, 20231221*. https://doi.org/10.1177/09632719231214309

Viljoen, A., & Wiskerke, J. (Eds.). (2012). *Sustainable food planning: Evolving theory and practice*. Wageningen Academic Publishers.

Viviano, F. (2017). *How the Netherlands feeds the world*. National Geographic. https://www.nationalgeographic.com/magazine/2017/09/holland-agriculture-sustainable-farming/

Wackernagel, M. (1994). *Ecological footprint and appropriated carrying capacity: A tool for planning toward sustainability*. PhD thesis, The University of British Columbia.
Waldheim, C. (2010, November). Notes toward a history of Agrarian urbanism. *Places Journal*. Accessed April 3, 2012, from https://placesjournal.org/article/notes-toward-a-history-of-agrarian-urbanism/
Ward, N. (1993). The agricultural treadmill and the rural environment in the post-productivist era. *Sociologia Ruralis, 33*, 348–364.
Wiedmann, T., Lenzen, M., Keyßer, L. T., & Steinberger, J. K. (2020). Scientists' warning on affluence. *Nature Communications, 11*(3107), 1–10. https://doi.org/10.1038/s41467-020-16941-y
WEF. (2017). *Shaping the future of global food systems: A scenarios analysis*. World Economic Forum.
Weirick, J. (2021, March 22). After a decade of lost opportunity to fix NSW's planning mess, here's a model for success. *Sydney Morning Herald*. https://www.smh.com.au/national/nsw/after-a-decade-of-lost-opportunity-to-fix-nsw-s-planning-mess-here-s-a-model-for-success-20210319-p57caq.html
Weller, R. (2009). *Boom town 2050*. UWA Press.
Weller, R., & Bolleter, J. (2013). *Made in Australia: The future of Australian cities*. UWA Publishing.
Wilkinson, J. (2011). *Agriculture in the Sydney region: Historical and current perspectives, e-brief 2/2011*. NSW Parliamentary Library Research Service. Accessed May 12, 2018, from https://www.parliament.nsw.gov.au/researchpapers/Documents/agriculture-in-the-sydneyregion-historical-and-/Agriculture%20in%20the%20Sydney%20region%202.pdf
Willis, A. M. (2007). From peri-urban to unknown territory. *Design Philosophy Papers, 5*, 79–90.
Winston, D. (1957). *Sydney's great experiment: The progress of the Cumberland County plan*. Halstead Press.
Wynne, L., Ruoso, L.-E., Cordell, D., & Jacobs, B. (2020). 'Locationally disadvantaged': Planning governmentalities and peri-urban agricultural futures. *Australian Geographer, 51*, 377–397.
Zeunert, J. (2013). Challenging assumptions in urban restoration ecology. *Landscape Journal, 32*(2), 231–242.
Zeunert, J. (2017). *Landscape architecture and environmental sustainability: Creating positive change through design*. London/New York.
Zeunert, J. (2018a). Dimensions of urban agriculture. In J. Zeunert & T. Waterman (Eds.), *Routledge handbook of landscape and food* (pp. 160–184). Routledge.

Zeunert, J. (2018b). Challenges in agricultural sustainability and resilience: Towards regenerative practice. In J. Zeunert & T. Waterman (Eds.), *Routledge handbook of landscape and food* (pp. 231–252). Routledge.
Zeunert, J. (2018c). *Multidimensional sustainability, landscape and food, and public urban agriculture in landscape architecture*. PhD Thesis. Deakin University.
Zeunert, J., & Freestone, R. (2022). From rural lands to agribusiness precincts: Agriculture in metropolitan Sydney 1948-2018. In C. Brisotto & F. Lemes de Oliveira (Eds.), *Re-imaging resilient food landscapes - perspectives from planning history* (pp. 247–272). Springer.
Zeunert, J. (2024a). Bridging rural and urban disconnections: Spatial graphic explorations of Australia's livestock landscapes. In R. Pernice & B. Chen (Eds.), *Urban regeneration and rural revitalization in Australia and China* (pp. 110–143). Routledge.
Zeunert, J. (2024b). States of Australia's agri-environment: Visual extractions from degraded landscapes. *Visual Studies,* 39(4), 609-621. https://doi.org/10.1080/1472586X.2023.2294004

8

Then and Now

This chapter presents a visual essay of before and after imagery of Sydney's metropolitan food landscapes, accompanied by select quotes. As Peter Spearitt has observed,

> There is always something brutal about the way cities expand into their hinterland. Anyone who has seen native forest flattened for housing blocks or even an old orchard carved up knows that the coming of new houses is often tinged with regret for the former landscape.[1]

Figure 8.1 shows the Cowpastures from J. Hassell's farm (1825–1828) (top), and via a view from the Northern Road at Oran Park facing south-east in September 2019 (bottom) (Fig. 8.1). In addition to general images of interest (Figs. 8.1, 8.9, 8.10 and 8.16), it includes three main themes: *Government Farms* (Figs. 8.2, 8.3, 8.4, 8.5 and 8.6); *Estates* (Figs. 8.7 and 8.8), and *Planning Rhetoric* (Figs. 8.11, 8.12, 8.13, 8.14 and 8.15). The latter juxtaposes extracts from Sydney's official metropolitan planning documents with photographic evidence of built outcomes.

J. Zeunert, A. Daroy, *Sydney's Food Landscapes*,
https://doi.org/10.1007/978-981-96-0710-5_8

Fig. 8.1 The Cow Pastures in 1825–1828 (top), and 2019 from the Northern Road at Oran Park facing south-east (bottom). Sources: Mitchell Library, State Library of New South Wales and Joshua Zeunert

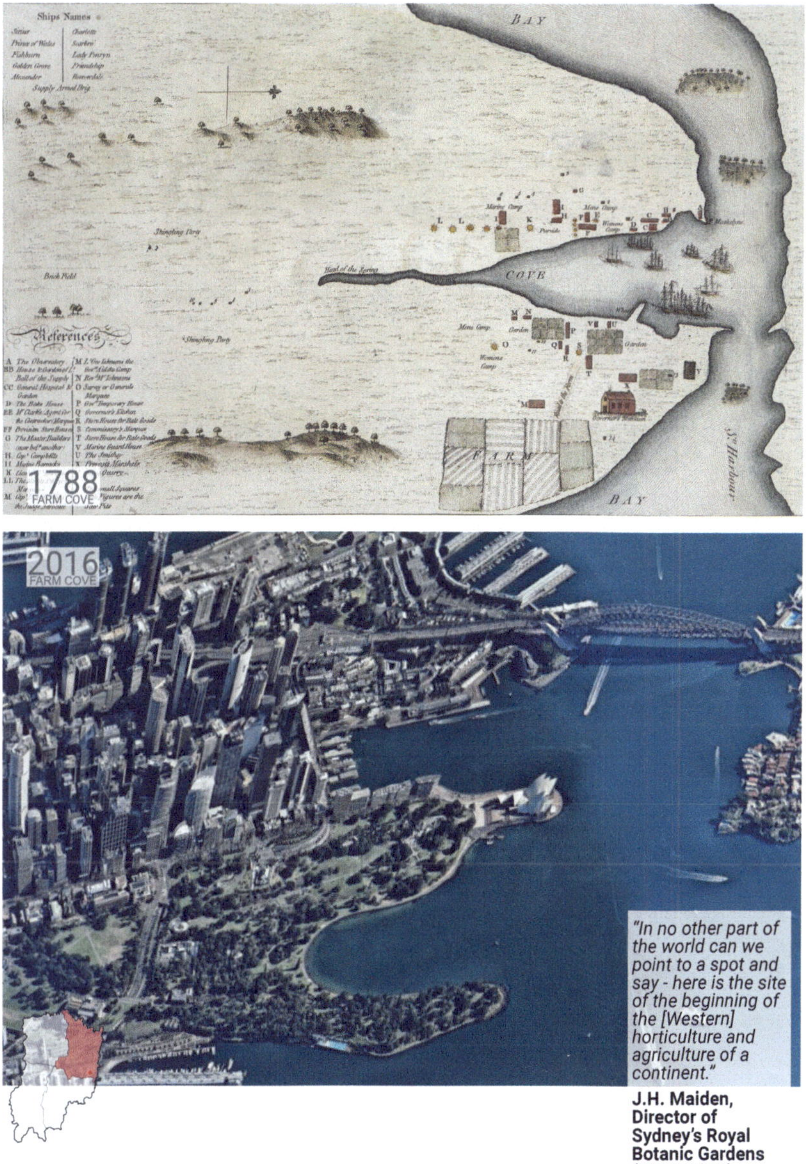

Fig. 8.2 Sydney Cove 1788 (top) and 2016 (bottom). Sources: Francis Fowkes, State Library of New South Wales and Nearmap

Fig. 8.3 Government Farm in Parramatta c. 1791 (top), 2021 (bottom). Sources: Alamy and Joshua Zeunert

Fig. 8.4 Toongabbie Government Farm c. 1795 (top), 2021 (bottom). Sources: David Collins, State Library of New South Wales and Joshua Zeunert

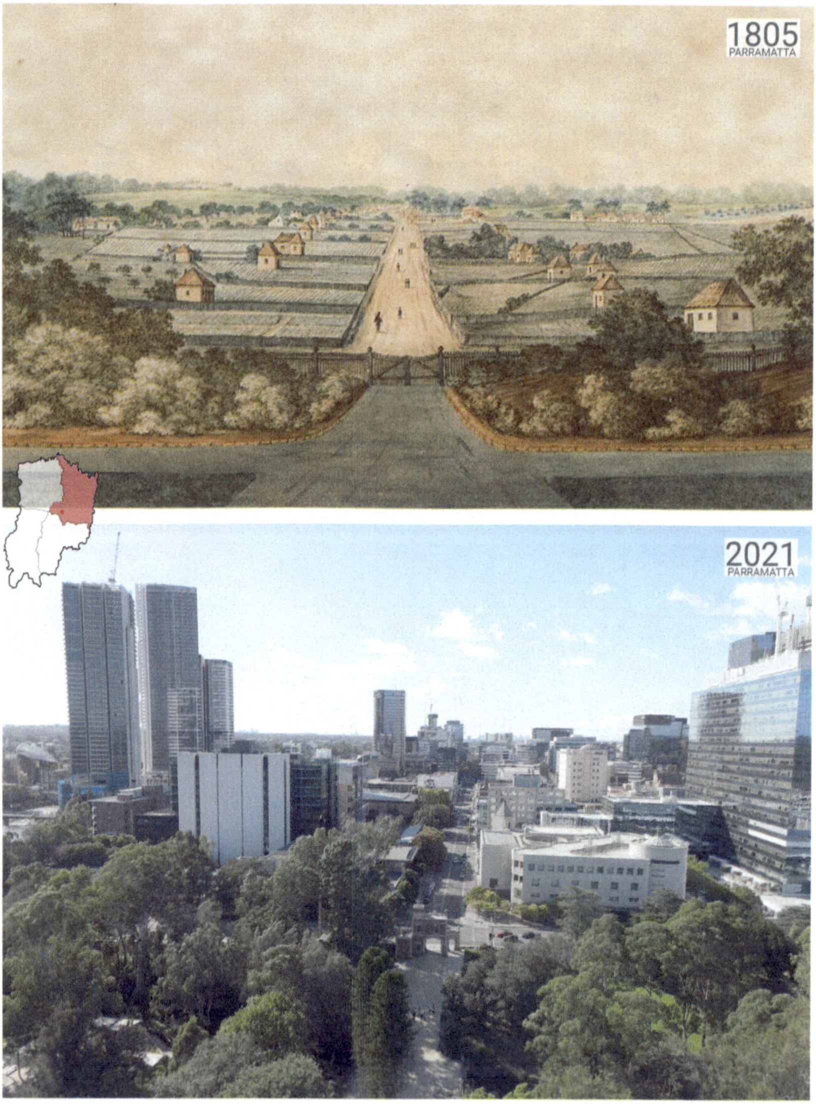

Fig. 8.5 Parramatta looking east from Government House Gates c. 1804–1805 (top) and 2021 (bottom). Sources: George William Evans, Caroline Simpson Library, Museums of History NSW and Joshua Zeunert

Fig. 8.6 Former Government Farm Castle Hill, c. 1806 (top) and 2021 (bottom). Sources: John Lewin, Mitchell Library State Library of New South Wales, and Joshua Zeunert

Government Farms

Sydney Cove's settlement was depicted by transported convict Francis Fowkes on April 16, 1788, less than three months after the British landing (Fig. 8.2, top). Although cartographically 'skewed', Fowkes' image is strikingly different from the 2016 Nearmap aerial satellite image (bottom), revealing Sydney's immense transformation only 228 years later.[2]

The painting in Fig. 8.3 (top) depicts the cultivated fields at Government Farm in Parramatta near 'The Crescent', c. 1791. Farm buildings, a brick kiln and fields have been hewn from the Cumberland plain woodland encircling them, bisected by the bridged Parramatta River. Rows of dots

Fig. 8.7 Rose Hill c. 1823 (top) and 2021 (Parramatta) (bottom). Sources: Joseph Lycett, State Library of New South Wales and Joshua Zeunert

and lines over pale washes of colour depict crops in the fields. In the far right, a man appears to be threshing wheat in a barn. Jumping to 2021 (bottom), the former Government Farm (Paramatta Park) forms part of a World Heritage listed landscape, replete with a recently redeveloped colosseum (the 30,000 capacity Western Sydney Stadium). While there are numerous clues to the site's past agricultural uses in names and faded signs, no active agricultural activities or plantings exist on the site. Thus, there is failure to express the fact that the key site and its immediate region were utterly crucial to local food provision and rendering the colony, and subsequent city of Sydney and Australian nation, a viable venture.

Given Farm Cove's failure, Australia's second successful Government Farm after Parramatta was at Toongabbie, engraved in Fig. 8.4 c. 1795 (top). It has since been largely lost to obscurity, as visible in 2021 (bottom). It operated from 1791 to 1817 as a form of convict labour-based public agriculture and played a critical role in feeding the struggling early colony, especially in the 'hungry years' of 1791 to 1794. It also held government livestock (cattle and sheep) before they were removed in 1813 ahead of its closure in 1817 that stemmed from Governor Macquarie's reappraisal of government agriculture.[3] Its agricultural heritage has washed away along with any interpretative signage. A neglected dirt 'settlers walk' pathway traverses the site along the creek bank, as the last vestige of an otherwise invisible history. A large factory occupies the southern portion (previously a large Chinese-Australian market garden). Suburbs lie to the north, estranged from a meaningful interface with the river. The site is bisected by buzzing power lines stemming from a substantial electricity substation and a major road to the west (A40, NW Transitway). Attempts to reinstate the indigenous vegetation community are visible and signposted. Albeit, this element of its landscape history is somewhat thwarted by the river, which carries and deposits exotic species, rubbish and elevated nutrient levels onto its rich banks that once supported wheat, maize and barley. The farm helped the fledgling colony escape starvation during its 'hungry years' (see *Period One*). The site is now not only a novel ecosystem, but is emblematic of the mish-mash of competing contemporary land uses, agendas and need for infrastructures to supply the insatiable appetite of the urban metabolism.

Fig. 8.8 Bungarribee 1858 (top) and 2021 (bottom). Sources: Joseph Fowles, Mitchell Library State Library of New South Wales and Joshua Zeunert

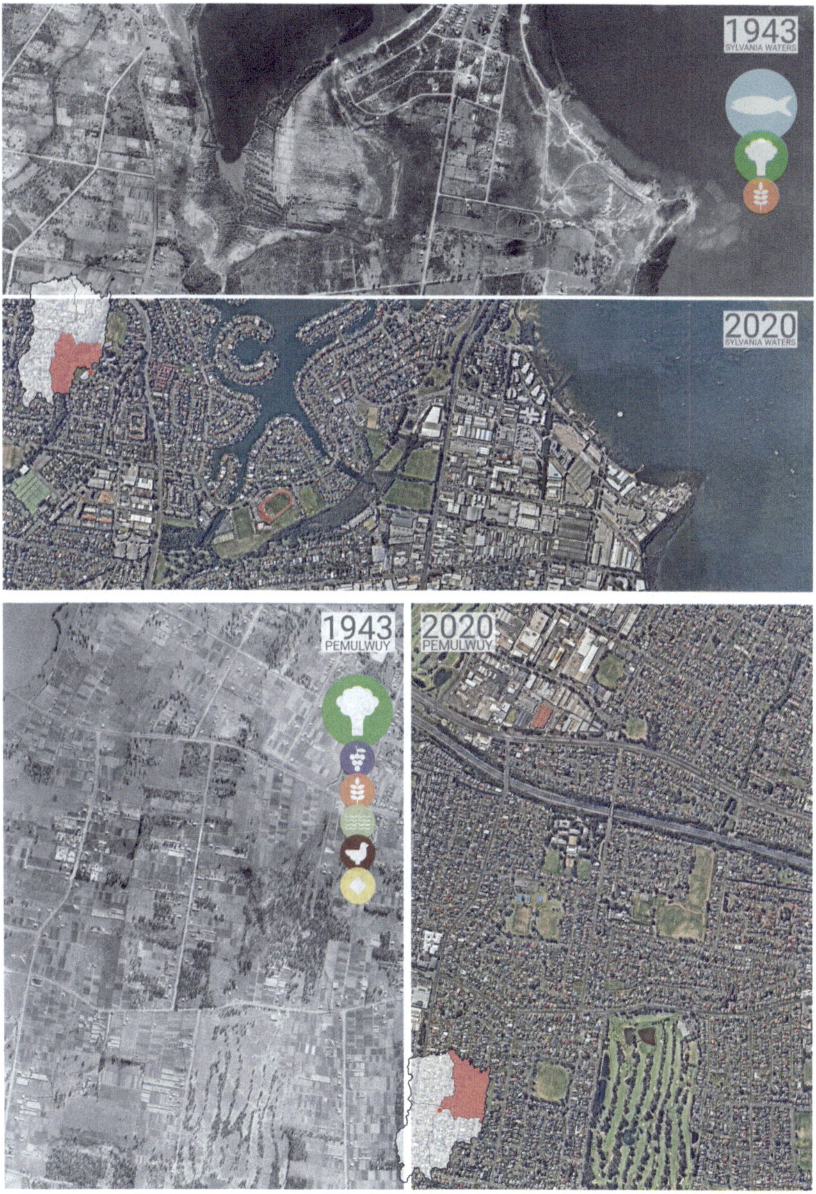

Fig. 8.9 Sylvania Waters 1943 (top) and 2020 (middle), Pemulwuy 1943 (bottom left) and 2020 (bottom right). Sources: Spatial Services NSW (© State of New South Wales (Spatial Services, a business unit of the Department of Customer Service NSW)) and Nearmap

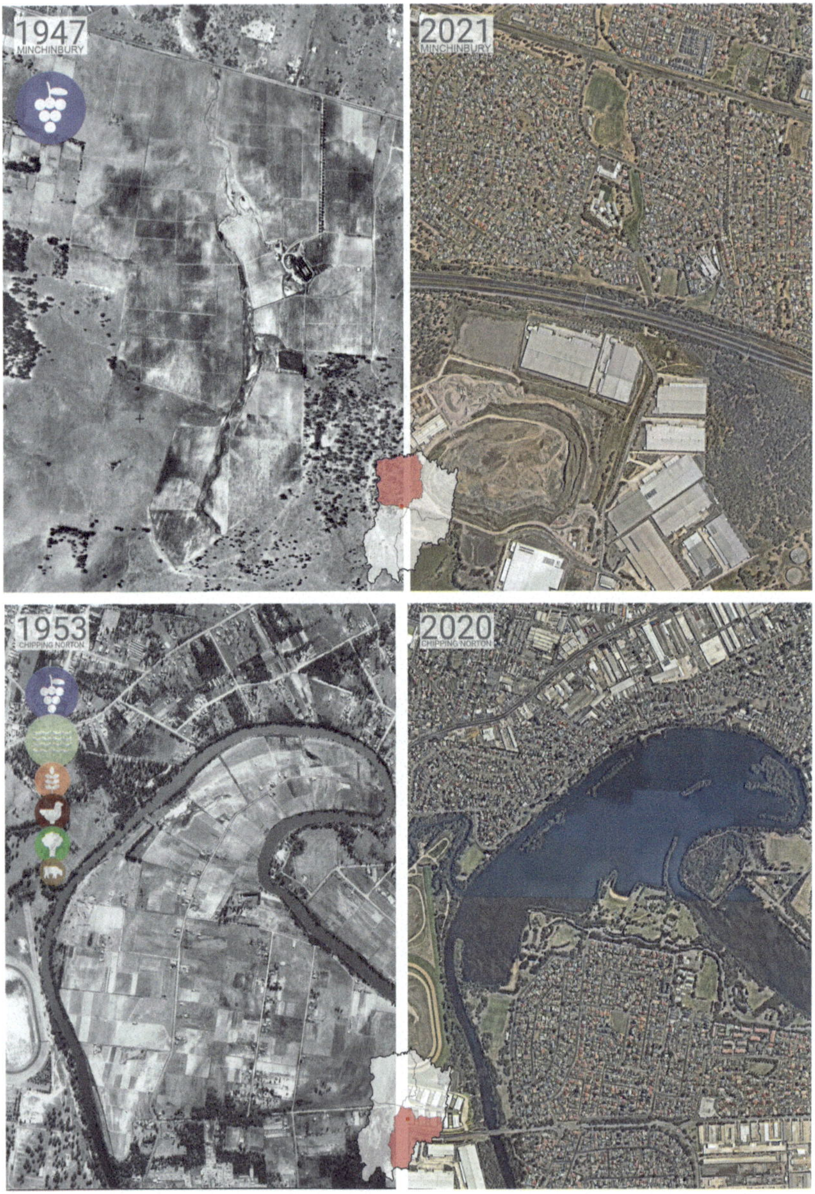

Fig. 8.10 Minchinbury 1947 (top left) and 2021 (top right), Chipping Norton 1953 (bottom left) and 2020 (bottom right). Sources: Spatial Services NSW and Nearmap

The Parramatta region was the birthplace of Australian agricultural success. Its fertility was exploited relatively quickly through extractive cropping and poor farming techniques. Some agriculturalists successfully transitioned to fruit production and particularly citrus, and the region became renowned for its oranges. Despite its marked transformation from 1805 to present day (Fig. 8.5), three key agricultural remnants exist in the towering skyline's shadows: Australia's first successful Government Farm in Parramatta Park, a cottage on the site of Australia's first successful individual farmer (James Ruse) and Elizabeth Farm—the first site of the formidable Macarthur Empire (Fig. 8.7). The Macarthurs were instrumental in Australia's agricultural establishment and successes, spanning numerous facets including pastoralism, viticulture and orcharding (see *Period One*).

Castle Hill Government Farm commenced in 1801 and its soil was described as the "best and most productive kind".[4] Eucalyptus stumps reveal their sacrifice to agriculture in the c. 1806 depiction (Fig. 8.6, top). The 'Blue Gum High Forest'[5] was clear-felled and burnt by convict labourers to aid cropping (visible) and running stock. The farm was extensive at 14,000 hectares.[6] Convicts were housed in stone barracks, which later became the norm for their accommodation.[7] The site is likely best known for its 1804 convict uprising and subsequent week of martial law in the colony. Governor Macquarie, arriving in 1810, reappraised the colony's system of government farms and ostensibly closed the farm in 1811, converting it to an asylum. Much later, after 1900, an orcharding phase ensued during which the south-west corner of the site was put to use for fruit growing. The 2021 aerial image (bottom) shows its evolution to an archetypal Australian park, consisting of native eucalyptus trees and irrigated turf.

Estates

A Joseph Lycett hand-coloured engraving (Fig. 8.7, top) depicts one of Australia's most significant agricultural sites, Elizabeth Farm. Titled the 'residence of John Macarthur Esq.', it was sketched in 1821, completed in England and published in 1825.[8] Macarthur knew the importance of

Fig. 8.11 Yobarnie, North Richmond 1975, 1991, 2010 and 2023. Sources: Spatial Services NSW, Nearmap and Joshua Zeunert

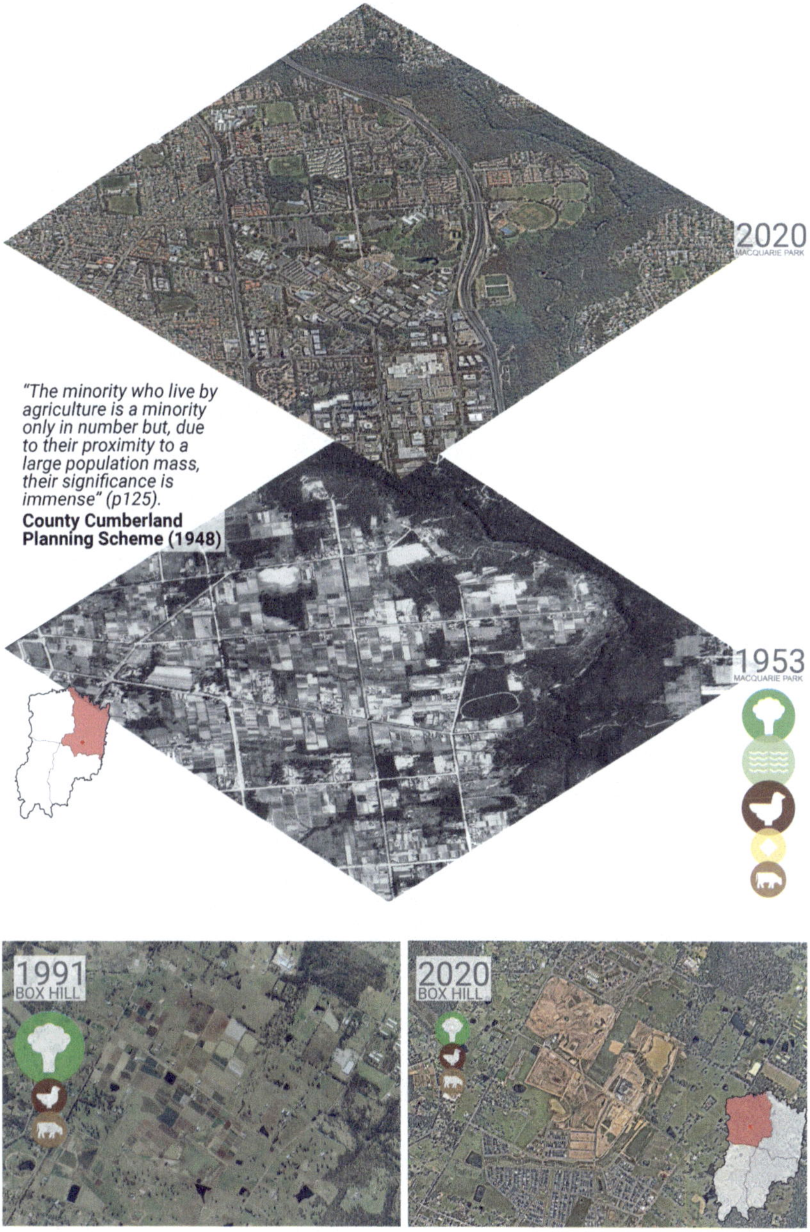

Fig. 8.12 Macquarie Park 1953 (middle) 2020 (top), Box Hill 1991 (bottom left) and 2020 (bottom right). Sources: Spatial Services NSW, Nearmap and Joshua Zeunert

Fig. 8.13 Warriewood 1961 (top left) and 2020 (top right), Pitt Town 1991 (bottom left) and 2020 (bottom right). Sources: Spatial Services NSW and Nearmap

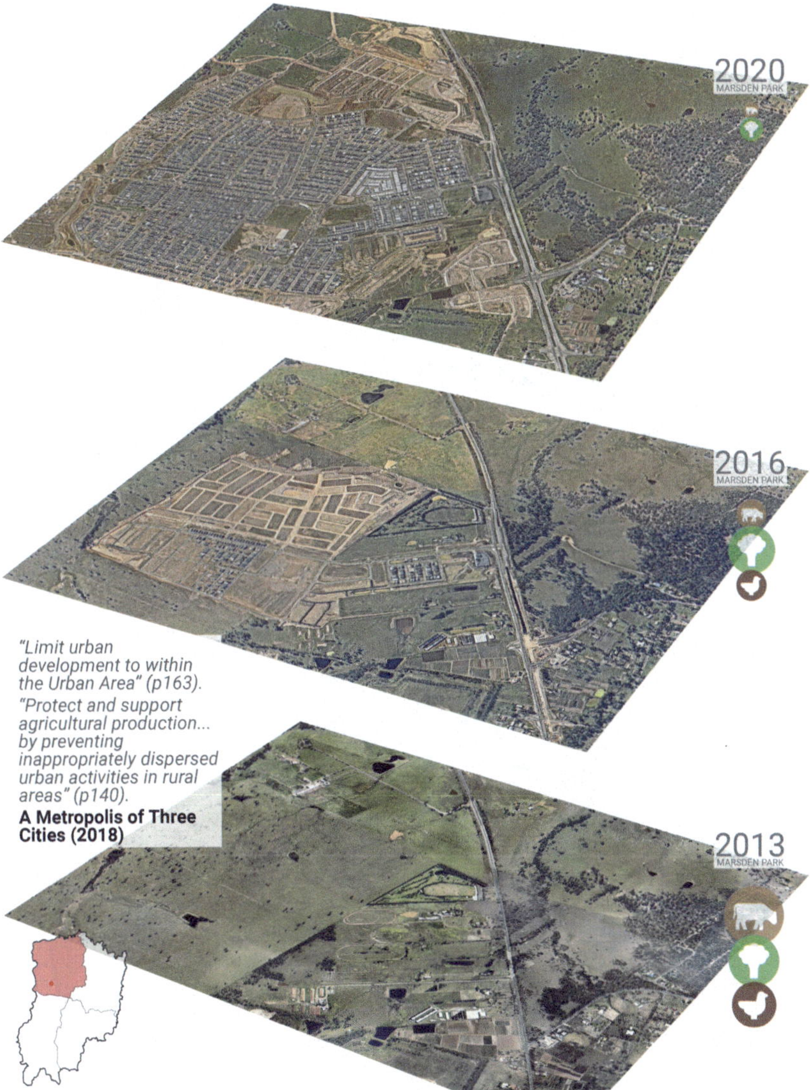

Fig. 8.14 Marsden Park 2013, 2016 and 2020. Sources: Nearmap

Fig. 8.15 Kellyville 2014, 2017 and 2020 (left), Arndell Park 2017 (top right) and 2020 (bottom right). Sources: Nearmap

Fig. 8.16 Luddenham 2015 (top) and 2021 (bottom). Sources: Nearmap

self-marketing to reach the uppermost social and political echelons. Lycett effectively conveyed the idyll of a gentleman's pastoral estate. The artwork displays enclosed, tamed and maintained fenced fields; a stately abode; servants quarters and (presumably) John and Elizabeth wandering their grounds—albeit on the opposite side of the Parramatta River to provide a foreground. The Macarthurs' pioneering enterprises spear-headed both at and from Elizabeth Farm exceeded the monocultural pastoral depiction. Convict labour land clearance facilitated sheep, cattle, horses, viticulture and the acclimatisation of plants and economic crops such as olives and vines. Yet Elizabeth Farm would prove diminutive compared to the scale of their subsequent Camden Park Estate (see *Period One: Grazing*). The 2021 image centres on Elizabeth Farm, with the cottage remaining (not visible) but lands long since extensively subdivided.

Bungarribee settler history was catalysed by an extensive 2000-acre land grant in 1823. Views from the 1858 depiction (Fig. 8.8, top) feature its farm buildings and the subsequently demolished (in 1957) homestead, along with all-important paddocks. These fields were once an important English horse breeding stud for the notorious East India Company, and for racing. The 1847–1852 watercolour is titled 'Honourable East India Company's Depot'. The long-horned cattle shown were motifs and a sideshow to the larger horse stud. An alternative 1847–1852 watercolour switched the horses to the foreground and cattle to the background.[9] The artistic style of the 1858 painting is more romantic but less expressive of a British pastoral idyll. Bungarribee's remnants are located in the now Western Sydney Parklands (bottom), shown in 2021 as extensively circled by encroaching suburbia.

General Images

The bays at the mouth of the Georges River were once home to thriving oyster leases. A 1943 image reveals geometric oyster farms laid around Shell Point and extending into Woolooware Bay (Fig. 8.9, top). Commencing in the 1960s, the extensive oyster beds were destroyed and the entire bay was completely remade into suburban marina Sylvania Waters, shown in 2020 (middle).[10] Pemulwuy is also depicted in 1943 (bottom left), once part of an extensive region of settler agriculture centring on Prospect Hill and its rich basaltic soils. In 2020 (bottom right) the area has long lost its use for food production, replaced by residential, recreation and light industrial land uses.

Minchinbury (see *Periods Three and Four*) was Sydney's most iconic winery and its extensive vineyards are visible in 1947 (Fig. 8.10, top left) when the then Penfolds operation had around 100 hectares under production. In 2021 (top right), little remained of the institution that served its sparkling wine to Queen Elizabeth II in 1954 beyond street names and a neglected olive grove. A once-productive bend of the Georges River (bottom left, 1953) was chiefly displaced by significant sand mining activities from the 1950s, particularly visible in 1960s and 1970s aerial imagery. The Chipping Norton Lake Authority was formed in 1977 and tasked with controlling sand extraction and developing the area into a recreational lake and associated parkland. While most agriculture also mines soil fertility, short-term resource exploitation through mining represents the erosion of longer-term food security and sustainability. This overall land use arc is a recurring theme of this book, marked here by a transition from Cobragul Country to agricultural land, then resource extraction and finally urban development and recreation.

Planning Rhetoric

P.A. Yeoman's former property 'Yobarnie' was an internationally significant agricultural landscape due to its pioneering 'keyline' design for water management and retention, as shown in Fig. 8.11 and documented in his 1965 book *Water for Every Farm.* The farm was cumulatively replaced by ubiquitous urban sprawl over the decades. The site was an anomaly within Australian private agriculture, offering a more civic minded educational and heritage resource that pioneered on-site water management. This occurred before Adelaide's 1960s municipal wetland systems,[11] Peter Andrew's farm water management techniques,[12] water sensitive urban design[13] and sponge-city concepts.[14] Like other notable sites of loss, its parks and streets are sadly now only token namesakes amidst ubiquitous sprawl.

Macquarie Park and the wider area surrounding Macquarie University was once a hub for food production, as evident in the 1953 aerial image (Fig. 8.12, centre). Box Hill's transformation is more recent, as evident in the change between 1991 (bottom left) and 2020 (bottom right).

Warriewood was once known as 'Glass City' due to its expanse of greenhouses producing fruits and vegetables (as well as some poultry in the region) shown in 1961 (top left) as a then peri-urban district of Sydney. It set the scene for later city-scale greenhouse districts like the 'plastic city' of Almeria in Spain and 'glass city' Westland in The Netherlands.[15] By 2020, all but a few remnants had vanished (top right). Figure 8.13 (bottom left) shows fruit and vegetable production, fodder cropping and agricultural infrastructures in Pitt Town in 1991. The same site is shown in 2020 (bottom right) under significantly encroaching sub-urbanisation, while its locality contains Australia's oldest still operating farms (though only just at the time of writing).[16]

Photographs of Marsden Park, in 2013 (bottom), 2016 (middle) and 2020 (top), reveal its rapid suburbanisation from grazing, market gardening and poultry, despite being located about 50 kilometres (north-west) of Sydney's city centre.[17] Its name derives from the Reverend Samuel Marsden who was a key colourful figure in Sydney's and Australia's early agricultural development (see *Period One*) (Fig. 8.14).

The three images of Kellyville, from 2014, 2017 and 2020 (left), reveal the rapid encroachment upon a poultry farm, making cohabitation challenging. The region was once significant for Sydney's agriculture. Top right shows an operating market garden in Arndell Park in 2017 that had disappeared under light industrial sheds in 2020 (bottom right) (Fig. 8.15).

Until forced to vacate for Sydney's second international airport in late 2015, Vicary's was Sydney's oldest continually operating winery.[18] The top image shows the winery (Fig. 8.16, top centre) and adjacent grape vines at Luddenham, along with Leppington Pastoral Company's irrigated pastures and dairy cattle (bottom). The bottom photograph reveals their obliteration by a much wider sea of earthworks in October 2021 to make way for the Western Sydney Airport.

Notes

1. Spearritt, 2000: 31.
2. Maiden, cited in Christie, 1988: 19. See obituary for Maiden via Cambage, 1926.
3. See Heritage NSW Database Number: 5061406, File number: 11/13622-1.
4. See Heritage NSW item: H01448.
5. Benson & Howell, 1990: 16–17.
6. The Hills Shire Council, n.d.
7. See Heritage NSW Database Number: 5061406, File number: 11/13622-1.
8. MHNSW, 2020.
9. There are also depictions from c. 1847–1852, see: https://archival.sl.nsw.gov.au/Details/archive/110339020
10. The suburb also hosted Australia's first reality TV show in 1992 of the same name.
11. See Zeunert, 2017: 60–61.
12. Andrews, 2006.
13. Zeunert, 2017: 100–110.
14. Daroy, 2018.
15. See Court & Zeunert, 2022; Zeunert, 2017: 32–33.
16. O'Connell, n.d.
17. In 2020, an approximately 600 square metre block of land in Marsden Park 53 kilometres from Sydney's CBD commanded $650,000.
18. Heritage NSW item: 5052646; Western Weekender, 2015.

References

Andrews, P. (2006). *Back from the brink: How Australia's landscape can be saved.* ABC Books.

Benson, D., & Howell, J. (1990). *Taken for granted: The bushland of Sydney and its suburbs*. Kangaroo Press.

Cambage, R. (1926). MR. J. H. Maiden, I.S.O., F.R.S. *Nature, 117*, 57–58. https://doi.org/10.1038/117057a0

Christie, M. (1988). *The Sydney markets 1788–1988*. Sydney Market Authority.

Court, T., & Zeunert, J. (2022). Novel horticultural adaptations to climate and their unique landscapes. In Y. Yang & A. Taufen, *Routledge handbook of sustainable cities and landscapes in the pacific rim* (pp. 164–179). Oxon. https://doi.org/10.4324/9781003033530-16

Daroy, A. (2018, February 28). In China, groundwater is dropping rapidly. Could Kongjian Yu's Sponge Cities prevent disaster? *Foreground*. https://www.foreground.com.au/agriculture-environment/kongjian-yu-sponge-cities/

MHNSW. (2020). *A turbulent past*. Accessed May 15, 2021, from https://mhnsw.au/stories/museum-stories/a-turbulent-past/

NSW Government. State Heritage inventory.

NSW Government. Database Number: 5061406, File number: 11/13622-1.

NSW Government. Item: 5052646.

O'Connell, J. (n.d.). *1794 Hawkesbury area settled*. https://australianfoodtimeline.com.au/hawkesbury-area/

Spearritt, P. (2000). *Sydney's century: A history*. UNSW Press.

The Hills Shire Council. (n.d.). *A brief history of the Shire*. Accessed April 26, 2020, from https://www.thehills.nsw.gov.au/Library/Library-e-Resources/Local-Studies-Family-History/A-Brief-History-of-the-Shire

Western Weekender. (2015). *Last call: Vicary's Winery to close after 92 years*. The Western Weekender. Accessed April 10, 2019, from https://westernweekender.com.au/2015/09/last-call-vicarys-winery-to-close-after-92-years/

Zeunert, J. (2017). *Landscape architecture and environmental sustainability: Creating positive change through design*. Bloomsbury.

9

2021+: Eating Itself

The harbour city's margins have spread steadily outward since invasion in 1788, consuming metropolitan agricultural land as they seep towards the outer limits of the city's topographic rim. In the process of feeding itself, Sydney has ultimately fed upon itself through commercial development's voracious consumption of life-supporting landscapes. Sydney's commercial food history is therefore also one of a lack of food planning—from its initial inception as an extractive economy to an increasingly precarious food future within an unsustainable growth model (Fig. 9.1).

As Richard Heinberg reminds us: "Food is energy" and "it takes energy to get food".[1] These two facts have together "always established the biological limits to the human population and always will".[2] Population growth has underpinned all stages of Sydney's development. Prior to occupation, the Sydney region supported approximately 2000–3000 peoples from Dharug, Kuringgai and Dharawal language groups. During *Period One* (1788–1850), British colonists violently dispossessed First Nations people and altered cultural food practices and land management systems,[3] clearing the way to generate commodity capital.[4] Imported practices such as mixed farming, grain cropping and livestock pastoralism

J. Zeunert, A. Daroy, *Sydney's Food Landscapes*,
https://doi.org/10.1007/978-981-96-0710-5_9

Fig. 9.1 Austral, 2022. Source: Joshua Zeunert

struggled initially but eventually established an agricultural blueprint still dominating the Australian continent today.

In *Period Two* (1851–1900), the discovery of gold in Wiradjuri Country (Orange) in 1851 ignited an international gold rush, while mechanisation and transport networks increased food production and distribution supporting a growing city. The colony of New South Wales' population increased nearly ten-fold, from 50,000 to 480,000.

Across *Period Three* (1901–1945) fossil-fuel driven industrialised processes supported international war efforts, creating and supplying new markets while employing an increasing proportion of Sydney's burgeoning population in manufacturing. The city's 481,830 residents at the time of Federation had more than tripled by 1945, reaching approximately 1.5 million. During *Period Four* (1946–1999), widespread verdant metropolitan agriculture was swept under decentralised suburban expansion. Catalysed by a post-World War II boom, Sydney's population increased from around 1.7 million to 4 million. By *Period Five* (2000–2020), neoliberal economics chased soaraway financial futures expanding free trade agreements into global markets, accompanied by long-chain and long-shadow supply logistics.

Sydney now requires colossal energy supplies to furnish the city with food from increasingly distant locations. In 2020, farmer Warren Waddell characterised local farming in Sydney as "finished",

> As land becomes more and more expensive, [local farming]—it's impossible to justify [...] we've moved from this term 'primary production' to what I would term 'rural enterprise', which is more like a thousand different land uses combining to generate a rural amenity that's quite enjoyable but has no relationship to primary production [...] We have a very valuable product and twice the average land holding of anyone around us, and it still would only pay one person full-time. [...] The agriculture industry in Australia is doing extremely well but it's all on scale and it's all based on a relationship with major retailers. What we have now around Sydney is just a combination of rural residential with hobbies that somehow align with different rural usage. Even in horticulture, the most active industries are turf, nurseries and flowers, none of which you can eat.[5]

Overwhelmingly, therefore, the city's history demonstrates a lack of sensitivity and attunement to place—towards Indigenous people, culture, food systems and ways of knowing,[6] and with respect to ecological symbiosis within Sydney's bioregion. This has also resulted in the failure to protect metropolitan agriculture as a meta-economic and fundamental land use, ultimately jeopardising its future food security. Sydney's spectacular natural beauty and its resilient remnant and regrowth urban bushland present the illusion of a more 'naturally attuned' city than is the case—the very resilience of region's tough sandstone ecologies almost defeated the early colonists before it could establish itself. The city has long since exceeded its carrying capacity, and its ecological footprint continues to trend in the wrong direction.[7] Consequently, the impacts of material processes required to support Sydney—extraction, production, manufacture and supply—are largely decoupled and outsourced to unseen places and flows. This final chapter reflects on metropolitan Sydney's past commercial food production and future in light of food and energy security, production capacity, planning and the shifts required in light of increasingly precarious climatic and resource futures.

Sydney's Commercial Agri*culture*

Sydney's commercial agriculture is bound within an economic imperative. From its outset, the colony of New South Wales was the castaway child of a rapidly industrialising British Empire at the inception of the Anthropocene.[8] Empires are built on the exploitation of people, land and resources and ultimately, such transgressions prove socially, politically, economically and ecologically temporary, given time. Since colonisation, Indigenous Care for Country has been replaced by what William Lines' terms 'taming the Great Southern Land' to create 'a camping ground for profit'.[9] The evisceration of ecological communities followed,[10] largely to facilitate settlement through Western agriculture and resource extraction. The ensuing fertility exhaustion and land degradation facilitated the transition to urban development through subdivision for maximised commercial gain.

Cultural and social structures and influences are complex. There were fundamental attempts to instil agriculture tracing back to Governor Phillip in *Period One*, persistence to establish commodity industries in *Period Two* and ongoing attempts for an agrarian nation in *Period Three*.[11] Civic and environmental values have also always been a part of metropolitan Sydney's fabric. From the symbiotic food systems of First Nations communities, to Government Farms, Commons and Governor Phillip's equitable distribution of rations among settlers and convicts during the nascent years and construction unions' environmental activism through 'green bans' in later years (see *Period Four*),[12] Sydney's food stories are nuanced.

Beneath the surface of civic virtue also lies frequent opportunism, inequality and outright greed. Injustices towards the Sydney region's First Nations peoples stands as a stark testament to this reality. Within the British colony, the transition from Governor Phillip to Lieutenant-Governor Grose marked a pivotal moment. The emergence of 'grants-for-mates' within the Rum Corps signalled a shift in power dynamics. Enter influential colonial figures like John Macarthur, whose image featured on Australia's former $2 note. Macarthur championed the archetype of the "industrious and vigilant master",[13] while simultaneously tactically manoeuvring political currents to create an intricate agribusiness empire within colonial parliament.[14] Sydney's most prosperous agriculturalists amassed vast tracts of land and in doing so, intergenerational wealth, leveraging their influence through agricultural expansion or the urban development of their former farmlands.

Land and governance arrangements were frequently skewed towards the ruling class and largely excluded smallholders. Support for land-based ventures was directed at large-scale commercial endeavours such as pastoralism (cattle and sheep), grain production and agricultural and processing activities at-scale. In the early colony, agriculture beyond subsistence or supplementary farming was regularly outcompeted by plum estates assisted by convict labour and government supply contracts. The effects of an embedded commodity production model is revealed in *Period Five*, whereby 72 per cent of Australia's agricultural output in 2017–2020 was exported.[15] Consequently, a lack of concession to, or subsidisation of, environmentally responsible small-scale agricultural operations has endured, reflecting a yield-effective but imposed, extractive and degradative agribusiness approach.[16]

Yet, amidst this landscape of elite privilege, the heart of Sydney's agricultural output remained rooted in resilient culturally and economically marginalised individuals and groups. During *Period One*, food production relied heavily on enforced convict labour and emancipists, supported at times by First Nations trade, knowledge or fishing. As the city evolved into *Periods Two* and *Three*, Chinese migrant-run market gardens emerged as caretakers of Sydney's commercial 'backyard vegetable garden', tending a key component of the city's sustenance. From *Periods Two* and *Three*, culturally diverse immigrants frequently assumed the role of agricultural labourers and heavy-lifters.[17] New immigrants often encountered a climate of social exclusion, evident in policies such as the *Immigration Restriction Act* or 'White Australia Policy' (1901–1958).[18] Despite cultural marginalisation, a wide range of Sydney's diverse cultural groups achieved success in smallholder farming.[19] Some eventually left the industry, some expanded or diversified, while others used agricultural production or the sale of lands as a means for themselves or their children to leapfrog into other ventures.[20]

Sydney's agricultural saga is thus a multifaceted tale—a blend of competing intentions, power dynamics and the resilience of those often relegated to the margins. Understanding this intricate interplay is essential for comprehending the city's past, present and its food futures,

> There is a very beautiful story to this area in south-west Sydney: all the migrant families—Italian, Maltese, Lebanese, Chinese, Cambodians, Vietnamese—came here from pretty much nothing. They've all worked very hard on the land and look what they have achieved over their lifetime. Why don't we have that sense of provenance towards our food, not just wine? [21]

Christina Kelman's above lament about the lack of value ascribed to regional agricultural identity invokes wider dynamics that can be traced to the colony's foundation and are entangled in ideological complexities.

Sydney's metropolitan food production history is also one multiple individual and family success stories defying economic climates of self-interest and opportunism,[22] frontier and cavalier behaviours and competitive productivism.[23]

Consequently, the processes sustaining Sydney, such as production, manufacture and supply, have been either pushed further into its margins, decoupled or altogether externalised to distant shadow places.[24] As such, resource drawdown, land degradation and pollution are less palpable. Ecological impacts are frequently difficult to discern or even intangible, with fleeting glimpses of extensive logistics across infrastructural networks of container ships, trucks, trains and planes.[25] Unlike metropolitan land set aside for biodiversity conservation, Sydney has not ascribed intrinsic or meta-economic value to its agricultural lands to protect them from development within its overall matrix of land use, nor their social and economic ecosystems. Instead, the city's trajectory has largely been one geared to "commercial peoples" as the "final products of the natural development of human society".[26]

In 2020, Sydney's two most financially viable food-producing activities were poultry (meat and eggs) and mushroom farming. Neither intensive modes resemble what we may think of as agrarian 'farming', taking place within spatially concentrated factory sheds rather than fields (Figs. 7.8 and 7.13). Thus they lack bucolic character still associated with food production landscapes, including decision-makers shaping Sydney's metropolitan agriculture such as the Western Sydney Parklands Trust.[27] As such, Australian agriculture's perception and its commercial realities are routinely out of step.[28] Meanwhile, nurseries/cut flowers/turf collectively generate more financial value than either eggs or mushrooms in Sydney's Metropolitan Rural Area, returning us to Waddell's earlier observation that Sydney's highest value primary production outputs do not contribute to the city's food security. This also mirrors Sydney's preference for ornamental rather than edible horticulture—suggesting the ongoing status of tired colonial aesthetics akin to Joseph Banks's legacy in *Period One*.[29]

Metropolitan Sydney's Food Security and Future Scenarios

Agriculture is historically intertwined with human settlement. A viable distance between food production and population is the lifeline to survival. Today's population numbers and their separation from proximate food production landscapes is historically unprecedented. Stable food supplies and ongoing food surpluses are relatively uncommon across history's multi-generational timeframes. Currently, Sydney and Australia's politicians, planners and key decision-makers assume a steady, ongoing food supply, with questionable resilience to accommodate events beyond surface-level cracks, such as crises, let alone collapse.

Food supply security is a crucial component of social and environmental sustainability, as evident in decades of discourse surrounding planning environmentally just futures.[30] According to the World Economic Forum (WEF), the "future of global food systems is uncertain".[31] Within Australia, "domestically, food and water security are, if sensible to covet in the short term, more uncertain than they have ever been for the long term".[32] Concerns stem from a multiplicity of reasons but fundamentally from finite supplies of fossil resources, particularly crude oil, natural gas (and its use for fertiliser) and phosphate rock underpinning all aspects of contemporary food systems.[33] Fresh water and the condition of soils and supporting ecosystems for agriculture are additionally high areas of concern.[34] A 2010 report, for example, highlighted that London's food system contains only three or four days of available food.[35] The duration of Sydney's stores are likely to be similarly short, as made evident during the periods of bare shelved supermarkets during the Covid-19 pandemic. These reflect the inherent precarities of a 'just-in-time' food supply system.[36] While the electrification of grids and supply chains is developing, the Australian island's global isolation presents geographically immutable logistical, technical, embodied energy, cost and operational barriers regardless of fuel sources utilised for global transportation.

Sydney has experienced tasters of scarcity during 2019–2022, including moments of the aforementioned pandemic, floods and bushfires. State government planning continues to focus on population growth and expanding the city's spatial area, reducing proximate agricultural land and thus arable land per capita and local food supply security.[37] Compounding the issue, federal and state governments push 'competitive productivism' geared to large producers, rather than local scales of supply.[38] Consequently, Sydney is increasingly reliant on food sourced from further and further afield. Such food systems logistics must also navigate Sydney's bottleneck geography constraining its supply route flows. Lifecycle studies have shown that international air transport consumes more fossil fuel energy than the air-freighted food nutritionally provides, resulting in net energy loss food systems and drawdown of planetary biocapacity.[39] As such they are the embodiment of intergenerationally unsustainable food systems, yet Sydney's recent agribusiness precinct intends to grow food locally to ship it overseas to the highest bidder (see *Period Five*).

Currently, decision-makers and metropolitan planners are betting on consistent and perpetual food supplies with enough resilience within production and infrastructural systems to accommodate shocks. Future scenarios depend on the severity of potential stresses and characteristics of production and supply chains. These span geopolitics; global trade and diplomacy:[40] food sovereignty;[41] climate change[42] and related events;[43] surface water and groundwater supplies; salinity and pollution; population growth;[44] fossil fuels and finite resource availability;[45] soil degradation, erosion and health;[46] ecosystem provision and resilience and transportation and supply chain resilience. Other challenges, including bushfires, cyclones, floods or global pandemics, can present more immediate shocks and shorter-term influences on the usual production and supply rhythms. These can occur at varying scales, from local and regional to national and global. The larger the scale, the more interruptive and significant the ramifications.

By externalising its supporting and provisioning systems, Sydney faces increased vulnerabilities to issues beyond its sphere of influence and control. There appears to be no ongoing role for metropolitan agricultural production to help increase resilience of the city's food supply system. Consequently, Sydney's 'planned' approach consigns its residents to what the WEF characterises as 'unchecked consumption'—namely, a "world of high GDP growth with high environmental cost".[47] In medium- and long-term scenario planning, this is most likely to lead to 'collapse' due to drawdown of supporting ecosystems and environments provisioning resources, and conflicts and competition for remnants.[48]

Numerous questions arise in the event of a food crisis. To what extent could Sydney financially overpower other international and national markets to channel-in flows of essentials?[49] How long would it take to repurpose ornamental production areas (turf, cut flowers, ornamental plant nurseries) and public landscapes to food production? Moreover, how much suitable and uncontaminated land area is available for repurposing, in the wake of decades of planned urban sprawl? What skills and knowledge exist to achieve successful metropolitan food production? What percentage of Sydney's food supply could be locally supplied, if under intensive horticultural production? What transportation networks and distances would be viable for food supply networks? Would rural Australian agribusinesses—including those owned by foreign nationals—forsake potentially more lucrative export markets to feed national and local markets?[50]

> We are slaves in the sense that we depend for our daily survival upon an expand-or-expire agroindustrial empire—a crackpot machine—that the specialists cannot comprehend and the managers cannot manage. Which is, furthermore, devouring world resources at an exponential rate.[51].

While the two world wars answered to some of these scenario questions stemming from scarcity and a significant need for self-reliance, circumstances were distinctly different from present. Not only was Sydney's population considerably lower (around 720,000[52] in WWI and 1.5 million in WWII), but the per capita availability of agricultural production land was considerably higher. Food production, animal husbandry and butchery, food processing and storage skills were considerably more widespread than in today's more highly urbanised and consumer-oriented society. Extensive local industries and manufacturing could produce infrastructures and tools, and fossil fuels, especially quality crude oil, were in comparative abundance to underpin life's daily rhythms. The past four decades of neoliberal governance has regularly proved to be held hostage to short-term interests of powerful corporate minorities and their influence over media and lobby groups, rather than supporting Sydney's wider population's best interests and intergenerational equity.

While Sydney's historic environmental movements saved and nurtured precious remnants of indigenous plant and animal webs in hybrid and novel ecosystems, the city continues to fail to protect its precious agricultural lands and those with the skills to successfully tend to them. This invokes William Cronon's astute observation that "the dream of an unworked natural landscape is very much the fantasy of people who have never themselves had to work the land to make a living".[53] Sydney's history reveals an arc of resistance to work on the land by a considerable proportion, and in doing so, separates culture and nature, action and impact, supply and consumption.

Only a real-life crisis would reveal just how much land could be turned over to food production.[54] A time-lag occurs in agricultural production systems between sourcing seeds, cuttings and rootstocks, germination, sowing and harvesting—months for the quickest foodstuffs (such as leafy greens) and multiple years for fruit and nut trees and most livestock. Even under a scenario of maximum intensive local food production, Sydney would rely considerably on food supplied from afar, particularly for crops (cereals and grains), non-poultry meat, nuts and most fruit. This has been case to varying degrees throughout Sydney's colonial history but now faces an unprecedented risk due to its continuously increasing population and urban footprint (Fig. 2.7).

Planning and Policy Pathways

Inspired by Sydney's first planning scheme in 1951, soon-to-be New South Wales Premier Cahill's words that the "rural areas of the county [of Cumberland] will be preserved not only for vital food production and as a countryside to serve the metropolis, but also for soil conservation, irrigation, afforestation" have long since been forsaken. Aside from the visionary but flawed County Cumberland Council's 1948 scheme (see *Period Four*), metropolitan agriculture has been either token or entirely absent from Sydney's metropolitan planning,[55]

> At the Royal Australian Planning Institute Conference at Penrith in Western Sydney in 1993, in response to a question about planning for agriculture in the Sydney region, a planning official replied: 'There is no place for agriculture in the Sydney region. Agriculture belongs over the (Great Dividing) Range and any agricultural land is land awaiting higher economic development'.[56]

Intergenerationally strategic planning needs to be decoupled from neoliberalism's econocentrism to situate human culture within planetary, national and local carrying capacities and equity measures of ecological justice. Sydney's most environmentally damaging outcomes have occurred under the watch of 75 years of official planning. Ecological architect Paul Downton characterises planners as "far from being advocates" but "more typically soothsayers of informed complacency" reiterating "conventional wisdom".[57] Downton's own shadow plans for Adelaide, meanwhile, offer strategic planning capabilities at a metropolitan-scale for spatially articulated visions of intergenerationally sustainable futures.[58]

Metropolitan planning requires considering multigenerational timescales that underpin processes taking decades to implement, with legacies that remain inscribed in hard landscapes for centuries and even millennia. Sydney's population growth and urban development are filling Sydney's finite topographical basin and quashing remnant life-supporting landscape systems with built typologies of questionable value. Remnant productive land is being reduced to fragmented follies (Fig. 9.2). Development is taking place in fertile and productive flood plains simultaneously risking residential damage. Low density suburban form is creating poor microclimates and escalating urban heat.[59] Investment in the Western Sydney Airport's agribusiness precinct focuses on export rather than local food supply, while further urbanising the Cowpastures district, which catalysed Australian pastoralism (see *Period One*).

Windfalls from state and local government-endorsed agricultural rezoning for urban development have led to widespread speculative land banking across western Sydney. Venture capitalists, investment firms and even crowdfunding campaigns (including certain farming families) purchase agricultural land as a medium-term commercial land redevelopment investment. Land regularly sits fallow or underutilised, waiting for land

Fig. 9.2 Aerial images of market gardening in Catherine Field, Freemans Reach and Maroota in 2022. Source: Joshua Zeunert

rezonings. This contributes to the impression that the agricultural land is degraded, supporting its rezoning. It is a largely irreversible process, as larger agricultural land parcels become subdivided into tens, hundreds and even thousands of individual titles and strata bodies.[60] Furthermore, land rezoning regularly consigns remaining agricultural operations to land-use conflicts stemming from urban encroachment (Figs. 7.3, 7.4 and 8.15). More widely, rezoned and subdivided lands collectively constrain Sydney's ongoing and future food supply security in a post-carbon world.[61]

Sydney has similarly witnessed farmers subdividing or altogether selling their land to developers seeking to rezone the land from agricultural to urban land uses, frequently to build new residential or mixed-use developments. The most economically savvy rezone and develop the land themselves in a vertically integrated process.[62] Concurrently, pressures on land values[63] from urban encroachment, combined with price pressures exerted by larger regional agricultural production scales in cheaper locations mean that most forms of Sydney's agricultural production struggle to remain economically viable—at least as long as transportation energy sources remain cheap.[64]

The Greater Sydney Commission's (GSC) metropolitan strategy—*A Metropolis of Three Cities* (2018)—took the approach of dividing the city into three 'cities'. This deemphasised the traditional radial focus on a singular core and effectively drew attention away from Sydney's expanding girth. The GSC's plans to continue to expand the city with 725,000 proposed new dwellings between 2018 and 2036,[65] including relaxed provisions for greenfield developments, principally in the 'Western Parkland City' (WPC).[66] Despite a change of government in 2023 and the dissolution of the Greater Cities Commission, the 2018 *Metropolis of Three Cities* remains the official metropolitan plan at the time of writing in early 2024.

The rush to fill Sydney's topographic basin with 8–8.9 million people by 2061 contains the hallmarks of a 'geography of nowhere'[67] displayed in twentieth-century North American suburbanism.[68] Sydney's future vision continues to emphasise suburbia[69] as lamented by architect Robin Boyd in 1960[70] and more recently by comedian Tim Ross in relation to 'McMansion' housing.[71] Implications, however, are much more serious than mere aesthetic preferences. For much of Sydney, growth correlates to suburban sprawl, non-independently certified multi-story buildings, generic big-box retail, traffic congestion and expensive toll roads. Spatially pervasive suburban forms now consist of low-density, increasingly large houses on smaller and smaller land parcels. Designs frequently demonstrate poor passive solar orientation and comprised energy efficiency,[72] use unrecyclable and toxic materials and finishes constituting 'monstrous hybridity'.[73] Australia's low-density and low-quality suburbanism supported by finite-fossil-fuel-dependent lifestyles hence evokes what James H. Kunster characterised in the American context as the "greatest misallocation of resources in the history of the world"[74]. Permaculture co-originator David Holmgren's notion of 'retrosuburba'[75] observes that these spatial arrangements offer limited potential to be retrofitted for sustainable living and/or suburban food production in an energy descent future.[76]

Suburbanisation increases the urban heat-island effect, hard surfaces deny groundwater recharge and intensify storm runoff and flood events, arterial roads reduce air quality, while bushfire buffer zones, social and aesthetic amenity decrease (Fig. 9.3). These effects and their systemic costs are not taken into account by the greenfield property development lobby's advocacy for peri-urban land releases for detached housing as the panacea to high property prices,[77] a notion systematically dispelled by Cameron Murray in *The Great Housing Hijack*.[78]

The cultural, environmental and intergenerational values of Sydney's metropolitan agriculture must be considered beyond economics, supported by protection mechanisms in perpetuity and enshrined in legislation. Former US President Bill Clinton, a key perpetuator of neoliberal ideology during *Period Four*, purportedly regrets agricultural marginalisation,

> We need the World Bank, we need the IMF, we need all the big foundations, we need all the governments to admit that for 30 years we all blew it, including me, when I was president. We blew it. We were wrong to believe that food is like some other product in international trade. And we all have to go back to a more environmentally responsible, sustainable form of agriculture. We should go back to a policy of maximum food self-sufficiency.[79]

A 1999 report found that agriculture was becoming marginalised in the Sydney Basin during *Period Four*.[80] Marginalisation was, however, constant under the prevailing pattern of Sydney's urban growth. By the end of *Period Four*, thousands of commercial agricultural producers altogether evaporated, as evidenced through historic aerial photography and

Fig. 9.3 Oran Park 2019 (top) and 2021 (bottom). Source: Joshua Zeunert

comparison of *Period Four* and *Period Five*'s maps (Figs. 9.4 and 2.7). The incremental erosion of larger commercial farms also saw 'rural lifestyle living' allotments become the main form of peri-urban land use. Thus, the number of allotments per hectare increased, while overall agricultural output in proportion to Sydney's population size significantly fell.

Almost two decades later, a 2017 report commissioned by the New South Wales Department of Planning and the Greater Sydney Commission and completed by a consultant team led by 'AgEconPlus' critically appraised 'values' in the wider metropolitan rural area to inform metropolitan planning.[81] The report lists agriculture as one of 12 values but assesses it in terms of economic and spatial indicators that are not applied to the other 11 values. It therefore ascribes meta-economic values to two-thirds (at least eight) of its other recognised values: biodiversity, water and air quality, scenic, recreation, rural lifestyle, towns and villages and heritage. The report further uses selective evidence of New South Wales' land quality to devalue Sydney's agricultural land to advocate for further development. It fails to contextualise that no land in the state meets the 'Class 1' agricultural land classification, that a mere 1.8 per cent of the state is 'Class 2', only a small amount of 'Class 3' land exists east of the Great Dividing Range and that Sydney has a large proportion of 'Class 4' land. It also fails to recognise that many of Sydney's historic and current market gardens have thrived on Class 4, 5 and 6 land.[82] The report also does not acknowledge that speculative land banking stemming from rezoning agricultural land for urban development significantly contributes to the 'minimal use' classification prevalent in peri-urban Sydney. Nor does it state that minimal use agricultural land further stems from urban encroachment and subsequent land use conflicts with agriculture. In short, the report effectively dismisses metropolitan agriculture's value to the metropolis, paving the way for further land development.

Over its half-century legacy, Sydney's metropolitan planning has failed to influence power structures to support the medium- and long-term interests of its population. If left to continue its current trajectory, Sydney's agricultural capacity and its metropolitan area will reach the geographic limit of the topographic basin. Change seems "hopelessly utopian in the face of Sydney's property market, even without the spectre of corruption thrown in for good measure".[83]

Sydney's Agricultural and Food Futures

> When all our prime agricultural land has houses on it, how are we going to feed ourselves?[84]

This book has focused on revealing Sydney's past metropolitan agriculture (Figs. 9.4, 9.5 and 9.6). To answer the question of future food production viability, further research must determine,

1. How much food *does* Sydney produce?
2. How much food *could* Sydney produce?
3. To what extent might this feed its growing population?

An earlier study offers a useful projection. The Sydney Food Futures project (SFF) (2015–2016) calculated how much food the 'Sydney Basin' produced in the year 2011. The project also forecast food supply in 2031 based on four future scenarios: urban sprawl; agriculture prioritised and intensive agriculture.[85] It is crucial to note, though, that the SFF included an enormous area within its calculations of 'Sydney': the extensive 'Sydney Peri Urban Network of Councils' (SPUN) boundary area[86] combined with all metropolitan Sydney local government areas (LGAs).[87] The project termed this the 'Sydney basin'[88] but the SPUN area differs from both the Sydney Basin Bioregion and the Sydney Geological Basin.[89] For clarity, SFF's area is shown in Fig. 2.8, in comparison to this book's boundary.

The SFF's study exceeds metropolitan Sydney's boundaries, extending 227–345 kilometres south of the CBD (via road depending on route) to Durras North at southern Shoalhaven LGA just north of Batemans Bay. At 19,710 square kilometres, its area is almost double that of the Greater Sydney Commission[90] and nearly four times larger than the 5325 square kilometres used in this book.[91] As such, SFF findings calculate considerably increased amounts of 'Sydney's' food production than within the city's immediate metropolitan area. Even so, findings were damning,

> According to the 2015–2016 Sydney Food Futures report [...] Sydney can currently [in 2011] only supply 20 percent of its food needs. In 2031, the report estimates Sydney will be able to provide only six per cent of the city's

> food needs. By 2040, Sydney would have grown to a megacity of over six million people and have needed to build another 750,000 homes—mostly by taking the last productive farmland left in the Sydney basin. By then, it is projected that Sydney will be able to provide less than three per cent of its food needs.[92]

The SFF scenario of '*intensive agriculture*' highlights that food production could be increased to a potential eight-fold rise from 2011 production levels. This was based on spatially expanding existing food production within its large boundary[93] rather than increasing the on-site intensity of individual production approaches (for example, more factory farming, feedlots, biointensive farming[94] and replication of, for example, the Netherlands' intensive approach to horticultural production).[95] In other words, further production increases are possible given a pressing need and planning for the implementation of intensified food production, albeit likely to result in land use conflicts from intensive approaches and their proximities to dwellings (odours, dust, nutrient runoff, water, air and soil pollution). Also, much of the SFF area is not local to Sydney and reliant on trucking and freight movements.

To cite social theorist Murray Bookchin: "the assumption that what currently exists must necessarily exist is the acid that corrodes all visionary thinking".[96] Reversing Sydney's trajectory given the expansive timeframes required for metropolitan planning—let alone instilling a sustainable or ideally, ecologically regenerative, approach to living and governance—requires progressive ecological thinking. A more sustainable food supply system necessitates stemming urban encroachment on agricultural land, condensing built form and reducing the consumption of resource intensive livestock products.[97] More substantially, it requires directing population growth into new and regional centres[98], or departing from population growth to ethical degrowth.

Due to its already considerable population and the intensity of Sydneysiders' diets, the 5325 square kilometre area of Sydney used in this book (Figs. 2.1 and 2.8) possesses a very limited capacity to supply the city's food needs, even if scaled up and intensified. Nutrient-dense, fresh local produce, however, contributes to Sydney's food needs beyond numerical measures. Does the fact that local food production constitutes a small overall proportion in the food supply negate metropolitan food production altogether? Should proximate metropolitan agricultural land increase its agricultural (meta-economic) value in the face of increasing climate and resource uncertainties in the twenty-first century and beyond? Two key ways to increase local food production's contribution to feeding the population are through increased production intensity[99] and greater shifts towards plant-based diets as the most ecologically efficient and just forms of food production.[100]

A resilient metropolitan agricultural industry celebrates and supports local production. Following a four-year study into Sydney's urban agriculture in 2010, Merson et al. recommended that local and state governments must intervene in land use policies, regulate activities and subsidise Sydney's metropolitan agriculture.[101] They emphasised the role of agricultural production landscapes as a bushfire buffer zone between Sydney's urban area and its encircling bushland protected areas (Fig. 2.11), including the World Heritage area of the Blue Mountains that were decimated by bushfires in 2019–2020.

Greater economic support for smaller agricultural enterprises is further needed at federal, state and local government levels for ongoing employment and to sustain widespread smallholder agriculture. Redirecting fossil fuel and carbon-intensive industry subsidies offers one example of how this shift from an ecologically destructive to more sustainable model could be catalysed. Planning protection is further needed to ensure that Sydney's best agricultural lands are harnessed for food production and not merely for uses such as turf and amenity horticulture. This is especially critical in Class 3 high-production-value lands, such as those adjoining the Hawkesbury River flats near Windsor.

A seismic shift is required to protect remaining agricultural lands. This could be pursued in the manner of the conservation movement's successful establishment of National Parks,[102] with agricultural lands protected from development and rezoning.[103] As with conservation areas, a method of financial compensation or voluntary land donation would need to accompany the scheme. These protected lands could be utilised for environmentally and socially responsible commercial food production,

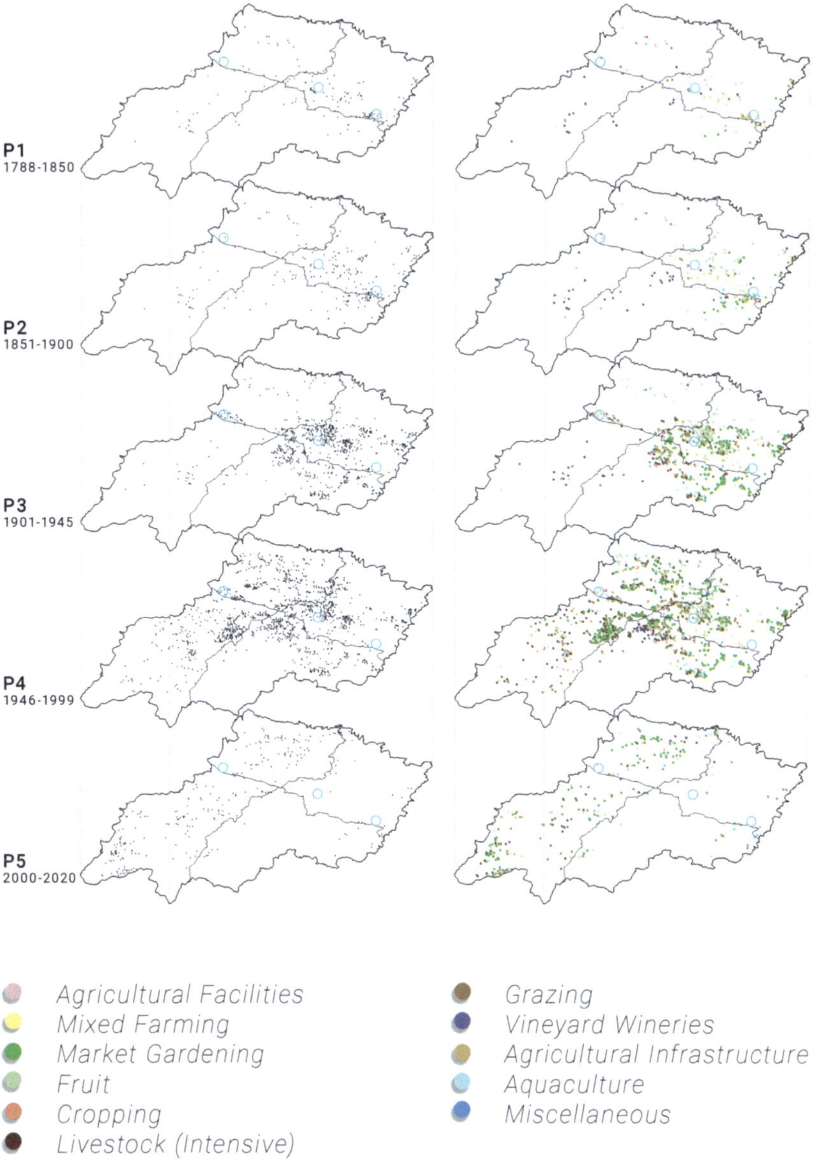

Fig. 9.4 Five Map Database Timeframes Compared: 3D. Each of the five timeframes is shown by total entries (in black, left column) and by 11 colours relating to the food landscape typologies (right column). Source: Joshua Zeunert, Josh Gowers and Stephanie Stankiewicz

offering more equitable access and supporting sustainable practices. It is essential that this occur in concert with First Nations consultancy and management, applying ecological knowledge systems, and regenerative land management practices together with food production.

Closing Thoughts

Few cities in the world can match the beauty of Sydney's azure harbour, beaches and golden sandstone geology. The glittering global metropolis belies its hungry engine room sustained by perpetual urban and financial growth, insatiably devouring resources becoming more and more obfuscated within distant and imperceivable shadow places and logistics networks. The city has removed most of its heavy industries and working landscapes. The processes and resources sustaining the city are not visible as they are, for instance, in neighbouring Newcastle and Wollongong's coal and steel industries. Sydney's dazzling surface hence conceals a web of outsourced landscape violence.

By focusing on food systems this book seeks to more meaningfully connect the everyday ritual of eating with food's origins. While not yet altogether lost, Sydney's food landscapes' trajectory continues to worsen and policy appetites for urban growth remain unrelentingly voracious. As we write this conclusion at the beginning of 2024, it seems that the fate of Sydney's local food systems is terminal. This is not a foregone conclusion and with the right combined and concerted commitment, change may be possible. Overwhelmingly, though, this book serves as a case study warning for unsustainable local food planning for other cities. It is a 'what not to do', unless a commercial and degradative 'survival of the richest' is the desired endgame. In many senses, this research into Sydney's lost food landscapes invokes Glenn Albrecht's notion of "solastalgia", namely, melancholia from the negative transformation of a loved home environment.[104] In this vein, we can only begin to comprehend the vast solastalgia felt by Australia's and Sydney's First Nations peoples since the Union Jack first forced its way into Sydney's sandstone land.

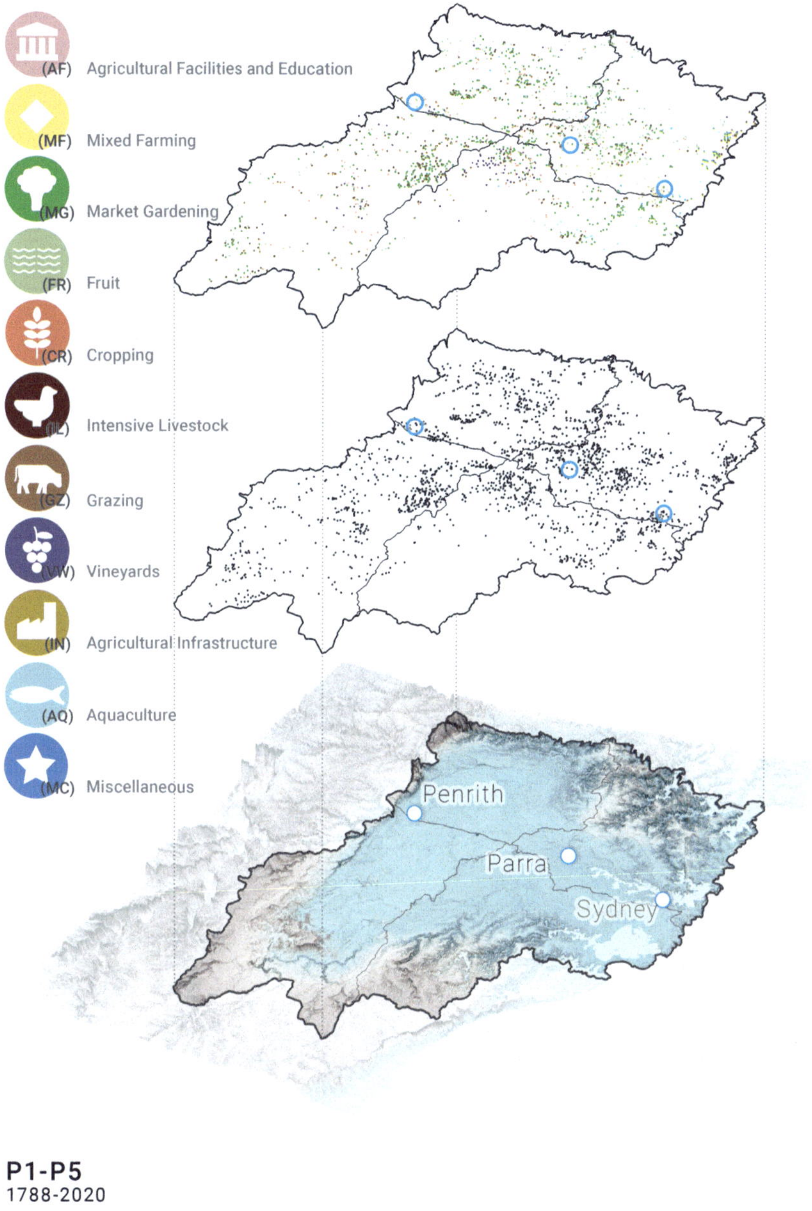

Fig. 9.5 Map Database Food Landscapes Compiled: 3D. The drawing depicts all of Sydney's documented food landscape database entries 1788–2020 by their coloured food landscape typologies (top) and in total (middle, black). Source: Joshua Zeunert, Josh Gowers and Stephanie Stankiewicz

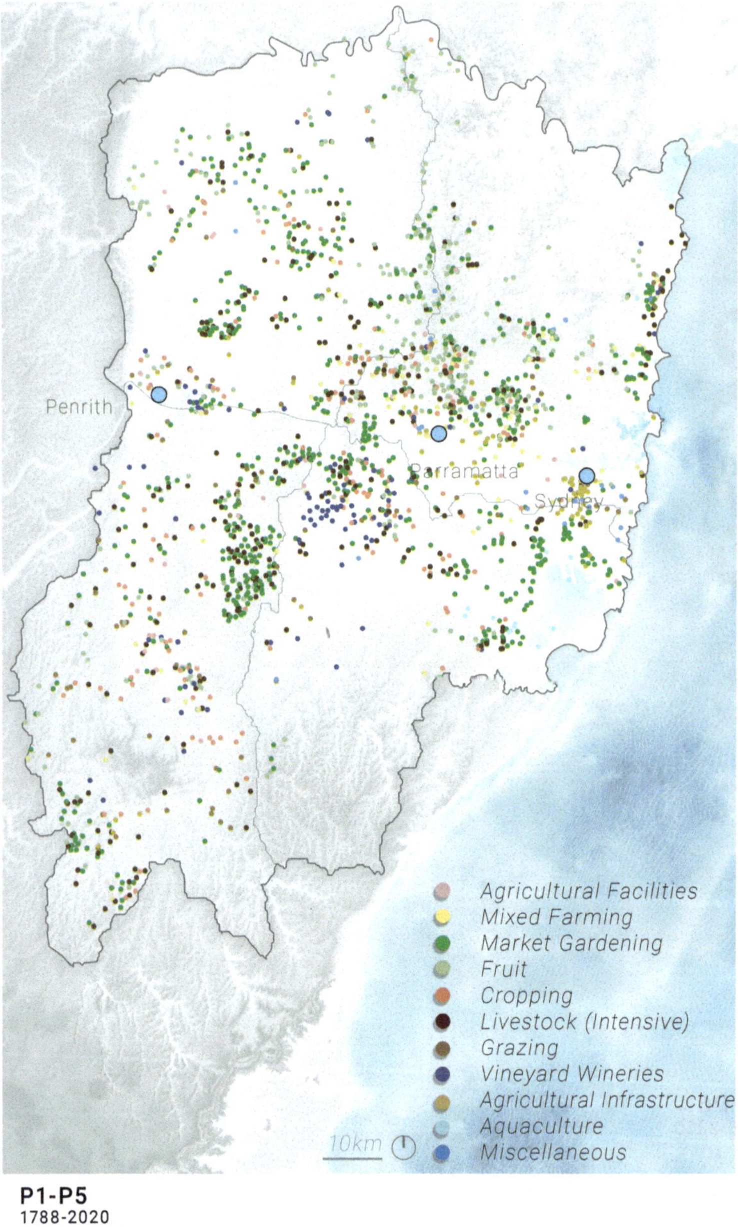

Fig. 9.6 The map shows all of Sydney's documented food landscape database entries 1788–2020 by their coloured food landscape typologies. Source: Joshua Zeunert, Josh Gowers and Stephanie Stankiewicz

Notes

1. Heinberg, 2005.
2. Heinberg, 2005
3. Ma Rhea, 2017.
4. See Lines, 1991.
5. Warren Waddell as quoted by Lehmann, 2020.
6. Seddon, 1972.
7. See Rees, 1992; Wackernagel, 1994 and https://www.footprintnetwork.org/
8. Crutzen & Stoemer, 2000.
9. Lines, 1991.
10. Benson & Howell, 1990.
11. Carruthers' extensive 'million [white] farms' campaign in the early twentieth–century is a notable example, see *Period Three*.
12. Mundey, 1981.
13. Cited in Birmingham, 1999: 283.
14. See *Period One* and Birmingham, 1999: 320.
15. ABARES, 2022.
16. Zeunert, 2023, 2024a, b.
17. Much like Australia's contemporary horticultural industry's seasonal labour workforce, which was embroiled in calls for a Royal Commission in 2020 and 2021.
18. Perera & Pugliese, 1998.
19. See Christie, 1988; James, 2016.
20. See Christie, 1988.
21. Kelman, cited in Lehmann, 2020.
22. See for example Birmingham, 1999; Lines, 1991. More recent examples involved the resignation of NSW Premier Gladys Berejiklian and Deputy Premier John Barilaro amidst corruption scandals in 2021.
23. See Thompson, 2007: 18 for a simple overview of value systems, specifically *egocentric (self-centered)*; *homo/anthropocentric (human-centred)*; *biocentric (life/living beings-cantered)* and *ecocentric (ecologically centered)*.
24. Plumwood, 2008.
25. Zeunert, 2024d.
26. McIntyre, 2009: 4–5.
27. See *Period Five's* discussion of the Western Sydney Parklands Trust and its gearing to rural 'character' for agriculture.
28. Zeunert, 2023, 2024a–c.

29. See Clarke, 2017: 8. The Australian Bureau of Statistics (ABS) collates nurseries/cut flowers/turf in one classification.
30. And others, such as Holmgren, 2002, a process which he termed 'energy descent' as the most civilised approach of four future scenarios.
31. WEF, 2017: 4.
32. Beilin, 2015: 79.
33. See for example Barrett & Odum, 2000; Kunstler, 2006; Steel, 2008; Roberts, 2008; Cribb, 2010, 2017; Zeunert, 2018.
34. Zeunert, 2023, 2024a, b.
35. LA and GLA, 2010.
36. See Roberts, 2008; Gaballa and Abraham, 2008; Zeunert & Waterman, 2018.
37. Zeunert & Freestone, 2022. Also see Fig. 2.7.
38. Dibden & Cocklin, 2005; Willis, 2007: 83; Poole, 2022; Zeunert, 2024b.
39. Zeunert, 2018.
40. Marshall, 2019.
41. Kelly, 2018.
42. See summary in Zeunert, 2018: 240.
43. See for example Cribb, 2010; Roberts, 2008; Kunstler, 2006.
44. See summary in Zeunert, 2018: 240.
45. See summary in Zeunert, 2018: 240.
46. Masters, 2019.
47. WEF, 2017: 5.
48. Holmgren, 2009.
49. Furthermore, to what extent would money retain its value? See Harari, 2015.
50. Australia exported 72 per cent of its total agricultural output by value from 2017 to 2020 (ABARES, 2022).
51. Abbey in Cooley, 1994: 21.
52. https://nswanzaccentenary.records.nsw.gov.au/on-the-homefront/sydney-1914/
53. Cronon, 1996: 80.
54. As occurred to some extent during the world wars, particularly WWII with federal Government impetus from 1942. Although some public spaces were retooled for food production (e.g. parks) this campaign targeted home gardeners in an age in which people had larger yards in which to grow food. See for example: https://www.abc.net.au/news/2020-03-25/the-return-of-the-world-war-victory-garden/12085190

55. See Zeunert & Freestone, 2022 and Clarke, 2017.
56. Mason & Knowd, 2010: 64.
57. Downton, 2009: 481.
58. See Downton, 2009: 314–315 or Zeunert, 2017: 298. His 1996 shadow plans included consolidated and increased density satellite settlements linked by efficient sustainable public transport systems; rehabilitation and restoration of rivers, creeks and the marine environments accounting for to sea level rise; revegetation of lands to support native animal and bird populations; and local food production.
59. See for example: https://www.theguardian.com/australia-news/2021/mar/08/extremely-dangerous-how-much-of-the-heat-can-western-sydney-bear
60. Detroit in the US, for example, has seen abandoned metropolitan land repurposed for agriculture, although this must navigate land contamination.
61. Heinberg & Lerch, 2010.
62. Such as the Perich, Baiada and Ingham families, namely Leppington Pastoral Company (Perich), Celestino (Baiada) and Ingham Property.
63. House values in Sydney are regularly internationally chart-topping.
64. Namely petrol and diesel via crude oil. NSW's energy production is about 90 per cent coal-generated, so even if greater Sydney transitions to electric vehicles, these equate to coal-powered electric transportation until NSW's energy grid transitions to renewable energy.
65. GSC, 2018: 10, 48, 58. See *Period Five.*
66. As outlined and critiqued in *Period Four* and particularly *Period Five.*
67. See Kunstler's 1993 book of the same name.
68. Countless work has been produced on this topic but an excellent starting point is the urban sociologist Jane Jacobs, a still revered urban sociologist who wrote the iconic *The Death and Life of Great American Cities* (1961), containing still relevant principles robust city-making. Also see: http://www.altimeterfilms.com/citizen-jane-battle-for-the-city
69. Albeit less so than most other Australian's cities. See for example Weller, 2009; Weller & Bolleter, 2013.
70. Boyd, 1961.
71. See Tim Ross in ABC's 'Streets of Your Town' (2016).
72. See e.g. https://theconversation.com/the-great-australian-dream-new-homes-in-planned-estates-may-not-be-built-to-withstand-heatwaves-166266

73. McDonough & Bruangart, 2002, 2013.
74. See: https://www.ted.com/talks/james_howard_kunstler_the_ghastly_tragedy_of_the_suburbs/transcript
75. Holmgren, 2018.
76. Holmgren, 2009.
77. Yet an approximately 600 square metre block of land 53 kilometres from Sydney's CBD commands $650,000 in 2020 based on, for example, blocks for sale in 'Marsden Park'.
78. Murray, 2024.
79. Bill Clinton, United Nations World Food Day Speech (October 16, 2008), cited in Cebon, 2010: 30 in Zeunert, 2017: 174.
80. Cited in CCPCoC, 2009: 3.
81. Clarke, 2017.
82. Zeunert et al., 2025.
83. Ihlein, 2014.
84. Sydney dairy farmer Gavin Moore, as quoted by Lehmann, 2020.
85. Unfortunately the interactive scenario maps are no longer functioning on their website: http://www.sydneyfoodfutures.net/interactive-maps/#
86. SPUN were the client of the Sydney Food Futures project.
87. Pers. comm. Laura Wynne 29 April 2021. This area is shown in Fig. 2.8 and here: http://www.sydneyfoodfutures.net/interactive-maps/#
88. The 'Sydney Basin' is formally the geological basin (Fig. 2.8) but frequently used loosely and inconsistently.
89. See for example Lehmann, 2020; Carey & James, 2018; and Pearson, 2018.
90. The Greater Sydney Commission refer to 'Greater Sydney' as a distinct entity, but never clearly show what this is (it appears to be the area shown in GSC, 2018: 12–13).

    - An overall 'Region Boundary', which all areas fall within (GSC, 2018: 157). It is likely but unclear if this is 'Greater Sydney'. Sydney extends predominantly westward and north-west and we have calculated it as approximately 10,330 square kilometres (Fig. 2.8);
    - Numerous sub-regions within the overall region: including a "Metropolitan Rural Area" (GSC, 2018: 144, 157) (approximately 2750 square kilometres +/-); "Protected Natural Area" (GSC, 2018: 144) (approximately 5530 square kilometres +/-); and the "urban area" (GSC, 2018: 144) (approximately 2050 square kilometres +/-),

which is limited to built-up areas and those zoned/planned for such. The 'urban area' is treated in a comparatively trim way to that of 'Greater Sydney', seemingly inferring a further capacity to continue sprawling and consuming its hinterland with urban development.

91. This equates to the residents of Shoalhaven, Kiama and Wingecarribee Council areas being part of (peri-urban) Sydney and integral to its food supply bowl.
92. Pearson, 2018.
93. Pers. comm. Laura Wynne 29 April 2021.
94. See Jeavons, 2006.
95. See Viviano, 2017; Court & Zeunert, 2022. This is planned for the agribusiness precinct for the under construction Western Sydney Airport, see Zeunert & Freestone, 2022.
96. Bookchin, 1995: 248.
97. Pimentel & Pimentel, 2008; Oppenlander, 2013; Poore & Nemecek, 2018; Springmann et al., 2018; Zeunert, 2018, 2024b; Scarborough et al., 2023.
98. Weller and Bolleter, 2013.
99. Biointensive agriculture, technologically (and energy intensive) production approaches such as those widely deployed in the Netherlands, mushroom farming, intensive poultry, aquaculture and feedlots, for example, exhibit spatially efficient metropolitan footprints but require significantly ranging additional (embodied) lands in order to support their immediate spatial footprint—their embodied footprint or lifecycle.
100. Australian livestock consume two-thirds to three-quarters of domestic grain and contemporary grainfed livestock production systems that constitute over half of Australia's livestock sector result in net energy loss food systems (Zeunert, 2024b). Also see Jeavons, 2006; Pimentel & Pimentel, 2008; Oppenlander, 2013; Viviano, 2017; Poore & Nemecek, 2018; Springmann et al., 2018; Zeunert, 2018, 2024a, c and Scarborough et al., 2023.
101. Merson et al., 2010: 84.
102. James, 2013.
103. Ensuring continuous and responsible productive usage under agricultural ownership, lease and/or agreements.
104. Albrecht, 2005: 44–59.

References

ABARES. (2022). *Snapshot of Australian Agriculture 2022.* Australian Bureau of Agricultural and Resource Economics and Sciences: Australian Government Department of Agriculture, Fisheries and Forestry. Accessed May 1, 2022, from https://www.awe.gov.au/abares/products/insights/snapshot-of-australian-agriculture-2022

Albrecht, G. (2005). 'Solastalgia': A new concept in health and identity. *PAN: Philosophy, Activism, Nature, 3*, 41. https://doi.org/10.4225/03/584f410704696

Barrett, G., & Odum, E. (2000). The twenty-first century: The world at carrying capacity. *BioScience, 50*, 363–368.

Beilin, R. (2015). Agriculture, water, and nationhood: Reboot or rewrite? *Georgetown Journal of International Affairs, 16*(2), 77–90. https://www.jstor.org/stable/43773698

Benson, D., & Howell, J. (1990). *Taken for granted: The bushland of Sydney and its suburbs.* Kangaroo Press.

Birmingham, J. (1999). *Leviathan: The unauthorised biography of Sydney.* Knopf.

Bookchin, M. (1995). *From urbanization to cities: Toward a new politics of citizenship.* Cassell.

Boyd, R. (1961). *The Australian ugliness.* Cheshire.

Carey, R., & James, S. (2018). Peri-urban agriculture in Australia: Pressure on the urban fringe. In J. Zeunert & T. Waterman (Eds.), *Routledge handbook of landscape and food* (pp. 213–227). Routledge.

CCPCoC. (2009). *Economic profile of agriculture on the Central Coast Plateau, NSW.* Central Coast Plateau Chamber of Commerce. NSW Government Industry and Investment. Accessed November 3, 2020, from https://s3.ap-southeast-2.amazonaws.com/dpe-files-production/s3fs-public/dpp/213581/Final%20Farming%20Economic%20Profile%20Report%20Dec%2009.pdf

Christie, M. (1988). *The Sydney markets 1788–1988.* Sydney Market Authority.

Clarke, M. (2017). *Values of the metropolitan rural area of the Greater Sydney Region.* AgEconPlus Consulting. Accessed December 10, 2018, from https://gsc-public-1.s3.amazonaws.com/s3fs-public/Values_of_the_Metropolitan_Rural_Area_of_the_Greater_Sydney_Region_(Ag_Econ_Plus).pdf

Cooley, J. (1994). *Earthly words: Essays on contemporary American nature and environmental writers.* University of Michigan Press.

Court, T., & Zeunert, J. (2022). Novel horticultural adaptations to climate and their unique landscapes. In Y. Yang & A. Taufen (Eds.), *Routledge handbook of sustainable cities and landscapes in the Pacific Rim* (pp. 164–179). Routledge. https://doi.org/10.4324/9781003033530-16

Cribb, J. (2010). *The coming famine: The global food crisis and what we can do to avoid it*. University of California Press.

Cribb, J. (2017). *Surviving the 21st century: Humanity's ten great challenges and how we can overcome them*. Springer.

Cronon, W. (1996). The trouble with wilderness; or, getting back to the wrong nature. In W. Cronon (Ed.), *Uncommon ground: Rethinking the human place in nature* (pp. 69–90). W.W. Norton.

Crutzen, P. J., & Stoermer, E. F. (2000). The "anthropocene". *Global Change Newsletter, 41*, 17.

Dibden, J., & Cocklin, C. (2005). *Sustainability and change in rural Australia*. University of New South Wales Press.

Downton, P. (2009). *Ecopolis: Architecture and cities for a changing climate*. CSIRO Publishing.

Gaballa, S., & Abraham, A. (2008). *Food miles in Australia: A preliminary study of Melbourne, Victoria*. Centre for Education and Research in Environmental Strategies.

GSC. (2018). *Greater Sydney Region plan: A metropolis of three cities*. Greater Sydney Commission: NSW Government. https://www.greater.sydney/publications

Harari, Y. N. (2015). *Sapiens: A brief history of humankind*. Harper.

Heinberg, R. (2005, June 23–25). *What will we eat as the oil runs out?* Conference Paper, FEASTA Conference, Dublin, Ireland.

Heinberg, R., & Lerch, D. (2010). *The post carbon reader: Managing the 21st century's sustainability crises*. Watershed Media.

Holmgren, D. (2002). *Permaculture: Principles & pathways beyond sustainability*. Holmgren Design Services.

Holmgren, D. (2009). *Future scenarios: How communities can adapt to peak oil and climate change*. Chelsea Green.

Holmgren, D. (2018). *RetroSuburbia: The downshifter's guide to a resilient future*. Melliodora Publishing.

Ihlein, L. (2014). *A heritage vision for sustainable housing goes to ICAC*. The Conversation. https://theconversation.com/a-heritage-vision-for-sustainable-housing-goes-to-icac-21616

Jacobs, J. (1961). *The death and life of Great American cities*. Random House.

James, P. (2013). *Cosmopolitan conversationalists: Greening modern Sydney*. Australian Scholarly.

James, S. (2016). *Farming on the fringe: Peri-urban agriculture, cultural diversity and sustainability in Sydney*. Springer International Publishing.

Jeavons, J. (2006). *How to grow more vegetables: And fruits, nuts, berries, grains, and other crops than you ever thought possible on less land than you can imagine*. Ten Speed Press.

Kelly, M. (2018). Food sovereignty. In J. Zeunert & T. Waterman (Eds.), *Routledge handbook of landscape and food* (pp. 475–486). Routledge.

Kunstler, J. (1993). *The geography of nowhere: The rise and decline of America's man-made landscape*. Simon & Schuster.

Kunstler, J. (2006). *The long emergency: Surviving the converging catastrophes of the 21st century*. Atlantic Books.

LA and GLA. (2010). *Cultivating the capital: Food growing and the planning system in London*. London Assembly and Greater London Authority.

Lehmann, M. (2020, June 13). Farming on Sydney's fringe; can it survive urban sprawl? *The Weekend Australian Magazine*. https://www.theaustralian.com.au/weekend-australian-magazine/farming-on-sydneys-fringe-can-it-survive-urban-sprawl/news-story/048b857628c86161f482319238dc9bad

Lines, W. (1991). *Taming the great south land: A history of the conquest of nature in Australia*. University of Georgia Press.

Ma Rhea, Z. (2017). *Frontiers of taste: Food sovereignty, sustainability and indigenous-settler relations in Australia*. Springer.

Marshall, T. (2019). *Prisoners of geography: Ten maps that tell you everything you need to know about global politics*. Elliott and Thompson Limited.

Mason, D., & Knowd, I. (2010). The emergence of urban agriculture: Sydney, Australia. *International Journal of Agricultural Sustainability, 8*(1–2), 62–71.

Masters, N. (2019). *For the love of soil: Strategies to regenerate our food production systems*. Printable Reality.

McDonough, W., & Braungart, M. (2002). *Cradle to cradle*. North Point Press.

McDonough, W., & Braungart, M. (2013). *The upcycle: Beyond sustainability—designing for abundance*. North Point Press.

McIntyre, J. (2009). Not rich and not British: Philip Schaeffer, 'Failed' Colonial Farmer [online]. *Journal of Australian Colonial History, 11*, 1–20. https://search.informit.com.au/documentSummary;dn=997044012583652;res=IELHSS

Merson, J., Attwater, R., Ampt, P., Wildman, H., & Chapple, R. (2010). The challenges to urban agriculture in the Sydney basin and lower Blue Mountains region of Australia. *International Journal of Agricultural Sustainability, 8*, 72–85.

Mundey, J. (1981). *Green bans and beyond*. Angus & Robertson.
Murray, C. (2024). *The great housing hijack: The hoaxes and myths keeping prices high for renters and buyers in Australia*. Allen & Unwin.
Oppenlander, R. A. (2013). *Food choice and sustainability: Why buying local eating less meat and taking baby steps won't work*. Langdon Street Press.
Pearson, S. (2018, July 5). *A fresh set of eyes: We're well placed to satisfy Sydney's hunger*. Western Advocate. Accessed May 28, 2020, from https://www.westernadvocate.com.au/story/5509201/a-fresh-set-of-eyes-were-well-placed-to-satisfy-sydneys-hunger/
Perera, S., & Pugliese, J. (1998). Detoxifying Australia? *Migration Action, 20*(2), 4–18.
Pers. comm. Laura Wynne 29 April 2021.
Poole, B. (2022, April 20). Get big or Get out. *SBS The Feed*. Accessed 22 June 2022. https://www.sbs.com.au/news/the-feed/article/get-big-or-get-out-trying-to-survive-in-australias-agricultureindustry/k9otk6j7y
Poore, J., & Nemecek, T. (2018). Reducing food's environmental impacts through producers and consumers. *Science, 360*, 987–992.
Pimentel, D., & Pimentel, M. (2008). *Food, energy, and society*. CRC Press.
Plumwood, V. (2008). Shadow places and the politics of dwelling. *Australian Humanities Review, 44*. https://australianhumanitiesreview.org/2008/03/01/shadow-places-and-the-politics-of-dwelling/
Rees, W. (1992). Ecological footprints and appropriated carrying capacity: What urban economics leaves out. *Environment and Urbanization, 4*, 121–130.
Roberts, P. (2008). *The end of food*. Houghton Mifflin.
Scarborough, P., Clark, M., Cobiac, L., et al. (2023). Vegans, vegetarians, fish-eaters and meat-eaters in the UK show discrepant environmental impacts. *Nature Food*. https://doi.org/10.1038/s43016-023-00795-w
Seddon, G. (1972). *Sense of place*. UWA Press.
Springmann, M., Clark, M., Mason-D'Croz, D., et al. (2018). Options for keeping the food system within environmental limits. *Nature, 562*, 519–525. https://doi.org/10.1038/s41586-018-0594-0
Steel, C. (2008). *Hungry city: How food shapes our lives*. Random House.
Thompson, I. (2007). The ethics of sustainability. In J. Benson & M. Roe (Eds.), *Landscape and sustainability* (2nd ed.). Taylor & Francis.
Viviano, F. (2017). *How the Netherlands feeds the world*. National Geographic. https://www.nationalgeographic.com/magazine/2017/09/holland-agriculture-sustainable-farming/

Wackernagel, M. (1994). *Ecological footprint and appropriated carrying capacity: A tool for planning toward sustainability*. PhD thesis, The University of British Columbia.

WEF. (2017). *Shaping the future of global food systems: A scenarios analysis*. World Economic Forum.

Weller, R. (2009). *Boom town 2050*. UWA Press.

Weller, R., & Bolleter, J. (2013). *Made in Australia: The future of Australian cities*. UWA Publishing.

Willis, A. M. (2007). From peri-urban to unknown territory. *Design Philosophy Papers, 5*, 79–90.

Zeunert, J. (2017). *Landscape architecture and environmental sustainability: Creating positive change through design*. Bloomsbury.

Zeunert, J. (2018). Challenges in agricultural sustainability and resilience: Towards regenerative practice. In J. Zeunert & T. Waterman (Eds.), *Routledge handbook of landscape and food* (pp. 231–252). Routledge.

Zeunert, J. (2023). *Food | Landscapes Australia*. https://www.foodlandscapes.com.au/. https://doi.org/10.26190/s0kk-8x57

Zeunert, J. (2024a). States of Australia's agri-environment: Visual extractions from degraded landscapes. *Visual Studies,* 39(4), 609–621. https://doi.org/10.1080/1472586X.2023.2294004

Zeunert, J. (2024b). Bridging rural and urban disconnections: Spatial graphic explorations of Australia's livestock landscapes. In R. Pernice & B. Chen (Eds.), *Urban regeneration and rural revitalization in Australia and China* (pp. 110–143). Routledge.

Zeunert, J. (2024c). Agriculture's shadow connections: Tracing heavy industries to food landscapes. In M. Lozanovska & U. de Jong (Eds.), *Architecture vacancy: Pasts and futures of Geelong, Australia* (pp. 68–75). Cambridge Scholars.

Zeunert, J. (2024d). An Australian agri-industrial landscape sublime. *Visual Communication, 24*(4). https://doi.org/10.1177/14703572241258895

Zeunert, J., & Waterman, T. (Eds.). (2018). *Routledge handbook of landscape and food*. Routledge.

Zeunert, J., & Freestone, R. (2022). From rural lands to agribusiness precincts: Agriculture in metropolitan Sydney 1948–2018. In C. Brisotto & F. Lemes de Oliveira (Eds.), *Re-imaging resilient food landscapes—Perspectives from planning history* (pp. 247–272). Springer.

Zeunert, J., Hawken, S., & Gowers, J. (2025). Visualising and valuing urban agriculture for land use planning: A critical GIS analysis of Sydney and neighbouring regions. *Land, 14*(4), 854. https://doi.org/10.3390/land14040854

Index[1]

[1] Note: Page numbers followed by 'n' refer to notes.

J. Zeunert, A. Daroy, *Sydney's Food Landscapes*,
https://doi.org/10.1007/978-981-96-0710-5

The manufacturer's authorised representative in the EU is Springer Nature Customer Service Centre GmbH, Europaplatz 3, 69115 Heidelberg, Germany. If you have any concerns regarding our products, please contact ProductSafety@springernature.com

Printed and bound by CPI Group (UK) Ltd, Croydon, CR0 4YY
01/10/2026
02225913-0008